A–Z of Digital Research Methods

This accessible, alphabetical guide provides concise insights into a variety of digital research methods, incorporating introductory knowledge with practical application and further research implications. *A–Z of Digital Research Methods* provides a pathway through the often-confusing digital research landscape, while also addressing theoretical, ethical and legal issues that may accompany each methodology.

Dawson outlines 60 chapters on a wide range of qualitative and quantitative digital research methods, including textual, numerical, geographical and audio-visual methods. This book includes reflection questions, useful resources and key texts to encourage readers to fully engage with the methods and build a competent understanding of the benefits, disadvantages and appropriate usages of each method.

A–Z of Digital Research Methods is the perfect introduction for any student or researcher interested in digital research methods for social and computer sciences.

Catherine Dawson is a freelance researcher and writer specialising in the use and teaching of research methods. She has taught research methods courses at universities in the UK, completed a variety of research projects using qualitative, quantitative and mixed methods approaches, and written extensively on research methods and techniques.

A–Z of Digital Research Methods

Catherine Dawson

Routledge
Taylor & Francis Group

LONDON AND NEW YORK

First edition published 2020
by Routledge
2 Park Square, Milton Park, Abingdon, Oxon, OX14 4RN

and by Routledge
52 Vanderbilt Avenue, New York, NY 10017

Routledge is an imprint of the Taylor & Francis Group, an informa business

British Library Cataloguing-in-Publication Data
A catalogue record for this book is available from the British Library

Library of Congress Cataloging-in-Publication Data
Names: Dawson, Catherine, author.
Title: A-Z of digital research methods / Catherine Dawson.
Other titles: A to Z of digital research methods
Description: Abingdon, Oxon ; New York, NY : Routledge, 2019. |
Includes bibliographical references and index.
Identifiers: LCCN 2019009327 (print) | LCCN 2019016155 (ebook) |
ISBN 9781351044677 (eBook) | ISBN 9781138486799 (hardback) |
ISBN 9781138486805 (pbk.)
Subjects: LCSH: Research–Data processing. | Research–Methodology. |
Computer simulation.
Classification: LCC Q180.55.E4 (ebook) | LCC Q180.55.E4 D39 2019
(print) | DDC 001.4/20285–dc23
LC record available at https://lccn.loc.gov/2019009327

ISBN: 978-1-138-48679-9 (hbk)
ISBN: 978-1-138-48680-5 (pbk)
ISBN: 978-1-351-04467-7 (ebk)

Typeset in Melior and Bliss
by Swales & Willis, Exeter, Devon, UK

Author Biography

Dr Catherine Dawson has worked as a researcher, tutor and trainer for almost 30 years in universities, colleges and the private sector in the UK. She has designed and taught research methods courses for undergraduate and postgraduate students and has developed and delivered bespoke research methods training sessions to employees at all levels in the private sector. She has also carried out a variety of research projects using qualitative, quantitative and mixed methods approaches and has published a number of papers and books on research methods and techniques. Catherine has drawn on this experience to develop and produce the *A–Z of Digital Research Methods*, which provides an accessible, comprehensive and user-friendly guide for anyone interested in finding out more about digital research methods.

Contents

Contents

Contents

Introduction

The A–Z of Digital Research Methods provides an introduction to a wide variety of digital research methods including numerical, geographical, textual, audio and visual methods. The term 'digital' is used as an umbrella term to describe research methods that use computer-based products and solutions (or electronic technologies) for data collection and analysis, including online, mobile, location and sensor-based technologies. 'Research methods' are the tools and techniques used to collect and analyse data. Methodology (the guideline system or framework used to solve a problem) is not included except in cases where the boundary between method and methodology are blurred or indistinct (digital and online ethnography, for example). The book includes new methods and techniques that have developed together with relevant digital technology (big data analytics, machine learning and online analytical processing, for example), and traditional methods that have been modified, changed or aided by digital technology, even though the method in itself may have a long pre-digital history (data visualisation, interviews, questionnaires and social network analysis, for example). Both qualitative and quantitative approaches are covered, including naturalistic approaches, tightly-designed experiments and surveys and the collection of various forms of non-reactive data.

The book is aimed at researchers, educators and students working primarily within the social sciences, education and humanities, but will also be of interest to those within a number of other disciplines and fields of study including management and business, health and nursing, information sciences, human geography and urban planning. It will be useful for early-career researchers planning and moving forward with their thesis or research project; experienced researchers updating their knowledge about digital research methods and approaches; early-career research methods tutors designing a new research methods course/module and adding to reading lists; existing research methods tutors updating their knowledge about digital methods; and under-graduate and postgraduate students thinking about and choosing appropriate research methods for their project, dissertation or thesis.

The simple, alphabetically-ordered structure of the book enables the reader to read the book from start to finish (useful for students who know very little about digital research methods, perhaps who are trying to choose methods for their dissertation or thesis), or to dip in and out of the book to find out about a specific method (useful for tutors who need to add new content to their course and reading lists, for example). The alphabetical structure also encourages the reader to approach the book without being restricted or constrained by the qualitative and quantitative divide, thus encouraging them to consider mixed or hybrid methods or think about alternative (or perhaps less obvious or 'non-traditional') approaches that will help them to answer their research question.

Each entry in the book contains four simple categories for ease of reference: overview, questions for reflection, useful resources and key texts. The overview provides concise information about the method, including information, where relevant, about history, development, epistemological, theoretical and methodo-logical insight and practical information about how, when and why the method is used. The 'questions for reflection' section asks pertinent questions that help the reader to think more deeply about the method. These are divided into three categories: 1) epistemology, theoretical perspective and methodology; 2) ethics, morals and legal issues; and 3) practicalities. The questions ask the reader to consider whether the particular method is suitable, given theoretical and meth-odological standpoint; highlight pros and cons; point to possible pitfalls and address practicalities associated with the method. These questions help to stimulate deeper thought and can be used by tutors for group or class discussion or to help develop assignments, for example. The 'useful resources' section includes relevant websites, tools, videos, apps and/or online courses and the 'key texts' section provides a list of useful books, papers and journal articles, to guide and facilitate further inquiry.

The book is an introductory text that assumes no prior knowledge of each topic covered. As such, it is of use to those new to digital research methods and those who need to update their knowledge or find information about digital research methods that they have not yet come across. Questions are posed to stimulate thought and reflection, which help the reader to think more deeply about each method and work out whether it would be suitable for their project, dissertation or teaching. It is important to note, however, that this book provides an introduction: students or researchers will not be able to complete their research project purely from reading this book. Instead, it provides enough information about suitable methods that will help to facilitate choices and point toward relevant texts and resources for further inquiry.

CHAPTER 1

Agent-based modelling and simulation

Overview

Agent-based modelling and simulation (ABMS) is a method that enables researchers to create a computer model and simulation of active entities and their behaviour and interaction with each other and their environment. These interacting, autonomous and adaptive agents can be individuals, households, groups, organisations, vehicles, equipment, products or cells, in social and evolutionary settings, for example. ABMS is a specific type of computer model and simulation: other types and a general discussion of modelling and simulation can be found in Chapter 7. There are different terms that are used to describe the same, or similar techniques, and these include agent-based computational modelling, agent-based modelling (ABM), agent-based simulation modelling, agent-based social simulation (ABSS) and agent-based simulation. It is important to note that models provide representations whereas simulations use models (or simulate the outcomes of models) for study and analysis: ABMS is a term that covers both.

ABMS enables researchers to build models of systems from the bottom up (micro to macro), with the aim of producing simulations in which patterns, structures and behaviours emerge from agent interaction. Models and simulations can be exploratory, descriptive and predictive: they can be used to provide insight into behaviour and decision-making, make predictions about future behaviour and trends or help to analyse, validate or explain data collected from other sources. Models and simulations can be relational, dynamic, responsive and adaptive. For example, agents:

- respond to the actions of others;
- respond to environmental stimuli;

- influence each other;

- learn from each other;

- learn from their experiences;

- adapt their behaviour as a result of other agents' behaviour;

- adapt their behaviour to suit their environment.

Researchers from a variety of disciplines and areas of study use ABMS including sociology and social psychology (Chattoe-Brown, 2014; Conte and Paolucci, 2014; Eberlen et al., 2017), geography (Millington and Wainwright, 2017), health and medicine (Auchincloss and Garcia, 2015), economics (Caiani et al., 2016), politics (Fieldhouse et al., 2016), the sports sciences (Lauren et al., 2013); the environmental sciences (Kerridge et al., 2001; Sun and Taplin, 2018) and the computer sciences (Abar et al., 2017). Examples of research projects that have used ABMS include a study that adapts principles of developmental biology and agent-based modelling for automated urban residential layout design (Sun and Taplin, 2018); research into pedestrian flow and movement (Kerridge et al., 2001); research into the interaction between the development of creative industries and urban spatial structure (Liu and Silva, 2018); research that helps to predict rates of burglary (Malleson et al., 2009); a study to model, simulate and test tactics in the sport of rugby union (Lauren et al., 2013); and research into voter turnout (Fieldhouse et al., 2016).

If you are interested in finding out more about ABMS, and using it for your research, a good reference to begin with is Silverman et al. (2018), which is an open access book that provides in-depth coverage of methodological issues and complexities associated with ABM and the social sciences. A useful reference for those working within, or studying, economics is Hamill and Gilbert (2016), which provides a practical introduction and history to ABM methods and techniques. Another is Caiani et al. (2016), which provides a practical guide and basic toolkit that highlights practical steps in model building and is aimed at undergraduates, postgraduates and lecturers in economics. A useful reference for those working within geography (and who are interested in mixed methods approaches) is Millington and Wainwright (2017: 68) who discuss 'mixed qualitative-simulation methods that iterate back-and-forth between "thick" (qualitative) and "thin" (simulation) approaches and between the theory and data they produce'. Auchincloss and Garcia (2015) provide a brief introduction to carrying out a simple agent-based model in the field of urban health research. Chapter 7 provides a useful overview of computer modelling and simulation and contains additional references and relevant questions for reflection. If you are interested in predictive modelling, more information can be found in Chapter 45.

Questions for reflection

Epistemology, theoretical perspective and methodology

- Miller (2015: 175) proposes critical realism as a philosophical perspective to understand, orient and clarify the nature and purpose of agent-based modelling research. Does this perspective have resonance with your research and, if so, in what way? How might this perspective help you to evaluate, validate and assess models?

- Do you intend to use agent-based modelling as a standalone research method, or do you intend to adopt a mixed methods approach? Is it possible to integrate diverse forms of data (and interdisciplinary data) with agent-based modelling? Chattoe-Brown (2014) believes so, illustrating why and how from a sociological perspective, and Millington and Wainwright (2017) discuss mixed method approaches from a geographical perspective.

- How might ABMS be used to complement and improve traditional research practices? Eberlen et al. (2017) will help you to reflect on this question in relation to social psychology.

- Can phenomena emerging from agent-based models be explained entirely by individual behaviour? Silverman et al. (2018) provide a comprehensive discussion on this and other methodological considerations.

- Do models represent the real world, or are they a researcher's interpretation of the real world?

- What are the strengths and weaknesses of ABMS? Conte and Paolucci (2014) will help you to address this question in relation to computational social science and Eberlen et al. (2017) discuss these issues in relation to social psychology.

Ethics, morals and legal issues

- Is it possible that modelling can be to the detriment of individuals? Can model outcomes lead to unethical or inappropriate action that can cause harm to individuals? Can individuals be singled out for action, based on models? What happens when predictions are based on past behaviour that may have changed? Can individuals correct model inputs?

- Have data been volunteered specifically for modelling purposes?

- Is it possible that individuals could be identifiable from models?

- Millington and Wainwright (2017: 83) ask a pertinent question that needs to be considered if you intend to use ABMS: 'how might new-found understandings by individuals about their agency be turned back to geographers to understand the role of agent-based simulation modelling itself as an agent of social change?'

Practicalities

- How will you go about building your model? Jackson et al. (2017: 391–93) provide a seven-step guide to creating your own model:

  ○ Step 1: what are your world's dimensions?

  ○ Step 2: how do agents meet?

  ○ Step 3: how do agents behave?

  ○ Step 4: what is the payoff?

  ○ Step 5: how do agents change?

  ○ Step 6: how long does your world last?

  ○ Step 7: what do you want to learn from your world?

- Do you know which is the most appropriate agent-based modelling and simulation toolkit for your research? How do you intend to choose software and tools? A concise characterisation of 85 agent-based toolkits is provided by Abar et al. (2017).

- How accurate is your model? How important is accuracy (when action is to be taken, or decisions made, based on your model outcomes, for example?)

- How do you intend to verify and validate your model (ensuring the model works correctly and ensuring the right model has been built, for example)?

Useful resources

There are a wide variety of agent-based modelling and simulation software and digital tools available. A few examples available at time of writing are given below (in alphabetical order).

- Adaptive Modeler (www.altreva.com);

- AnyLogic (www.anylogic.com);

- Ascape (http://ascape.sourceforge.net);

- Behaviour Composer (http://m.modelling4all.org);

- Cougaar (www.cougaarsoftware.com);

- GAMA (https://gama-platform.github.io);

- JADE (http://jade.tilab.com);

- NetLogo (https://ccl.northwestern.edu/netlogo);

- OpenStarLogo (http://web.mit.edu/mitstep/openstarlogo/index.html);

- Repast Suite (https://repast.github.io);

- StarLogo TNG (https://education.mit.edu/portfolio_page/starlogo-tng);

- Swarm (www.swarm.org/wiki/Swarm_main_page).

The Society for Modeling and Simulation International (http://scs.org) and the European Council for Modelling and Simulation (www.scs-europe.net) provide details of conferences, workshops, publications and resources for those interested in computer modelling and simulation (see Chapter 7 for more information about these organisations and for additional tools and software that can be used for computer modelling and simulation).

Key texts

Abar, S., Theodoropoulos, G., Lemarinier, P. and O'Hare, G. (2017) 'Agent Based Modelling and Simulation Tools: A Review of the State-of-Art Software', *Computer Science Review*, 24, 13–33, May 2017, 10.1016/j.cosrev.2017.03.001.
Auchincloss, A. and Garcia, L. (2015) 'Brief Introductory Guide to Agent-Based Modeling and an Illustration from Urban Health Research', *Cadernos De Saude Publica*, 31(1), 65–78, 10.1590/0102-311X00051615.
Caiani, A., Russo, A., Palestrini, A. and Gallegati, M. (eds.) (2016) *Economics with Heterogeneous Interacting Agents: A Practical Guide to Agent-Based Modeling*. Cham: Springer.
Chattoe-Brown, E. (2013) 'Why Sociology Should Use Agent Based Modelling', *Sociological Research Online*, 18(3), 1–11, first published August 31, 2013, 10.5153/sro.3055.
Chattoe-Brown, E. (2014) 'Using Agent Based Modelling to Integrate Data on Attitude Change', *Sociological Research Online*, 19(1), 1–16, first published March 5, 2014, 10.5153/sro.3315.
Conte, R. and Paolucci, M. (2014) 'On Agent-Based Modeling and Computational Social Science', *Frontiers in Psychology*, 5(668), first published July 14, 2014, 10.3389/fpsyg.2014.00668.
Eberlen, J., Scholz, G. and Gagliolo, M. (2017) 'Simulate This! An Introduction to Agent-Based Models and Their Power to Improve Your Research Practice', *International Review of Social Psychology*, 30(1), 149–160. 10.5334/irsp.115.

Fieldhouse, E., Lessard-Phillips, L. and Edmonds, B. (2016) 'Cascade or Echo Chamber? A Complex Agent-Based Simulation of Voter Turnout', *Party Politics*, 22(2), 241–56, first published October 4, 2015, 10.1177/1354068815605671.

Hamill, L. and Gilbert, N. (2016) *Agent-Based Modelling in Economics*. Chichester: John Wiley & Sons Ltd.

Jackson, J., Rand, D., Lewis, K., Norton, M. and Gray, K. (2017) 'Agent-Based Modeling: A Guide for Social Psychologists', *Social Psychological and Personality Science*, 8(4), 387–95, first published March 13, 2017, 10.1177/1948550617691100.

Kerridge, J., Hine, J. and Wigan, M. (2001) 'Agent-Based Modelling of Pedestrian Movements: The Questions that Need to Be Asked and Answered', *Environment and Planning B: Urban Analytics and City Science*, 28(3), 327–41, first published June 1, 2001, 10.1068/b2696.

Lauren, M., Quarrie, K. and Galligan, D. (2013) 'Insights from the Application of an Agent-Based Computer Simulation as a Coaching Tool for Top-Level Rugby Union', *International Journal of Sports Science & Coaching*, 8(3), 493–504, first published September 1, 2013, 10.1260/1747-9541.8.3.493.

Liu, H. and Silva, E. (2018) 'Examining the Dynamics of the Interaction between the Development of Creative Industries and Urban Spatial Structure by Agent-Based Modelling: A Case Study of Nanjing, China', *Urban Studies*, 55(5), 1013–32, first published January 25, 2017, 10.1177/0042098016686493.

Malleson, N., Evans, A. and Jenkins, T. (2009) 'An Agent-Based Model of Burglary', *Environment and Planning B: Urban Analytics and City Science*, 36(6), 1103–23, first published January 1, 2009, 10.1068/b35071.

Miller, K. (2015) 'Agent-Based Modeling and Organization Studies: A Critical Realist Perspective', *Organization Studies*, 36(2), 175–96, first published December 3, 2014, 10.1177/0170840614556921.

Millington, J. and Wainwright, J. (2017) 'Mixed Qualitative-Simulation Methods: Understanding Geography through Thick and Thin', *Progress in Human Geography*, 41(1), 68–88, first published February 4, 2016, 10.1177/0309132515627021.

Silverman, E., Courgeau, D., Franck, R., Bijak, J., Hilton, J., Noble, J. and Bryden, J. (2018) *Methodological Investigations in Agent-Based Modelling: With Applications for the Social Sciences*. Cham: Springer Open.

Sun, Y. and Taplin, J. (2018) 'Adapting Principles of Developmental Biology and Agent-Based Modelling for Automated Urban Residential Layout Design', *Environment and Planning B: Urban Analytics and City Science*, first published January 31, 2017, 10.1177/2399808317690156.

CHAPTER 2

Audio analysis

Overview

Audio analysis is a term that is used to describe the study of audio content. There are different types, terms and/or methods of audio analysis that are used within different disciplines and fields of study and for different purposes. Examples include (in alphabetical order):

- Acoustic analysis: this can be acoustic analysis of voice (used by researchers in health and medicine, psychology and linguistics, for example) and acoustic analysis of sound (used by researchers in engineering, biology, ecology and textile design, for example). Acoustic analysis is performed by inspecting visualised speech or sound and can involve waveform analysis, phonetic analysis, periodicity analysis and intensity analysis, for example. Anikin and Lima (2018) provide an example of this type of analysis in their paper on perceptual and acoustic differences between authentic and acted nonverbal emotional vocalisations.

- Audio diary analysis: audio diaries, as a means of data collection, are used by researchers approaching their work from a variety of disciplines and fields of study, including the social sciences, psychology, education, arts and humanities, anthropology and health and medicine. Various types of analysis can take place, depending on epistemology, theoretical perspective and methodology (narrative analysis, content analysis or discourse analysis, for example). More information about audio diaries, along with useful references, can be obtained from Chapter 30.

- Audio event and sound recognition/analysis: this is for home automations, safety and security systems and surveillance systems. It is of interest to researchers working in business, retail, security, policing and criminology, for example. Audio sensors and software can be used to recognise, analyse and react to sounds and events such as windows smashing, smoke alarms, voices of intruders, anomaly detection or babies crying, for example. More information about sensor-based methods can be found in Chapter 49.

- Multimodal analysis: this involves the annotation, analysis and interpretation of social semiotic patterns and choices from language, text, image and audio resources (timing, frequency and volume of utterances, for example). It is a method of analysis used by researchers working with the social sciences, psychology, education and health and medicine. Sasamoto et al. (2017) provide an example of this type of analysis in their paper on Japanese TV programmes. Multimodal discourse analysis is one type of multimodal analysis (the world is understood as multimodal: different semiotic systems and their interactions are considered). Dash et al. (2016) provide an example of this type of analysis in their study of Indian TV commercials.

- Music information retrieval: this involves the process of retrieving information from music to categorise, manipulate, analyse and/or create music. It is of interest to researchers working in musicology, music composition, acoustics, information sciences, neuroscience and computer science, for example. It can involve processes such as instrument recognition, recognising the sequence of notes being played, music transcription (notating a piece or sound) and genre categorisation or recognition, for example. Kızrak and Bolat (2017) provide an example of this type of analysis in their research on classical Turkish music.

- Psychoacoustic analysis: this involves the analysis of sound and the effect that it has on physiological perception and stress. It considers the relationship between sound and what a person hears (or their hearing perception) and is used by researchers working in the natural sciences, engineering, fabric and textile design, health and medicine and psychology, for example. Cho and Cho (2007) provide an example of this type of analysis in their research on fabric sound and sensations.

- Semantic audio analysis: this is a method of analysis that enables researchers to extract symbols or meaning from audio signals. It is closely related to music information retrieval within the study of music and is used to analyse and manipulate music in an intuitive way (considering single notes, instruments or genre, for example). This type of analysis is used by researchers approaching their work from a number of disciplines and fields of study including the social sciences, psychology, information sciences, computer sciences, music studies and engineering. Comprehensive information about this type of analysis can be found in the doctoral thesis produced by Fazekas (2012).

- Sound analysis: this is the study of multiple frequencies and intensities in sound (and how they change over time). It can include analysing the spectral content of sound and analysing the pitch, format and intensity contours of sound, for example. This analysis method is used by researchers working in health and medicine, engineering, biology, ecology, linguistics, musicology, music composition and computer sciences, for example. Malindretos et al. (2014) illustrate how this type of analysis can be used in medical research.

- Sound scene and event analysis: sound scene analysis refers to the study of the entirety of sound within a given scene (sounds from various sources) and sound event analysis refers to a specific sound from a specific source. They can involve processes such as sound scene classification, audio tagging and audio event detection and are used by researchers approaching their work from a number of disciplines and fields of study including geography, biology, social sciences, engineering and computer sciences. Comprehensive information about sound scene and event analysis is provided by Virtanen et al. (2018).

- Speech analysis: this involves the analysis of voice and speech patterns of people. There are a wide variety of techniques that are covered by the umbrella term of speech analysis, and this can include recognition of patterns, structures and/or content of speech; analysis of the sequence of words within speech; analysis of intonation, rhythm, stress or emphasis within speech; recognition of the speaker; turn taking within speech; and identification of the speaker (in situations where more than one person is speaking), for example. It is used in HR analytics (see Chapter 22), business analytics (Chapter 4) and by researchers working in the social sciences, health and medicine, psychology, education and politics, for example. Loo et al. (2016) provide an example of this type of analysis in their research on learning a second language.

If you are interested in finding out more about audio analysis for your research it is important that you get to grips with the different techniques, methods and purposes outlined above. The references provided above will help you to do this, along with a consideration of the questions for reflection listed below.

There is a vast array of digital tools and software available for those interested in audio analysis: examples are given below. You may also be interested in the related field of video analysis, which is discussed in Chapter 55.

Questions for reflection

Epistemology, theoretical perspective and methodology

- What theoretical perspective and methodological framework will guide your audio analysis? As we have seen above, there are various approaches to audio analysis and the method(s) that you choose must fit within your theoretical perspective and methodological framework. For example, semantic audio analysis is an area of study within semiotics, which is a type of discourse analysis that focuses on how signs, symbols and expressions are interpreted, used and create meaning. Other areas of study within semiotics include syntactics, which looks at the relationship of signs, symbols and expressions to their formal structures, and pragmatics, which considers the relationships between signs, symbols and expressions and the effects that they have on people who use them. An understanding of such approaches will help you to clarify your methodological framework.

- What role does memory play in audio perception?

- How can you ensure a good balance between machine and human listening? Why is balance important?

- Virtanen et al. (2018: 8) point out that 'to date there is no established taxonomy for environmental sound events or scenes'. What impact, if any, might this observation have on your research? Is an established taxonomy necessary for your audio analysis?

Ethics, morals and legal issues

- How can you ensure anonymity when recording and analysing speech? Audio recordings of speech contain aspects of speech (intonation, tone and voice characteristics, for example) that may help to identify the speaker. Issues of anonymity are of particular importance if secondary analysis or reanalysis is carried out by researchers who were not involved in the original interview. Pätzold (2005) discusses technical procedures that can be used to modify the sound of the audio source so that the possibility of recognition can be reduced.

- How can you ensure against unauthorised release of audio data?

- How might constraints on human audio perception lead to error and bias?

- How can you ensure that all people being recorded know and understand the purpose of the recording, what will happen to the recording, who will hear it, how the recording will be stored and when it will be deleted, for example? The Oral History Society (www.ohs.org.uk/advice/ethical-and-legal) provides comprehensive advice about ethical and legal issues associated with recording voices.

Practicalities

- Do you have a good understanding of relevant hardware, software and digital tools? How accurate, flexible and scalable is the technology? What are the integration, calibration and certification processes? What service contract is available, if relevant? Some examples of software and tools are listed below. Sueur (2018) discusses the free and open-source software R and provides step-by-step examples and useful case studies for those who are interested in using this software for audio analysis.

- Does your recorder have a 'voice operated switch' or a 'voice operated exchange' (a switch used to turn a transmitter or recorder on when someone speaks and off when someone stops speaking, used to save storage space). Is speech loud enough to operate the switch? Will speech be lost when the switch operates?

- How might sound be affected by distance from microphone, or distance from audio sensor, in particular where there are many sounds created at various distances? How can you ensure that you obtain the best recording possible in such situations?

- What other factors might influence the quality of recording (sub-standard hardware, technical faults, vibrations and external noises, for example)?

Useful resources

There is a vast array of tools and software for audio analysis, with differing functions, capabilities, purposes and costs. Some are aimed at biologists who specialise in a particular animal (bats or birds, for example), some are aimed at those who want to view and analyse the contents of music audio files and others are aimed at researchers who are interested in calculating various psychoacoustic parameters or those wanting to use automatic speaker recognition software. Examples available at time of writing are given below, in alphabetical order.

- ai3 sound recognition technology (www.audioanalytic.com/software);
- ArtemiS suite for sound and vibration analysis (www.head-acoustics.com /eng/nvh_artemis_suite.htm);
- Audacity multi-track audio editor and recorder (www.audacityteam.org);
- Kaleidoscope Pro Analysis Software for recording, visualising and analysing bat sounds (www.wildlifeacoustics.com/products/kaleidoscope-software-ultrasonic);
- Praat for phonetics by computer (www.fon.hum.uva.nl/praat);
- Raven sound analysis software (http://ravensoundsoftware.com);
- seewave an R package for sound analysis and synthesis (http://rug.mnhn.fr /seewave);
- Sonic Visualiser for viewing and analysing the content of music audio files (www.sonicvisualiser.org);
- Sound Analysis Pro for the analysis of animal communication (http:// soundanalysispro.com);
- Soundecology R package for soundscape ecology (http://ljvillanueva .github.io/soundecology);
- Speech Analyzer for acoustic analysis of speech sounds (https://soft ware.sil.org/speech-analyzer);
- Wasp for recording, displaying and analysing speech (www.phon.ucl.ac.uk /resource/sfs/wasp.php);
- WaveSurfer for sound visualisation and manipulation (https://source forge.net/projects/wavesurfer).

Key texts

Alderete, J. and Davies, M. (2018) 'Investigating Perceptual Biases, Data Reliability, and Data Discovery in a Methodology for Collecting Speech Errors from Audio Recordings', *Language and Speech*, first published April 6, 2018, 10.1177/ 0023830918765012.
Anikin, A. and Lima, C. (2018) 'Perceptual and Acoustic Differences between Authentic and Acted Nonverbal Emotional Vocalizations', *Quarterly Journal of Experimental*

Psychology, 71(3), 622–41, first published January 9, 2017, 10.1080/17470218.2016.1270976.

Cho, J. and Cho, G. (2007) 'Determining the Psychoacoustic Parameters that Affect Subjective Sensation of Fabric Sounds at Given Sound Pressures', *Textile Research Journal*, 77(1), 29–37, first published January 1, 2007, 10.1177/0040517507074023.

Dash, A., Patnaik, P. and Suar, D. (2016) 'A Multimodal Discourse Analysis of Glocalization and Cultural Identity in Three Indian TV Commercials', *Discourse & Communication*, 10(3), 209–34, first published February 8, 2016, 10.1177/1750481315623892.

Fazekas, G. (2012) *Semantic Audio Analysis: Utilities and Applications*, PhD Thesis, Centre for Digital Music, School of Electronic Engineering and Computer Science, Queen Mary University of London.

Kızrak, M. and Bolat, B. (2017) 'A Musical Information Retrieval System for Classical Turkish Music Makams', *Simulation*, 93(9), 749–57, first published May 24, 2017, 10.1177/0037549717708615.

Loo, A., Chung, C. and Lam, A. (2016) 'Speech Analysis and Visual Image: Language Learning', *Gifted Education International*, 32(2), 100–12, first published April 16, 2014, 10.1177/0261429414526332.

Malindretos, P., Liaskos, C., Bamidis, P., Chryssogonidis, I., Lasaridis, A. and Nikolaidis, P. (2014) 'Computer Assisted Sound Analysis of Arteriovenous Fistula in Hemodialysis Patients', *The International Journal of Artificial Organs*, 37(2), 173–76, first published November 29, 2013, 10.5301/ijao.5000262.

Pätzold, H. (2005), 'Secondary Analysis of Audio Data. Technical Procedures for Virtual Anonymisation and Modification', *Forum Qualitative Sozialforschung/Forum: Qualitative Social Research*, 6(1), Art. 24, retrieved from www.qualitative-research.net/fqs-texte/1-05/05-1-24-e.htm.

Sasamoto, R., O'Hagan, M. and Doherty, S. (2017) 'Telop, Affect, and Media Design: A Multimodal Analysis of Japanese TV Programs', *Television & New Media*, 18(5), 427–40, first published November 17, 2016, 10.1177/1527476416677099.

Sueur, J. (2018) *Sound Analysis and Synthesis with R*. Cham: Springer.

Virtanen, T., Plumbley, M. and Ellis, D. (eds.) (2018) *Computational Analysis of Sound Scenes and Events*. Cham: Springer.

CHAPTER 3

Big data analytics

Overview

Big data analytics refers to the process of examining extremely large and complex data sets. These are referred to as 'big data' ('small data' refers to datasets of a manageable volume that are accessible, informative and actionable and 'open data' refers to datasets that are free to use, re-use, build on and redistribute, subject to stated conditions and licence). Some big data are structured: they are well-organised, with defined length and format that fit in rows and columns within a database. Other big data are unstructured, with no particular organisation or internal structure (plain text or streaming from social media, mobiles or digital sensors, for example). Some big data are semi-structured in that they combine features from both the above (email text combined with metadata, for example). All three types of data can be human or machine-generated.

Big data analytics involves capturing, extracting, examining, storing, linking, analysing and sharing data to discover patterns, highlight trends, gain deeper insight, identify opportunities and work out future directions. Big data analytics can involve individual datasets or data from multiple channels (social media, web, mobile or digital sensors, for example). It is covered by the umbrella term of data analytics (Chapter 10) and can involve data mining (Chapter 12), data visualisation tools and techniques (Chapter 13) and predictive modelling (Chapter 45). There are different types of big data analytics (descriptive, diagnostic, predictive, prescriptive, real-time, security, speech, voice, text and visual) and these are discussed in Chapter 10.

Big data analytics is used in a wide variety of fields of study and research including criminology (Chan and Bennett Moses, 2016); interdisciplinary research (Norder et al., 2018); hospitality and tourism (Xie and Fung, 2018); health and medicine (Wong, 2016); sports science (Zuccolotto et al., 2018); public policy and administration (Rogge et al., 2017); and business and management (Sanders, 2016). Examples of research projects that have used big data analytics include a study to assess the multidisciplinarity and interdisciplinarity of small group research (Norder et al., 2018) and scoring probabilities in basketball (Zuccolotto et al., 2018). Critiques of the use of big data analytics include a case study into corruption and government power in Australia (Galloway, 2017) and the impact of big data analytics on media production and distribution (Harper, 2017).

If you are interested in big data analytics for your research it is important that you become familiar with the different tools and software that are available to help you analyse data, the storage platforms that are available and sources of open data that can be used for research purposes. A wide range is available and the tools, platforms and sources that are relevant depend on your research topic, research question, methodology and the purpose of your research. Some examples are given below. There are also various free online university courses available for those who are new to big data analytics, which will help you to get to grips with tools and techniques (examples of courses can be found at www.edx.org/learn/big-data and www.coursera.org/specializations/big-data). An understanding of data mining (Chapter 12), data visualisation (Chapter 13), predictive modelling (Chapter 45) and machine learning (Chapter 29) is also important. Kitchin (2014), O'Neil (2017) and Richterich (2018) will help you to reflect on the use and abuse of big data analytics and think more about ethical issues associated with your research. If you are new to statistics for big data analytics, Anderson and Semmelroth (2015) provide a good introduction.

Questions for reflection

Epistemology, theoretical perspective and methodology

- Schöch (2013) believes that big data in the humanities is not the same as big data in the natural sciences or in economics. Do you agree with this? If so, how does the term 'big data' fit with your subject of study?

- Fuchs (2017: 47) argues that big data analytics is a form of digital positivism and that researchers should consider an alternative approach that 'combines critical social media theory, critical digital methods and critical-realist social media research ethics'. What relevance does this argument have to your own theoretical perspective and methodological standpoint?

- How can you ensure validity, reliability and authenticity of data? Is it possible to work with multiple sources of data to triangulate results (combining structured data analytics using statistics, models and algorithms to examine well-organised data with unstructured data analytics such as images, voice and text, for example)? Is it possible to combine big data with qualitative questions about perceptions, feelings and reasons for action, or are such approaches diametrically opposed? Mills (2017) provides an interesting discussion on these issues.

Ethics, morals and legal issues

- How do you intend to address the challenges and complexities surrounding individual privacy protection and privacy self-management? A comprehensive and illuminating discussion on these issues is provided by Baruh and Popescu (2017).

- How can you ensure that individuals have given informed consent for their data to be used? How can you ensure that datasets do not include information about individuals who have not given their consent for their data to be collected (in cases where data about individuals have been harvested from social networks, for example: see Chapter 53)?

- How might big data analytics perpetuate inequalities, lead to disadvantage or cause harm to users? O'Neil (2017) provides some interesting case studies and examples.

- Baruh and Popescu (2017: 592) state that 'the ideological power of the big data logic is to render the forces that shape decisions over individual lives both ubiquitous and unintelligible to the individual'. How can you address these concerns when adopting big data analytics?

- How will you address issues surrounding unequal big data (countries in which data systems are weak, restricted, non-existent or invisible to outsiders, for example)?

- How far can you trust big data analytics (and the data that are analysed)? How far can you trust inferences made and conclusions reached?

Practicalities

- What sources of data do you intend to use? How are you going to deal with massive amounts of data and with diverse and complex datasets that may be growing or expanding on a continual basis? A careful analysis of the quality, quantity, size, complexity and diversity of data will enable you to make decisions about analysis tools and software.

- How can you take account of variability (possible differences in data at different sub-set levels when using temporal and spatial data, for example)?

- Are data sources public or private access? How will you gain access? Are significant costs involved? Some examples of open datasets are listed below.

- Are you familiar with tools and techniques associated with big data analytics (whether, for example, they provide a complete or partial view of the data)? Are tools and software freely available within your organisation? Do you need additional training and, if so, is this provided by your organisation or available through tutorials or online courses? Some tools and software are listed below and in Chapters 10, 12 and 58: it is useful to visit some of the websites listed so that you can gain a deeper understanding of functions, capabilities and purpose.

- Do you have a good understanding of relevant analysis techniques? This can include, for example, categorical data analysis, cluster analysis (Chapter 5), exploratory data analysis, multivariate analysis, nonparametric analysis, psychometric analysis, regression and mixed-model analysis. Anderson and Semmelroth (2015) provide a useful introduction for those new to analysing big data.

Useful resources

Open datasets that are available for research purposes include (in alphabetical order):

- Australian open government data (https://data.gov.au);

- European Data Portal (www.europeandataportal.eu);

- Google Public Data Explorer (www.google.com/publicdata/directory);

- Open Data New Zealand (https://data.govt.nz);

- Open Science Data Cloud (www.opensciencedatacloud.org);

- UK Government's open datasets (https://data.gov.uk);
- US Government's open datasets (www.data.gov);
- World Bank Open Data (http://data.worldbank.org).

There are a variety of tools and software that can be used for big data analytics. These include (in alphabetical order):

- Jaspersoft BI Suite (www.jaspersoft.com);
- KNIME Analytics Platform (www.knime.com/knime-analytics-platform);
- OpenRefine (http://openrefine.org);
- Pentaho (www.hitachivantara.com/go/pentaho.html);
- Rapid Miner (https://rapidminer.com);
- Skytree (www.skytree.net);
- Talend (www.talend.com/products/talend-open-studio).

Additional tools and software that can be used for big data analytics are listed in Chapters 10, 12 and 58.

Storage (and organisation) platforms for big data include:

- Cassandra (http://cassandra.apache.org);
- CouchDB (http://couchdb.apache.org);
- Hadoop (http://hadoop.apache.org);
- mongoDB (www.mongodb.com);
- Neo4j (https://neo4j.com);
- Redis (https://redis.io).

The Analytics and Big Data Society (www.abdsociety.org) is a professional organisation in the US that supports the growth of analytics and big data through professional development. Event listings can be found on the website.

Rajapinta is a 'scientific association that advocates the social scientific study of ICT and ICT applications to social research in Finland'. There is a useful blog on the website by Juho Pääkkönen that discusses big data epistemology in the social sciences, and lists a number of papers that cover these issues from a variety of disciplines: https://rajapinta.co/2017/01/27/how-to-study-big-data-epistemology-in-the-social-sciences [accessed May 25, 2018]

The Linked Open Data cloud diagram can be accessed at https://lod-cloud.net. This image shows datasets that have been published in the Linked Data format (see Chapter 11 for more information).

The European Open Science Cloud (https://ec.europa.eu/research/open science/index.cfm?pg=open-science-cloud) aims to be up and running by 2020. It aims to make data FAIR (findable, accessible, interoperable and reusable) by bringing together and providing open access to global scientific data.

Key texts

Anderson, A. and Semmelroth, D. (2015) *Statistics for Big Data for Dummies*. Hoboken, NJ: John Wiley & Sons, Inc.

Baruh, L. and Popescu, M. (2017) 'Big Data Analytics and the Limits of Privacy Self-Management', *New Media & Society*, 19(4), 579–96, first published November 2, 2015, 10.1177/1461444815614001.

Chan, J. and Bennett Moses, L. (2016) 'Is Big Data Challenging Criminology?', *Theoretical Criminology*, 20(1), 21–39, first published May 19, 2015, 10.1177/1362480615586614.

Felt, M. (2016) 'Social Media and the Social Sciences: How Researchers Employ Big Data Analytics', *Big Data & Society*, first published April 29, 2016, 10.1177/2053951716645828.

Fuchs, C. (2017) 'From Digital Positivism and Administrative Big Data Analytics towards Critical Digital and Social Media Research!', *European Journal of Communication*, 32 (1), 37–49, first published January 8, 2017, 10.1177/0267323116682804.

Galloway, K. (2017) 'Big Data: A Case Study of Disruption and Government Power', *Alternative Law Journal*, 42(2), 89–95, first published September 18, 2017, 10.1177/1037969X17710612.

Harper, T. (2017) 'The Big Data Public and Its Problems: Big Data and the Structural Transformation of the Public Sphere', *New Media & Society*, 19(9), 1424–39, first published April 19, 2016, 10.1177/1461444816642167.

Kitchin, R. (2014) *The Data Revolution: Big Data, Open Data, Data Infrastructures and Their Consequences*. London: Sage.

Mills, K. (2017) 'What are the Threats and Potentials of Big Data for Qualitative Research?', *Qualitative Research*, first published November 30, 2017, 10.1177/1468794117743465.

Norder, K., Emich, K. and Sawhney, A. (2018) 'Evaluating the Interdisciplinary Mission of Small Group Research Using Computational Analytics', *Small Group Research*, 49(4), 391–408, first published February 22, 2018, 10.1177/1046496418755511.

O'Neil, C. (2017) *Weapons of Math Destruction: How Big Data Increases Inequality and Threatens Democracy*. London: Penguin.

Richterich, A. (2018) *The Big Data Agenda: Data Ethics and Critical Data Studies*. London: University of Westminster Press.

Rogge, N., Agasisti, T. and De Witte, K. (2017) 'Big Data and the Measurement of Public Organizations' Performance and Efficiency: The State-Of-The-Art', *Public Policy and Administration*, 32(4), 263–81, first published January 17, 2017, 10.1177/0952076716687355.

Sanders, R. (2016) 'How to Use Big Data to Drive Your Supply Chain', *California Management Review*, 58(3), 26–48, first published May 1, 2016, 10.1525/cmr.2016.58.3.26.

Schöch, C. (2013) 'Big? Smart? Clean? Messy? Data in the Humanities', *Journal of Digital Humanities*, 2(3), summer 2013, retrieved from http://journalofdigitalhumanities.org /2-3/big-smart-clean-messy-data-in-the-humanities.

Wong, K. (2016) 'A Novel Approach to Predict Core Residues on Cancer-Related DNA-Binding Domains', *Cancer Informatics*, first published June 2, 2016, 10.4137/CIN. S39366.

Xie, K. and Fung, S. K. (2018) 'The Effects of Reviewer Expertise on Future Reputation, Popularity, and Financial Performance of Hotels: Insights from Data-Analytics', *Journal of Hospitality & Tourism Research*, 42(8), 1187–209, first published December 4, 2017, 10.1177/1096348017744016.

Zuccolotto, P., Manisera, M. and Sandri, M. (2018) 'Big Data Analytics for Modeling Scoring Probability in Basketball: The Effect of Shooting under High-Pressure Conditions', *International Journal of Sports Science & Coaching*, 13(4), 569–89, first published November 6, 2017, 10.1177/1747954117737492.

CHAPTER 4

Business analytics

Overview

Business analytics refers to the process of examining an organisation's data to measure past and present performance to gain insight that will help with future planning and development. It includes various types of analytics including descriptive, predictive, prescriptive and diagnostic (see Chapter 10). Business analytics also refers to the set of skills, tools, software and statistical analysis techniques that are used to capture, explore, analyse and visualise data, along with models that can help to make predictions about future performance. It can help businesses to learn from past mistakes, refine their business strategy, develop and improve products or services, make comparisons with competitors, improve public relations and retain staff, for example. Business analytics can also draw on insights from customer analytics, which uses customer behaviour data to develop models and drive decisions concerning customer transactions, engagement, satisfaction, relation management, development and retention (this can also be referred to as customer relationships management analytics or CRM analytics). Bijmolt et al. (2010) provide a detailed discussion for those who are interested in analytics for customer engagement. A related method is marketing analytics, which seeks to use customer data to offer more value to customers, increase customer loyalty, enhance customer experience, drive marketing activities and improve company profits, for example. If you are interested in finding out more about marketing analytics, useful information is provided by Grigsby (2015) and Winston (2014), whereas a critical examination of marketing analytics is provided by Wedel and Kannan (2016).

Sources of data for business analytics can be internal (under the control of, and owned by, the business) and external (private or public data generated outside the business). Examples of data that are used for business analytics include:

- Internal data:
  - point of sale information;
  - transactional data (e.g. business purchases and shopping trends of customers);
  - finance data (e.g. cash flow reports and budget variance analyses);
  - customer relationship management (CRM) system information such as customer location and affiliation;
  - human resources data (see Chapter 22);
  - marketing data;
  - internal documents (e.g. word documents, PDFs and email);
  - digital and remote sensors (e.g. employee location and interaction patterns: see Chapter 49).

- External data:
  - public government data available from websites such as https://data.gov.uk, https://data.govt.nz, www.data.gov and https://data.gov.au;
  - customer surveys;
  - market research;
  - buyer behaviour trends;
  - social media analytics (see Chapter 52);
  - Google data sources such as Google Trends and Google Finance.

There are various digital tools and software packages available for business analytics and examples of these are listed below. It is useful to visit some of the websites listed so that you can get an idea of functions, capability, purpose and cost. Functions vary, depending on the software or tool that you choose but, in general, you will be able to:

- collect, prepare, organise and analyse data using a single tool;
- join diverse tables and charts from multiple sources;
- import, sort, filter, explore, modify and drill down data;

- combine the work of different analysts and share dashboards and metrics;
- run ad hoc and concurrent queries on changing data;
- search, manipulate and interact with data;
- detect anomalies and trends;
- create, deploy and share interactive visualisations and reports.

If you are interested in business analytics for your research project, a concise overview is provided by the Harvard Business Review (2018) and Evans (2017) provides a comprehensive guide to descriptive, predictive and prescriptive business analytics. Lall (2013) presents an interesting case study that illustrates how various business analytics techniques such as multivariate clustering, hierarchical clustering, regression, co-relation and factor analysis, can be used by telecom companies to understand customer segments and increase revenue. Mathias et al. (2011) provide a detailed case study of how business analytics can be used to support and enhance marketing, sales and business operations of companies requiring Risk Evaluation and Mitigation Strategy programmes to manage risks associated with drugs. Lam et al. (2017) discuss potential problems associated with analytics for frontline management, and illustrate how big data can be combined and integrated with small data to overcome such problems. A good understanding of data analytics (Chapter 10), big data analytics (Chapter 3), data mining (Chapter 12), data visualisation (Chapter 13) and predictive modelling (Chapter 45) are also important. Knowledge of social media analytics (Chapter 52) and web analytics (Chapter 58) will raise awareness of external sources that can be used for business analytics.

Questions for reflection

Epistemology, theoretical perspective and methodology

- What role does theory play in business analytics? Does analytics obviate the need for theory? A brief discussion that will help you to reflect on this issue is provided by Chiaburu (2016) who posits that theory, in analytics, 'is replaced by intuitions, hunches, direct observations, speculations, and real-time data'.

- How much can business analytics tell us about human behaviour? Can business analytics help us to identify behavioural causes? Are grounded-research, prior knowledge or hermeneutic sensibilities necessary for those undertaking business analytics? These issues are raised and discussed by

Baruh and Popescu (2017): their paper will help you to reflect on these questions.

- Lam et al. (2017) illustrate that big data should be combined with small data so as to avoid higher costs and reduced benefits that can occur when only big data are analysed. Do you agree with this argument and, if so, does your methodology allow for the combination and integration of big and small data when undertaking business analytics? Definitions of big, small and open data are provided in Chapter 3.

Ethics, morals and legal issues

- What value does business analytics add to businesses and society? How can it be of benefit?

- When engaging in business analytics, is there a mutually-beneficial trade in personal information, or does one party benefit above the other? If so, in what way? How can you address this issue in your research?

- How far can business analytics cause harm to individuals (privacy loss, use of e-scores that disadvantage individuals, hidden algorithms that perpetuate inequality or disadvantage, rational discrimination, risk identification and missed marketing opportunities, for example)? Baruh and Popescu (2017) discuss these issues and will help you to reflect on this question.

- How can you ensure that individuals have read, understood and agreed to privacy and data use policies? How will you address these issues when using open source data, for example?

- How are data handled within your chosen business? How are they procured, stored managed, used and disposed of? What data security procedures are in place? Who handles data within the business? What protections are in place to prevent the leak or misuse of sensitive data? Have all handlers signed confidentiality agreements? Are all relevant data governance and regulatory compliance commitments supported and enforced? Are data management policies credible, clear and transparent? More information about data protection and privacy can be obtained from the International Conference of Data Protection and Privacy Commissioners (details below).

- What impact might business analytics have on members of staff within a business? Will actual (or perceived) impact have an influence on engagement with and attitudes towards your research? More detailed questions for reflection about these issues can be found in Chapter 22.

Practicalities

- How can you ensure that you have access to all relevant data and that you have a rich variety of information that will yield meaningful insights? This is identified by MacMillan (2010) as one of six important keys to business analytics success. The other five are planning a caching strategy; adopting a common, multilingual business model; producing a single, common business model; establishing role-based security for handling highly sensitive information; and developing models collaboratively.

- When working with internal and external sources of data, who owns the data? Are data freely and equally available for your research? How might restrictions on access affect your research? Some businesses, for example, only collect and store data for internal use, whereas others collect and sell data for both internal and external use.

- How can you avoid making the wrong conclusions (patterns are not simply translated into conclusions, for example)? Avoiding over-reliance on software and cultivating skills of comprehension, interpretation and analysis will help you to avoid such problems.

- Do you intend to share your data once your project is complete (as supplementary information to a journal paper or through a data repository, for example)? The European Open Science Cloud (https://ec.europa.eu/research/openscience/index.cfm?pg=open-science-cloud) is set to be developed by 2020 to enable researchers, businesses and the public to access, share and reuse data.

Useful resources

Examples of tools and software that are available for business analytics at time of writing are given below (in alphabetical order). Other relevant tools and software can be found in Chapters 3, 10 and 22.

- Analance (https://analance.ducenit.com);
- Chartio (https://chartio.com);
- Clear Analytics (www.clearanalyticsbi.com);
- Domo (www.domo.com);
- GoodData (www.gooddata.com);
- iccube (www.iccube.com);

- InsightSqaured (www.insightsquared.com);

- Microsoft Power BI (https://powerbi.microsoft.com/en-us);

- MicroStrategy (www.microstrategy.com);

- OpenText suite of tools (www.opentext.com);

- Qlik Sense (www.qlik.com/us/products/qlik-sense);

- SAP Chrsytal Reports (www.crystalreports.com);

- TapClicks (www.tapclicks.com).

TIAS School for Business and Society, Netherlands, runs The Business Analytics LAB (www.tias.edu/en/knowledgeareas/area/business-analytics), which is 'an open platform for developing and sharing knowledge on topics that fuse (big) data, data-related analysis, and data-driven business'. Relevant blogs, articles and courses can be found on their website.

The International Conference of Data Protection and Privacy Commissioners (https://icdppc.org) is a global forum for data protection authorities. It 'seeks to provide leadership at international level in data protection and privacy'. The website provides details of conferences, events and documents relevant to data ethics.

Key texts

Baruh, L. and Popescu, M. (2017) 'Big Data Analytics and the Limits of Privacy Self-Management', *New Media & Society*, 19(4), 579–96, first published November 2, 2015, 10.1177/1461444815614001.

Bijmolt, T., Leeflang, P., Block, F., Eisenbeiss, M., Hardie, B., Lemmens, A. and Saffert, P. (2010) 'Analytics for Customer Engagement', *Journal of Service Research*, 13(3), 341–56, first published August 11, 2010, 10.1177/1094670510375603.

Chiaburu, D. (2016) 'Analytics: A Catalyst for Stagnant Science?', *Journal of Management Inquiry*, 25(1), 111–15, first published August 24, 2015, 10.1177/1056492615601342.

Evans, J. (2017) *Business Analytics*, 2nd edition. Harlow: Pearson Education Ltd.

Grigsby, M. (2015) *Marketing Analytics: A Practical Guide to Real Marketing Science*. London: Kogan Page.

Harvard Business Review. (2018) *HBR Guide to Data Analytics Basics for Managers*. Boston, MA: Harvard Business Review Press.

Lall, V. (2013) 'Application of Analytics to Help Telecom Companies Increase Revenue in Saturated Markets', *FIIB Business Review*, 2(3), 3–10, 10.1177/2455265820130301.

Lam, S., Sleep, S., Hennig-Thurau, T., Sridhar, S. and Saboo, A. (2017) 'Leveraging Frontline Employees' Small Data and Firm-Level Big Data in Frontline Management: An Absorptive Capacity Perspective', *Journal of Service Research*, 20(1), 12–28, first published November 17, 2016, 10.1177/1094670516679271.

Business analytics

MacMillan, L. (2010) 'Six Keys to Real-Time Analytics: How to Maximize Analytics Initiatives', *Business Information Review*, 27(3), 141–43, first published October 25, 2010, 10.1177/0266382110378219.

Mathias, A., Kessler, K. and Bhatnagar, S. (2011) 'Using Business Intelligence for Strategic Advantage in REMS', *Journal of Medical Marketing*, 11(1), 84–89, first published February 1, 2011, 10.1057/jmm.2010.40.

Wedel, M. and Kannan, P. (2016) 'Marketing Analytics for Data-Rich Environments', *Journal of Marketing*, 80(6), 97–121, first published November 1, 2016, 10.1509/jm.15.0413.

Winston, E. (2014) *Marketing Analytics: Data-Driven Techniques with Microsoft Excel.* Indianapolis, IN: John Wiley & Sons, Inc.

CHAPTER 5

Cluster analysis

Overview

Cluster analysis is a collection of data analysis techniques and procedures that help researchers to group, sort, classify and explore data to uncover or identify groups or structures within the data. Although cluster analysis has a long pre-digital history (see Everitt et al., 2011), the procedures and techniques have been included in this book because of technological developments in digital tools, software and visualisation packages, along with the increasing use of digital methods that use cluster analysis techniques and procedures, such as exploratory data mining (Chapter 12), data analytics (Chapter 10) and machine learning (Chapter 29).

Techniques and procedures that can be used for cluster analysis include (in alphabetical order):

- Biclustering: simultaneous clustering of the rows and columns of a data matrix. It can also be referred to as co-clustering, block clustering two-dimensional clustering or two-way clustering. See Dolnicar et al. (2012) for a discussion on this technique.

- Consensus clustering: a number of clusters from a dataset are examined to find a better fit. See Şenbabaoğlu et al. (2014) for an analysis and critique of this technique.

- Density-based spatial clustering: to filter out noise and outliers and discover clusters of arbitrary shape. See Li et al. (2015) for an example of this technique used together with mathematical morphology clustering.

- Fuzzy clustering: clustering data points have the potential to belong to multiple clusters (or more than one cluster). See Grekousis and

Hatzichristos (2013) for an example of a study that uses this clustering technique.

- Graph clustering: this can include between graph (clustering a set of graphs) and within-graph (clustering the nodes/edges of a single graph). See Hussain and Asghar (2018) for a discussion on graph-based clustering methods that are combined with other clustering procedures and techniques.

- Hierarchical clustering: a hierarchy of clusters is built using either a bottom-up approach (agglomerative) that combines clusters, or a top-down approach (divisive) that splits clusters. It is also known as nesting clustering. See Daie and Li (2016) for an example of a study that uses matrix-based hierarchical clustering.

- K-means clustering: an iterative partitioning method of cluster analysis where data are clustered based on their similarity (the number of clusters is known and specified within the parameters of the clustering algorithm). See Michalopoulou and Symeonaki (2017) for an example of a study that combines k-means clustering and fuzzy c-means clustering.

- Model-based clustering: a method in which certain models for clusters are used and best fit between data and the models is determined (the model defines clusters). Malsiner-Walli et al. (2018) discuss this clustering technique in their paper.

There are many other procedures, methods and techniques that are used for cluster analysis, including latent class cluster analysis, mixture model clustering, neural network-based clustering, overlapping clustering, partition clustering, partial clustering, pyramid clustering, simultaneous clustering, nearest neighbour clustering and mode (or mode seeking) clustering. More information about these methods, along with a wide variety of other techniques and procedures, can be obtained from Wierzchoń and Kłopotek (2018).

Cluster analysis is used by researchers working in a variety of disciplines and fields of study including economic development (Bach et al., 2018), psychiatry (Chai et al., 2013), health informatics (Loane et al., 2012), nursing research (Dunn et al., 2017), medicine (Ammann and Goodman, 2009), customer profiling (Ghuman and Singh Mann, 2018), education (Saenz et al., 2011) and welfare reform (Peck, 2005). Examples of research projects that have used cluster analysis include sensor-based research into movement of older residents between and within homes (Loane et al., 2012); a study into whether it is possible to identify male sexual offender subtypes (Ennis et al.,

2016); research into the impact of economic cost of violence on internationalisation (Bach et al., 2018); research into the factors associated with acute admission to inpatient psychiatric wards (Chai et al., 2013); and a study into student engagement in community colleges (Saenz et al., 2011).

If you are interested in finding out more about cluster analysis for your research, comprehensive information and guidance can be found in Everitt et al. (2011), who provide an interesting introduction to classifying and clustering from a number of different disciplinary perspectives, before going on to discuss specific methods, techniques and procedures. If you are a biologist, or interested in bioinformatics, Abu-Jamous et al. (2015) provide comprehensive information and advice about clustering methods, including hierarchical clustering, fuzzy clustering, neural network-based clustering, mixture model clustering and consensus clustering. If you are interested in cluster analysis in big data, Wierzchoń and Klopotek (2018) provide a detailed guide. More information about specific cluster analysis tools and techniques can be found in useful resources and keys texts, below. You might also find it of interest to obtain more information about data analytics (Chapter 10), data mining (Chapter 12), big data analytics (Chapter 3) and machine learning (Chapter 29).

Questions for reflection

Epistemology, theoretical perspective and methodology

- Do you intend to use cluster analysis as a standalone data analysis method, or use it together with other types of analysis (discriminant analysis, for example)?

- What is a cluster? How can you define a cluster (based on the nature of data and the desired results, for example)? How can you overcome ambiguities in definition?

- Do you intend to cluster cases or variables? This depends on your research question.

- How can you ensure that models do not over-fit data? Can generalisations be made?

- Do you have a good understanding of cluster validation techniques? This can include determining the cluster tendency, determining the correct number of clusters and comparing the results of a cluster analysis to externally known results, for example. Is it possible to develop objective criteria for correctness or clustering validity?

- How can you understand the goodness of a cluster? Measures that you could consider include similarity measures, dissimilarity measures, dissimilarity matrix, matching-based measures, entropy-based measures and pairwise measures, for example.

- How can you test the stability and robustness of clusters?

- How can you take account of cluster evolution and change?

- How can you quantify similarity (temporal, geographical and spatial, for example)?

Ethics, morals and legal issues

- How reliable are cluster analysis software packages, digital tools and algorithms? Are they free of faults and bias? How can you check reliability and accuracy (using additional software, for example)? Can you trust software, tools and algorithms?

- Do you have a good understanding of software licensing conditions and correct citing practices for any tools and software used for cluster analysis?

- Do you have a thorough understanding of issues surrounding privacy, security and the misuse or abuse of data? How can you ensure that your cluster analysis is performed within the bounds of privacy regulations and informed consent?

- How can you ensure that your analyses, or action taken as a result of your analyses, will not cause harm or distress, or affect people in adverse ways?

Practicalities

- Do you have a good understanding of the tools and software that are available for cluster analysis? Have you looked into issues such as extensibility, completeness, compatibility, performance and development (and problems that may arise with the release of new versions, for example)? Can tools and software be used with different programming languages on different operating systems? Can tools be used with several scripting languages across platforms? What tutorials are available?

- Do you have a good understanding of the stages, techniques or procedures required for your cluster analysis? This can include, for example:
  - choosing digital tools;
  - assessing visual approaches;
  - preparing and standardising data;
  - choosing cluster method(s);
  - forming or establishing clusters;
  - estimating the number of clusters;
  - interpreting clusters;
  - comparing clusters;
  - calculating or measuring distances;
  - linking clusters;
  - inspecting clusters;
  - validating clusters.

- How can you ensure that you provide enough information for others to understand how your cluster analysis was conducted and to allow for replication or evaluation of the quality of your research (clarity, transparency, rigour and sufficient detail, for example)?

Useful resources

There are a variety of tools and software packages available for cluster analysis. It is useful to visit some of the websites listed below (in alphabetical order) so that you can get an idea of purpose, functions, capabilities and costs.

- Apache Mahout (https://mahout.apache.org);
- DataMelt (https://jwork.org/dmelt);
- ELKI (https://elki-project.github.io);
- GNU Octave (www.gnu.org/software/octave);
- MATLAB (https://uk.mathworks.com/products/matlab.html);

- Orange (https://orange.biolab.si);

- Python-Scikit (http://scikit-learn.org/stable);

- R (https://cran.r-project.org);

- Rapid Miner (https://rapidminer.com);

- SPSS (www.ibm.com/analytics/spss-statistics-software);

- Weka (www.cs.waikato.ac.nz/ml/weka).

Key texts

Abu-Jamous, B., Fa, R. and Nandi, A. (2015) *Integrative Cluster Analysis in Bioinformatics.* Chichester: John Wiley & Sons, Ltd.

Ammann, L. and Goodman, S. (2009) 'Cluster Analysis for the Impact of Sickle Cell Disease on the Human Erythrocyte Protein Interactome', *Experimental Biology and Medicine*, 234(6), 703–11, first published June 1, 2009, 10.3181/0806-RM-211.

Bach, M., Dumičić, K., Jaković, B., Nikolić, H. and Žmuk, B. (2018) 'Exploring Impact of Economic Cost of Violence on Internationalization: Cluster Analysis Approach', *International Journal of Engineering Business Management*, first published May 10, 2018, 10.1177/1847979018771244.

Chai, Y., Wheeler, Z., Herbison, P., Gale, C. and Glue, P. (2013) 'Factors Associated with Hospitalization of Adult Psychiatric Patients: Cluster Analysis', *Australasian Psychiatry*, 21(2), 141–46, first published February 20, 2013, 10.1177/1039856213475682.

Daie, P. and Li, S. (2016) 'Matrix-Based Hierarchical Clustering for Developing Product Architecture', *Concurrent Engineering*, 24(2), 139–52, first published March 8, 10.1177/1063293X16635721.

Dolnicar, S., Kaiser, S., Lazarevski, K. and Leisch, F. (2012) 'Biclustering: Overcoming Data Dimensionality Problems in Market Segmentation', *Journal of Travel Research*, 51(1), 41–49, first published February 21, 2011, 10.1177/0047287510394192.

Dunn, H., Quinn, L., Corbridge, S., Eldeirawi, K., Kapella, M. and Collins, E. (2017) 'Cluster Analysis in Nursing Research: An Introduction, Historical Perspective, and Future Directions', *Western Journal of Nursing Research*, first published May 16, 2017, 10.1177/0193945917707705.

Ennis, L., Buro, K. and Jung, S. (2016) 'Identifying Male Sexual Offender Subtypes Using Cluster Analysis and the Static-2002R', *Sexual Abuse*, 28(5), 403–26, first published March 20, 2014, 10.1177/1079063214527481.

Everitt, B., Landau, S., Leese, M. and Stahl, D. (2011) *Cluster Analysis*, 5th edition. Chichester: John Wiley & Sons, Ltd.

Ghuman, M. and Singh Mann, B. (2018) 'Profiling Customers Based on Their Social Risk Perception: A Cluster Analysis Approach', *Metamorphosis*, 17(1), 41–52, first published May 8, 2018, 10.1177/0972622518768679.

Grekousis, G. and Hatzichristos, T. (2013) 'Fuzzy Clustering Analysis in Geomarketing Research', *Environment and Planning B: Urban Analytics and City Science*, 40(1), 95–116, first published January 1, 2013, 10.1068/b37137.

Hussain, I. and Asghar, S. (2018) 'DISC: Disambiguating Homonyms Using Graph Structural Clustering', *Journal of Information Science*, 44(6), 830–47, first published March 5, 2018, 10.1177/0165551518761011.

Li, S., Xu, Y., Sun, W., Yang, Z., Wang, L., Chai, M. and Wei, X. (2015) 'Driver Fixation Region Division–Oriented Clustering Method Based on the Density-Based Spatial Clustering of Applications with Noise and the Mathematical Morphology Clustering', *Advances in Mechanical Engineering*, first published October 29, 2015, 10.1177/1687814015612426.

Loane, J., O'Mullane, B., Bortz, B. and Knapp, R. (2012) 'Looking for Similarities in Movement between and within Homes Using Cluster Analysis', *Health Informatics Journal*, 18(3), 202–11, first published September 24, 2012, 10.1177/1460458212445501.

Malsiner-Walli, G., Pauger, D. and Wagner, H. (2018) 'Effect Fusion Using Model-Based Clustering', *Statistical Modelling*, 18(2), 175–96, first published January 15, 2018, 10.1177/1471082X17739058.

Michalopoulou, C. and Symeonaki, M. (2017) 'Improving Likert Scale Raw Scores Interpretability with K-Means Clustering', *Bulletin of Sociological Methodology/Bulletin De Méthodologie Sociologique*, 135(1), 101–09, first published June 26, 2017, 10.1177/0759106317710863.

Peck, L. (2005) 'Using Cluster Analysis in Program Evaluation', *Evaluation Review*, 29(2), 178–96, first published April 1, 2005, 10.1177/0193841X04266335.

Saenz, V., Hatch, D., Bukoski, B., Kim, S., Lee, K. and Valdez, P. (2011) 'Community College Student Engagement Patterns: A Typology Revealed Through Exploratory Cluster Analysis', *Community College Review*, 39(3), 235–67, first published August 23, 2011, 10.1177/0091552111416643.

Şenbabaoğlu, Y., Michailidis, G. and Li, J. (2014) 'Critical Limitations of Consensus Clustering in Class Discovery', *Scientific Reports*, 4(6207), first published August 27, 2014, 10.1038/srep06207.

Wierzchoń, S. and Klopotek, M. (2018) *Modern Algorithms of Cluster Analysis*. Cham: Springer.

CHAPTER 6

Coding and retrieval

Overview

The term 'coding and retrieval' refers to the processes involved in collecting together, sorting, marking and categorising qualitative and/or quantitative data for analysis. Coding has a long pre-digital history and is still undertaken manually (see Saldaña, 2016). However, it has been included in this book because it is used to transform data to facilitate computer-assisted data analysis (see Chapter 9 for information on computer-assisted qualitative data analysis software: CAQDAS). Also, coding and retrieval processes are used by those adopting a variety of digital research methods including ethno-mining (Chapter 18), digital ethnography (Chapter 14), online ethnography (Chapter 37), online interviews (Chapter 40), online focus groups (Chapter 39), online questionnaires (Chapter 43) and mobile phone interviews (Chapter 33).

Coding is the process of marking or categorising data (parts or passages of text, visual images, sound or numerical data) that cover the same issue, topic, theme or concept. These are marked with a name, letter, number or colour that is associated with a longer explanation of what the code means. There are different stages or levels of coding. These can include first-stage or first-level coding (main headings) and second-stage or second-level coding (subheadings), for example. Coding can be undertaken in a highly detailed and systematic way that enables researchers to count and quantify how many times the same issue, topic or theme is raised. Or it can be used to identify and categorize (classify or organize into categories) general themes that can guide further data collection, perhaps using thematic analysis techniques. In some cases researchers highlight concepts and categories and then concentrate on the conditions that have led to or influenced these concepts or categories.

Retrieval refers to the process of retrieving data that have already been coded (coded text or text associated with code patterns, for example: see below), and to the process of obtaining, collecting and gathering together data from computer systems, databases and datasets for exploration and analysis. This can be done using data mining techniques (Chapter 12), by presenting a set of criteria in a query using a specific query language (the demanded data are selected, or retrieved, by the Database Management System) or by utilising a platform's application programming interface (API), for example (see Murphy, 2017 for an example of retrieving data using Twitter's API). It can also include the retrieval of raw data presented in visual images (see Moeyaert et al., 2016 for an analysis and comparison of programs that use point and click procedures to retrieve raw data presented in graphs).

Coding and retrieval tools and software are available to enable researchers to perform both tasks quickly and efficiently on both structured data (numbers, dates and groups of words in databases and spreadsheets, for example) and unstructured data (documents, research memos, transcripts, web pages, images, video/audio and observations, for example). Although functions vary, depending on the chosen tool or software, in general you will be able to:

- create a master code list/code tree;
- develop structured codebooks;
- add code to text files;
- weight codes for mixed methods analysis;
- correct and change codes;
- select codes from lists;
- search and filter codes;
- sort code references by code group (in order);
- search for patterns among codes;
- retrieve coded text;
- retrieve lines of text associated with code patterns;
- compute frequency tables for codes;
- develop themes or concepts that can be represented numerically;
- co-ordinate and facilitate team-based coding and analysis (cloning and copying documents without codes for checking inter-coder reliability, for example);

- import all types of media (documents, images, audio, video, spreadsheets and questionnaires) for qualitative and quantitative data integration and analysis;

- export data in various formats for further analysis (qualitative or quantitative);

- generate outputs such as summaries, reports and visualisations.

Coding and retrieval methods are used by researchers in a wide variety of disciplines and fields of study including health and medicine (Mao and Bottorff, 2017; White et al., 2016), psychology (Murphy, 2017), the social sciences (Ose, 2016) and language and communication (Geisler, 2018). Examples of research projects that have used coding and retrieval methods include a study to explore factors influencing parents' decisions to adhere and persist with ADHD pharma-cotherapy in children (Ahmed et al., 2017); a study to help health professionals understand how people with multiple health conditions experience their illness (White et al., 2016); a study to explore the narratives of three self-identified female gamers, and to examine 'the use of qualitative data analysis software (QDAS) as a means of disrupting conventional research practices' (Le Blanc, 2017); and a study into unassisted smoking cessation among Chinese Canadian immigrants (Mao and Bottorff, 2017).

If you are interested in finding out more about coding and retrieval methods for your research, Saldaña (2016) provides a good starting point, covering both manual and CAQDAS coding, first and second cycle coding methods, coding and theory and common coding errors. If you are interested in finding out more about using Nvivo for your research, Bazeley and Jackson (2013) provide comprehensive advice, with chapters 4 and 5 of their book covering the basics of coding and creating a coding system. An understanding of CAQDAS (Chapter 9), data mining (Chapter 12), data conversion and collection (Chapter 11), data analytics (Chapter 10) and information retrieval (Chapter 23) will help you to think more about coding and retrieval methods for your research, as will the questions for reflection, given below.

Questions for reflection

Epistemology, theoretical perspective and methodology

- Have you made clear methodological decisions before you make coding and retrieval decisions? Methodological decisions guide and drive forward your research, enabling you to make decisions about whether to undertake inductive coding (open-ended and exploratory) or deductive coding (pre-defined codes), for example.

- Do you have a clear understanding of advantages and disadvantages of using software for coding and retrieval? Your response to this question depends on theoretical perspective and epistemological and methodological standpoint. For example, possible advantages can include enhancing transparency, improving replicability, illustrating convergence and highlighting difference, whereas possible disadvantages can include coding bias, error in retrieval (or partial retrieval) and incompatibility of format, tools or software.

- Do you have a clear analytical strategy and do you understand the analytical process through which you need to proceed to answer your research question? There are various stages of analysis in a research project: tools and software help with coding and retrieval, but what other analytical strategies and processes are required?

Ethics, morals and legal issues

- When using existing databases or datasets, can you trust the data? What do you know about the content of data (shape, form, volume, topic, collection methods and collection context, for example)? What procedures do you need to adopt to evaluate secondary sources (a consideration of assumptions, provenance and limitations, for example)?

- Can data be used ethically and legally for your intended purpose (and for reuse)? This is of particular importance when data are retrieved from unknown sources, or when using raw qualitative data, for example. Issues of informed consent and sharing, collecting and encrypting data need to be addressed. More information about these issues can be obtained from the UK Data Service (www.ukdataservice.ac.uk) and from the UK Data Archive (http://data-archive.ac.uk). You will also need to ensure that you understand and adhere to all database and software licences, and understand how these might vary between countries. More information about data analysis software licensing can be obtained from Chapter 9.

- What are the implications for traditional citation practices of rapid and powerful retrieval from primary and secondary sources? How can you ensure correct, systematic and detailed citation?

Practicalities

- What tools and software do you intend to use for coding and retrieval? What technical support, tutorials or online training is available? Geisler (2018: 217) points out that making choices 'requires much more than a simple review of software features using standard criteria such as cost and simplicity'. She goes on to suggest that an important consideration when making choices is 'language complexity', which is discussed in her paper.

- What methods, other than CAQDAS (Chapter 9), can you use to sort and structure large amounts of unstructured data? Ose (2016) presents a simple method that uses Microsoft Excel and Microsoft Word for those who find CAQDAS too 'advanced and sophisticated'.

- How can you be sure that you are coding and retrieving data from real users, rather than from automated bots, paid-for followers/users or marketing companies, for example? This is of particular relevance when working with social media data.

- When using APIs to retrieve data, what limitations are in place that may have an influence on retrieval processes and sharing results? Murphy (2017) discusses these limitations in terms of:
  - API restrictions on searches;
  - rate limiting that restricts the number of requests;
  - problems that can occur with changes introduced to online platforms as they evolve;
  - issues surrounding restrictions on publication.

Useful resources

There are a wide variety of digital tools and software packages available for coding and retrieval. Some of these are open source and free to use, whereas others have significant costs involved. Tools and software vary enormously and your choice will depend on your research topic, needs and purpose (viewing, coding, retrieving and visualising text; collecting, organising, transcribing, coding and analysing audio/video files; or retrieval/extraction of data points from image files, for example). The following examples are available at time of writing and are presented in alphabetical order.

- Atlas.ti (https://atlasti.com);
- DataThief III (https://datathief.org);
- Dedoose (www.dedoose.com);
- Excel (https://products.office.com/en-gb/excel);
- HyperResearch (www.researchware.com);
- MAXQDA (www.maxqda.com);
- NVivo (www.qsrinternational.com/nvivo/home);
- QDA Miner (https://provalisresearch.com/products/qualitative-data-analysis-software);
- Quirkos (www.quirkos.com/index.html);
- Saturate (www.saturateapp.com);
- Sonal (www.sonal-info.com);
- The Ethnograph 6.0 (www.qualisresearch.com);
- WebPlotDigitizer (https://automeris.io/WebPlotDigitizer).

The E-Qual app has been developed by Hilary McDermott and her team at Loughborough University. It is a free interactive iPad app available on the App Store. The aim is to demonstrate the basic concept of coding qualitative data: students are able to code a section of data and compare their coding with that of an experienced researcher. More information about the E-Qual app can be obtained from the Loughborough University teaching and learning blog: http://blog.lboro.ac.uk/teaching-learning [accessed July 10, 2018].

Key texts

Ahmed, R., Borst, J., Wei, Y. and Aslani, P. (2017) 'Parents' Perspectives about Factors Influencing Adherence to Pharmacotherapy for ADHD', *Journal of Attention Disorders*, 21(2), 91–99, first published August 30, 2013, 10.1177/1087054713499231.
Bazeley, P. and Jackson, K. (2013) *Qualitative Data Analysis with Nvivo*, 2nd edition. London: Sage.
Geisler, C. (2018) 'Coding for Language Complexity: The Interplay among Methodological Commitments, Tools, and Workflow in Writing Research', *Written Communication*, 35(2), 215–49, first published March 22, 2018, 10.1177/0741088317748590.
Le Blanc, A. (2017) 'Disruptive Meaning-Making: Qualitative Data Analysis Software and Postmodern Pastiche', *Qualitative Inquiry*, 23(10), 789–98, first published October 3, 2017, 10.1177/1077800417731087.

Coding and retrieval

Mao, A. and Bottorff, J. (2017) 'A Qualitative Study on Unassisted Smoking Cessation among Chinese Canadian Immigrants', *American Journal of Men's Health*, 11(6), 1703–12, first published January 27, 2016, 10.1177/1557988315627140.

Moeyaert, M., Maggin, D. and Verkuilen, J. (2016) 'Reliability, Validity, and Usability of Data Extraction Programs for Single-Case Research Designs', *Behavior Modification*, 40 (6), 874–900, first published April 28, 2016, 10.1177/0145445516645763.

Murphy, S. (2017) 'A Hands-On Guide to Conducting Psychological Research on Twitter', *Social Psychological and Personality Science*, 8(4), 396–412, first published June 15, 2017, 10.1177/1948550617697178.

Ose, S. (2016) 'Using Excel and Word to Structure Qualitative Data', *Journal of Applied Social Science*, 10(2), 147–62, first published August 19, 2016, 10.1177/1936724416664948.

Saldaña, J. (2016) *The Coding Manual for Qualitative Researchers*, 3rd edition. London: Sage.

White, C., Lentin, P. and Farnworth, L. (2016) 'Multimorbidity and the Process of Living with Ongoing Illness', *Chronic Illness*, 12(2), 83–97, first published October 13, 2015, 10.1177/1742395315610392.

CHAPTER 7

Computer modelling and simulation

Overview

Computer modelling and simulation (or modelling and simulation: M&S) refers to the process of building and manipulating computer-aided representations of artificial or natural systems or phenomena. A model is the computer-based mathematical, graphical or algorithmic representation, whereas a simulation is the process of running a model to describe, explain, analyse, predict or optimise behaviour within the system. Computer modelling and simulation can be used in situations where it is impossible, difficult, costly, unsafe, too risky or unethical to conduct real-world experiments. The process can help to solve real-world problems, advance knowledge, counter misleading or false information, provide useful insights, inform policy or decision-making and lead to improvements in health, education and training programmes, for example.

There are three broad classes of computer model and simulation: microscale (each individual or object is described with its particular attributes and behaviour); macroscale (an amalgamation or aggregation of details into categories); and multiscale (the simultaneous use of multiple models at different scales). There are different types of model and simulation within these broad classes, examples of which include (in alphabetical order):

- Agent-based models and simulations: rules that govern the behaviour of entities are specified, with their behaviour and interaction determining the transmission or flow between states. They are dynamic models that simulate the behaviour of a system over time and are used for spotting trends and patterns, and for making predictions. Agent-based modelling and simulation has become increasingly popular in recent years, and is now used in a wide variety of disciplines. Therefore, a separate entry has been included in this book (see Chapter 1).

- Compartmental models and simulations: entities with similar characteristics are grouped into compartments (or classes), with the modeller specifying how they flow from one compartment to another. Models can be single-compartment models or multi-compartment models. Equations describing this flow can be deterministic (the same starting conditions will always give the same results) or stochastic (random factors are involved in the model and variability of model outcomes can be measured).

- Decision-analytical models and simulations: these enable modellers to synthesise two or more sources of evidence and visualise a sequence of events that can occur following alternative decisions or actions. They enable modellers to make decisions based on the outcomes of different scenarios and/or choose the path of optimum outcome.

- Discrete event models and simulations: in these models the fundamental components are states (a model value that is changed by events) and events (something that happens at an instant in time). The time of the next event can be calculated after each event with the simulation moving on to that time. It is a discreet event because it occurs instantaneously (as opposed to continuously or gradually) in time.

- Fixed-time models and simulations: these differ from the above in that the simulation is updated in fixed time periods. It enables modellers to determine which events have occurred since the last time-step and update the state of the model.

Computer modelling and simulation is used in a number of disciplines and fields of study including systems and software engineering, geography, biology, climatology, meteorology, geology, hydrology, physics, health and medicine, sports science, the social sciences, operations management, business, military science and defence studies. Examples of computer modelling and simulation research include a model of driver behaviour that integrates psychological and physiological factors affecting driver performance (Elkosantini and Darmoul, 2018); a model to develop a greenhouse control system in the real-world (Kim et al., 2017); research into the suitability of specific modelling techniques in meeting the competing metrics for sustainable operations management (Fakhimi et al., 2016); modelling and simulating to optimise the operation and maintenance strategy for offshore wind turbines (Santos et al., 2015); modelling and simulation of active noise control in a small room (Aslan and Paurobally, 2018);

a modelling and simulation framework for value-based healthcare systems (Traoré et al., 2018); modelling and simulation of landmine and improvised explosive device detection (Johnson and Ali, 2015) and modelling pedestrian behaviour (Al-Habashna and Wainer, 2016).

If you are interested in finding out more about computer modelling and simulation, a good starting point is Winsberg (2010). This book discusses philosophical, theoretical and methodological issues pertaining to the use of computer simulations in science, addressing questions such as how much do we know about the limits and possibilities of computer simulation? How do simulations compare to traditional experiments? Are they reliable? Edmonds and Meyer (2013) provide interesting reading for those approaching their work from the social sciences, with a collection of papers presented under the headings of methodology, mechanisms and applications. If you are interested in the 'problem solving methodology called modelling and simulation', Birta and Arbez (2013) provide a comprehensive and thought-provoking discussion and include sections on discrete event dynamic systems and continuous time dynamic system domains. Al-Begain and Bargiela (2016) provide a collection of conference papers from the European Council for Modelling and Simulation, which gives broad coverage of the topic for those approaching from a variety of disciplines. If you are interested in agent-based modelling and simulation more information is provided in Chapter 1 and if you are interested in predictive modelling more information is provided in Chapter 45.

Questions for reflection

Epistemology, theoretical perspective and methodology

- What theoretical basis or construct do you intend to use to formulate your model? Winsberg (2010) and Birta and Arbez (2013) will help you to address this question.

- Vallverdú (2014: 7) points out that providing an exact definition of computer modelling and simulation is difficult as definitions vary, according to discipline and area of expertise. Can you produce a clear and distinct definition, based on your subject of study and your research question?

- Are you interested in quantitative modelling and simulation, qualitative modelling and simulation or combined/integrated/mixed modelling and simulation?

- Can modelling and simulation 'be used in ways that are complementary to existing interpretative, heuristic and dialogic approaches' (Millington and Wainwright, 2017: 69)?

- Vallverdú (2014: 11) argues that 'simulations are not bad quality copies, but special cases of the construction of meaning' and that, if this argument is accepted 'we must conclude that they are as real as other entities we might analyze'. Do you agree with this argument and what relevance might it have to your research?

- To what extent is a model and simulation an approximation of what it purports to represent or emulate?

Ethics, morals and legal issues

- How can you ensure thorough and unbiased interpretation and evaluation of models and simulations?

- How can you ensure modelling and simulation activities will not cause harm to individuals, animals or the environment, for present and future generations? How can you ensure that action taken as a result of modelling and simulation will not have a detrimental impact on others?

- How can you ensure that you are sensitive to the needs and values of end users (those who may be affected by models, but who have no influence, control and/or understanding of modelling processes)?

Practicalities

- What, exactly, do you want to model and simulate? Is dedicated software available, or will you need to undertake programming (or re-programming)? Do you have a good understanding of software and tools that are available and the required skills and understanding to use the tools?

- What methods of modelling and simulation do you intend to use? For example, Kim et al. (2017: 579) discuss the differences between data modelling and simulation modelling. Also, have you looked into different computational paradigms, or mathematical frameworks, for your research? This can include cellular automata (CA), genetic algorithms, neural networks, swarm intelligence and Monte Carlo method, for example.

- What system design assumptions have been made? Have assumptions been made to reduce simulation complexity? If so, are these assumptions reasonable? Can assumptions invalidate the results of the model?

- What are possible limitations of modelling and simulation techniques? How will you recognise, acknowledge and report such issues?

- Padilla et al. (2018) note that one of the key findings from their research is the identification of an over-reliance on informal methods for conceptualisation and verification in modelling and simulation. How can you ensure that you undertake 'rigorous and repeatable conceptualization, verification and validation processes' when modelling and simulating (Padilla et al., 2018: 500)?

Useful resources

There are various digital tools and software packages available for researchers who are interested in computer modelling and simulation. Some of these are aimed at specific fields of study and occupations (health care, human resources, economics, management, transportation, mining and supply chains, for example) and provide specific tools and techniques (process mapping, discreet event simulation, linear static analyses, modal analysis and 3D model building, for example). Others are aimed at the general market and provide an array of tools. Examples available at time of writing are given below, in alphabetical order.

- AnyLogic simulation software (www.anylogic.com);

- Insight Maker (https://insightmaker.com);

- MATLAB (https://uk.mathworks.com/products/matlab.html) and Simulink (https://uk.mathworks.com/products/simulink.html);

- Minsky (https://sourceforge.net/projects/minsky);

- Powersim Studio (www.powersim.com);

- Scilab (www.scilab.org);

- Simio Simulation Software (www.simio.com/software/simulation-software.php);

- Simmer (https://r-simmer.org);

- Simprocess (http://simprocess.com);

- SimPy (https://pypi.org/project/simpy);
- SmartPLS (www.smartpls.com).

SIMULATION is a peer-reviewed journal covering subjects such as

> the modelling and simulation of: computer networking and communications, high performance computers, real-time systems, mobile and intelligent agents, simulation software, and language design, system engineering and design, aerospace, traffic systems, microelectronics, robotics, mechatronics, and air traffic and chemistry, physics, biology, medicine, biomedicine, sociology, and cognition.

The Society for Modeling and Simulation International (http://scs.org) is 'dedicated to advancing the use of modeling & simulation to solve real-world problems; devoted to the advancement of simulation and allied computer arts in all fields; and committed to facilitating communication among professionals in the field of simulation'. You can find information about conferences, publications and membership on their website, along with a Code of Conduct for 'simulationists'.

The European Council for Modelling and Simulation (www.scs-europe.net) is 'an independent forum of European academics and practitioners dedicated to research, development, and applications of modelling and simulation'. You can find information about conferences, workshops and publications on their website.

Key texts

Al-Begain, K. and Bargiela, A. (eds.) (2016) *Seminal Contributions to Modelling and Simulation: 30 Years of the European Council of Modelling and Simulation.* Cham: Springer.

Al-Habashna, A. and Wainer, G. (2016), 'Modeling Pedestrian Behavior with Cell-DEVS: Theory and Applications', *SIMULATION*, 92(2), 117–39, first published December 31, 2015, 10.1177/0037549715624146.

Aslan, F. and Paurobally, R. (2018), 'Modelling and Simulation of Active Noise Control in a Small Room', *Journal of Vibration and Control*, 24(3), 607–18, first published May 6, 2016, 10.1177/1077546316647572.

Birta, L. and Arbez, G. (2013) *Modelling and Simulation: Exploring Dynamic System Behaviour*, 2nd edition. London: Springer.

Edmonds, B. and Meyer, R. (eds.) (2013) *Simulating Social Complexity: A Handbook.* Heidelberg: Springer.

Elkosantini, S. and Darmoul, S. (2018) 'A New Framework for the Computer Modelling and Simulation of Car Driver Behavior', *SIMULATION*, first published January 24, 2018, 10.1177/0037549717748747.

Fakhimi, M., Mustafee, N. and Stergioulas, L. (2016), 'An Investigation into Modeling and Simulation Approaches for Sustainable Operations Management', *SIMULATION*, 92 (10), 907–19, first published August 30, 2016, 10.1177/0037549716662533.

Johnson, D. and Ali, A. (2015), 'Modeling and Simulation of Landmine and Improvised Explosive Device Detection with Multiple Loops', *The Journal of Defense Modeling and Simulation*, 12(3), 257–71, first published September 11, 2012, 10.1177/1548512912457886.

Kim, B., Kang, B., Choi, S. and Kim, T. (2017), 'Data Modeling versus Simulation Modeling in the Big Data Era: Case Study of a Greenhouse Control System', *SIMULATION*, 93(7), 579–94, first published June 9, 2017, 10.1177/0037549717692866.

Millington, J. and Wainwright, J. (2017), 'Mixed Qualitative-Simulation Methods: Understanding Geography through Thick and Thin', *Progress in Human Geography*, 41(1), 68–88, first published February 4, 2016, 10.1177/0309132515627021.

Padilla, J., Diallo, S., Lynch, C. and Gore, R. (2018), 'Observations on the Practice and Profession of Modeling and Simulation: A Survey Approach', *SIMULATION*, 94(6), 493–506, first published October 24, 2017, 10.1177/0037549717737159.

Santos, F., Teixeira, Â. and Soares, C. (2015), 'Modelling and Simulation of the Operation and Maintenance of Offshore Wind Turbines', *Proceedings of the Institution of Mechanical Engineers, Part O: Journal of Risk and Reliability*, 229(5), 385–93, first published June 8, 2015, 10.1177/1748006X15589209.

Traoré, M., Zacharewicz, G., Duboz, R. and Zeigler, B. (2018) 'Modeling and Simulation Framework for Value-Based Healthcare Systems', *SIMULATION*, first published June 4, 2018, 10.1177/0037549718776765.

Vallverdú, J. (2014) 'What are Simulations? An Epistemological Approach', *Procedia Technology*, 13, 6–15, open access, 10.1016/j.protcy.2014.02.003.

Winsberg, E. (2010) *Science in the Age of Computer Simulation*. Chicago, IL: University of Chicago Press.

CHAPTER 8

Computer-assisted interviewing

Overview

Computer-assisted interviewing is a term that is used to describe interviews that take place with the aid of a computer. This research method is used mostly in quantitative approaches, such as survey research, to collect data about thoughts, opinions, attitudes and behaviour. It enables researchers to undertake exploratory research, test theory and understand and describe a particular phenomenon, for example. Interviewees are asked a set of pre-defined questions (interviewer administered or self-administered) using methods such as questionnaires or structured interviews, assisted by computers. The method is sometimes used for qualitative approaches and in mixed methods research through using semi-structured or unstructured interview techniques, or by using both closed and open-ended questions in question-naires, for example. Computers assist in the research process by enabling a number of modes of communication such as voice, text or video (human or automated: see Chapter 51); helping with the construction and administration of interview questions; assisting with the input, recording and real-time analysis of answers (enabling further questions to be asked or questions to be skipped, for example); and enabling input errors, missing data or incon-sistencies to be detected and rectified as the interview proceeds.

Computer-assisted interviewing enables researchers to reach a wide audi-ence, across geographical boundaries and in places that may be hard to access. Software for laptops, mobiles and smartphones enables researchers to conduct interviews in the field where live internet access is not available. Devices can record survey answers offline, uploading them to a Web server when a connection becomes available. Devices can also use voice capture software to record answers in participant's own voices. This is useful when open questions have been asked or when it is difficult for an interviewer to input answers quickly enough. Recordings can be played back at a later stage for transcription, coding and analysis.

There are various types of computer-assisted interviewing. These include:

- computer-assisted personal interviewing (CAPI), which utilises computers to help interviewers conduct face-to-face interviews (Watson and Wilkins, 2015);

- mobile computer-assisted personal interviewing (mCAPI), which is a mobile form of CAPI, where an interviewer is present and uses mobile devices such as laptops, tablets or smartphones as part of the interviewing process (Poynter et al., 2014),

- computer-assisted self-interviewing (CASI) where interviewees complete the questionnaire themselves using PCs, laptops, tablets or mobile phones;

  ○ audio-CASI, or ACASI relates to questionnaires administered by phone (Couper et al., 2016);

  ○ video-CASI, or VCASI, relates to questionnaires that appear on screen;

- computer-assisted telephone interviewing (CATI), which is used to describe interviews that are administered over the phone (landline or mobile) by an interviewer who reads the questionnaire and inputs answers direct into the computer (Snidero et al., 2009);

- computer-assisted web interviewing (CAWI) where interviewees complete a questionnaire that is embedded into a website or blog, sent out as a bulk email to targeted groups (as an attachment or unique link), shared on social networks or as a QR Code (Quick Response Code) for respondents to scan with their mobile devices, for example (Chapter 43);

- online interviewing that use computer-mediated communication technology and is carried out over the internet (synchronous interviews that take place in real-time by video, webcam, text or instant messaging and asynchronous interviews that are not in real-time and can take place over an extended period of time by email, pre-recorded video, microblogs, blogs, wikis or discussion boards, for example: see Chapter 40).

Computer-assisted interviewing is used in a wide variety of disciplines including marketing and business, sociology, education, politics, health and medicine and geography. Examples of studies that have assessed and/or critiqued different types of computer-assisted interviewing include an assessment of the differences between live and digital interviewing in CASI, CAPI and CATI (Bronner and Kuijlen, 2007); the impact of CAPI on length of interview (Watson and Wilkins, 2015); an evaluation of the use of text-to-speech technology for ACASI (Couper et al., 2016); and a comparison of mobile phone-assisted personal interviewing and traditional paper surveys (van Heerden et al., 2014).

If you are interested in using computer-assisted interviewing as a research method it is important that you have a thorough understanding of interview techniques, questionnaire design and survey research. Comprehensive information can be found in Brinkmann and Kvale (2015), Fink (2017) and Gillham (2007). It is also important to find out more about sampling techniques: Blair and Blair (2015), Thompson (2012) and Daniel (2012) all provide useful information. Decisions about your research will also be aided by a good understanding of mobile phone interviews (Chapter 33), mobile phone surveys (Chapter 34), online interviews (Chapter 40), online questionnaires (Chapter 43) and smartphone questionnaires (Chapter 51).

Questions for reflection

Epistemology, theoretical perspective and methodology

- Do you understand how epistemological and theoretical assumptions have implications for methodology and methods? Chapter 40 discusses the different assumptions underpinning interviews conducted by those approaching their work from objective and subjective standpoints, for example.

- Do you intend to adopt a quantitative, qualitative or mixed methods approach? In quantitative research where the intention is to explain, predict or generalise to the whole research population, you must ensure that all people within the study population have a specifiable chance of being selected, which involves probability sampling. Since the sample serves as a model for the whole research population, it must be an accurate representation of this population. How will you ensure that this is the case? In qualitative research, however, description rather than generalisation is the goal: therefore, non-probability samples can be used (the sample is selected based on knowledge of the research problem). Comprehensive information about sampling techniques can be obtained from Blair and Blair (2015), Thompson (2012) and Daniel (2012).

- Which type of computer-assisted interviewing will be most appropriate, given your chosen topic and methodology? If your research is on a sensitive topic or covers illegal or undesirable behaviour, for example, would interviewees be reluctant to share information with an interviewer? Would CASI be more appropriate? Bronner and Kuijlen (2007) provide insight into these issues.

Ethics, morals and legal issues

- How will you address issues of respect and disclosure when you make contact for your interview? This could include:

  o identification of the researcher/person making the contact (displaying a caller number if using a mobile so that the interviewee can call to verify, if required);

  o identification of the university/funding body/organisation for which the researcher works;

  o information about the purpose of the contact and/or the nature of the research;

  o information about data collection processes, storage, management and sharing;

  o assurances that participation is voluntary;

  o explanation of what is meant by informed consent and obtaining informed consent (see Chapter 40);

  o guarantee of confidentiality and anonymity (only if this can be guaranteed: see Chapter 33);

  o information about financial compensation if costs are involved for the interviewee;

  o respect for time of contact (avoiding children's bed time when contacting parents, or avoiding working hours when contacting working people, for example);

  o polite termination if it is clear that the person does not wish to take part, cannot take part, is a vulnerable person or is a child (where permission has not been sought from a parent or guardian).

- How can you reduce interviewee burden, such as that produced by length of interview or complexity of questions? Do different types of computer-assisted interview have an influence on length of interview and the type and structure of questions? Watson and Wilkins (2015) will help you to reflect on appropriate length for interviews.

Practicalities

- How do you intend to overcome potential technological problems (battery failure, loss of internet connection or equipment malfunction, for example)? What back-up technology or contingency plans will be required?

- If interviewees are to answer their own questions, is the software that you intend to use responsive and adaptive, so that it is suitable for all mobile devices, laptops and PCs?

- What hosting services are available and are they in compliance with all regulations for data security? Will hosting take care of hardware, security, maintenance and software upgrades?

- Are you intending to conduct interviews yourself, put together an interview team or use a commercial survey/interview organisation to carry out your interviews? If you intend to use a commercial company, will they adhere to all rules and regulations involving contact (especially if using automated dialling), security and privacy? Böhme and Stöhr (2014) provide an interesting discussion about choosing interviewers and their workload.

- How will you decide on your sample size? This is the number of individuals or groups that are required to respond to achieve the required level of accuracy and refers to the final number of responses collected, rather than the number of individuals selected to provide responses. When sample sizes are calculated, the potential response rate will need to be taken into consideration, although response rates cannot be predicted. Methods of contacting interviewees will have an influence on response rates.

- When using CATI, how might interviewees who are contacted initially differ from those contacted by callback? Vicente and Marques (2017) discuss these issues in-depth and Vicente (2017) considers the effects of call attempts and time periods on call outcomes.

Useful resources

There are a wide variety of computer-assisted interviewing digital tools and software packages available for a diverse range of applications. A small selection of those available at time of writing is listed below.

- ACASI, CASI and CAPI software for research studies, developed by academic members of staff in the School of Medicine, Tufts University (http://acasi.tufts.edu);

- CATI, CAWI and CAPI software for market and social research, used by universities, public institutions and research institutes (www.softwarecati.com/eng);

- customer experience surveys, including tactical surveys, lifecycle surveys and relationship surveys through a variety of channels (www.confirmit.com);

- phone surveys for educational research and course evaluations (www.survoxinc.com/industries/education-survey-solutions);

- Rotator Survey 'Mobile interview' for applying CAPI questionnaires in the field (www.rotatorsurvey.com/en/rotator_mobile.html);

- Voxco Mobile Offline CAPI software for conducting personal interviews offline (www.voxco.com/survey-solutions/voxco-mobile-offline);

- web-based diagnostic tools for behavioural health research, administered by clinician interview or by self-report (https://telesage.com).

The Research Software Central database developed by Meaning Ltd (www.meaning.uk.com/your-resources/software-database) is a free database that enables you to search for market research and survey software by category, platform, name and/or manufacturer, and provides links to each website for further information about functions, hosting and costs.

The Association for Survey Computing (www.asc.org.uk) is a membership organisation for those using software and technology for survey research and statistics. Free access to conference proceedings and relevant journal papers is available on the website [accessed April 11, 2018].

Key texts

Blair, E. and Blair, J. (2015) *Applied Survey Sampling*. Thousand Oaks, CA: Sage.
Böhme, M. and Stöhr, T. (2014) 'Household Interview Duration Analysis in CAPI Survey Management', *Field Methods*, 26(4), 390–405, first published April 21, 2014, 10.1177/1525822X14528450.

Brinkmann, S. and Kvale, S. (2015) *InterViews: Learning the Craft of Qualitative Research Interviewing*, 3rd edition. Thousand Oaks, CA: Sage.

Bronner, F. and Kuijlen, T. (2007) 'The Live or Digital Interviewer – A Comparison between CASI, CAPI and CATI with respect to Differences in Response Behaviour', *International Journal of Market Research*, 49(2), 167–90, first published March 1, 2007, 10.1177/147078530704900204.

Couper, M., Berglund, P., Kirgis, N. and Buageila, S. (2016) 'Using Text-To-Speech (TTS) for Audio Computer-Assisted Self-Interviewing (ACASI)', *Field Methods*, 28(2), 95–111, first published December 29, 2014, 10.1177/1525822X14562350.

Daniel, J. (2012) *Sampling Essentials: Practical Guidelines for Making Sampling Choices.* Thousand Oaks, CA: Sage.

Fink, A. (2017) *How to Conduct Surveys: A Step-by-Step Guide*, 6th edition. Thousand Oaks, CA: Sage.

Gillham, B. (2007) *Developing a Questionnaire*, 2nd edition. London: Continuum.

Poynter, R., Williams, N. and York, S. (2014) *The Handbook of Mobile Market Research – Tools and Techniques for Market Researchers.* Chichester: Wiley.

Snidero, S., Zobec, F., Berchialla, P., Corradetti, R. and Gregori, D. (2009) 'Question Order and Interviewer Effects in CATI Scale-Up Surveys', *Sociological Methods & Research*, 38(2), 287–305, first published October 8, 2009, 10.1177/0049124109346163.

Thompson, S. (2012) *Sampling*, 3rd edition. Hoboken, NJ: John Wiley & Sons.

van Heerden, A., Norris, S., Tollman, S. and Richter, L. (2014) 'Collecting Health Research Data: Comparing Mobile Phone-Assisted Personal Interviewing to Paper-And-Pen Data Collection', *Field Methods*, 26(4), 307–21, first published January 27, 2014, 10.1177/1525822X13518184.

Vicente, P. (2017) 'Exploring Fieldwork Effects in a Mobile CATI Survey', *International Journal of Market Research*, 59(1), 57–76, first published January 1, 2017, 10.2501/IJMR-2016-054.

Vicente, P. and Marques, C. (2017) 'Do Initial Respondents Differ from Callback Respondents? Lessons from a Mobile CATI Survey', *Social Science Computer Review*, 35(5), 606–18, first published June 28, 2016, 10.1177/0894439316655975.

Watson, N. and Wilkins, R. (2015) 'Design Matters: The Impact of CAPI on Interview Length', *Field Methods*, 27(3), 244–64, first published May 6, 2015, 10.1177/1525822X15584538.

CHAPTER 9

Computer-assisted qualitative data analysis software

Overview

Computer-assisted qualitative data analysis software or CAQDAS is a term (and acronym) adopted by researchers to describe software that is used to facilitate qualitative approaches to data (it can also be referred to as computer-assisted qualitative data analysis: see the CAQDAS networking project at the University of Surrey, below). The term refers to software that breaks down text (and images, sound and video) into smaller units that can be indexed and treated as data for analysis. It includes a wide variety of qualitative data analysis software (QDAS) packages that help with the mechanical tasks of qualitative data analysis, which can include coding, memoing, annotating, grouping, linking, network-building, searching and retrieving, for example (see Chapter 6 for more information about coding and retrieval software and tools). It can also include software that enables collaborative and distributed qualitative data analysis (Chapter 36), software that supports quantitative integration for mixed methods approaches and data visualisation software that facilitates analysis by illustrating patterns and trends, or by helping researchers to visualise complex interrelations or networks (Chapter 13).

Various methodologies utilise CAQDAS, including grounded theory (Bringer et al., 2006), ethnography (see Chapters 14, 31 and 37) and phenomenology (Belser et al., 2017). It is used to aid and support the qualitative data analysis process, enabling researchers to undertake time-consuming and complex tasks in a shorter space of time. It is important to note, however, that CAQDAS packages do not work in a similar way to statistical packages, which analyse large amounts of data with little researcher input. Instead, the researcher is involved in all stages of the analysis process, using CAQDAS to aid and support this process as the research progresses (see questions for reflection, below). Methods of qualitative data analysis that can be aided and supported by CAQDAS include (in alphabetical order):

- comparative analysis that identifies, analyses and explains similarities across groups, societies, institutions and cultures;

- conceptual analysis that recognises, breaks down and analyses concepts;

- content analysis that identifies, codes and counts the presence of certain words, phrases, themes, characters, sentences, sounds or images;

- conversation analysis that focuses on interaction, contribution and sense-making during conversations;

- language analysis that identifies, counts and classifies words that relate to a particular emotion;

- recursive abstraction that summarises data in several steps (summarising the summary until an accurate final summary is produced);

- relational analysis that identifies and explores the relationship between concepts or themes;

- systematic analysis that identifies, collects and summarises empirical data;

- thematic analysis that identifies specific themes (clusters of linked categories that convey similar meanings);

- theoretical analysis that seeks to construct theory through the collection and analysis of data;

- transcription analysis that uses various analysis techniques to analyse transcripts (word counting, turn-taking, interaction, coding and classifying, for example).

CAQDAS is used by researchers undertaking qualitative research in a variety of disciplines, including the social sciences, education, health and medicine, business and the humanities. It can be used to collect, coordinate and analyse vast amounts of qualitative data from different team members in a project (Robins and Eisen, 2017) and facilitate researcher reflexive awareness (Woods et al., 2016b). It can also be used to assist with record-keeping and the organisation and management of qualitative data, provide an audit trail, enhance and demonstrate rigour and enable researchers to move beyond description to theory development (Bringer et al., 2006).

If you are interested in using CAQDAS for your research it is important that you choose the right software for your project (and for your methodology: see questions for reflection, below). Popular qualitative software packages include NVivo, ATLAS.ti, MAXqda, ANTHROPAC, LibreQDA, hyper RESEARCH, QDA Miner, Qualrus and Quirkos. Some data analysis software is easy-to-use, free and open source. Other software is freely available from institutions (IT staff will offer advice and guidance on its use). More information about making the right choices can be obtained from the CAQDAS networking project at the University of Surrey (see useful resources). You can also visit Predictive Analytics Today (www.predictiveanalyticstoday.com/top-qualitative-data-analysis-software) for reviews of popular software packages. Silver and Lewins (2014) provide a practical, user-friendly guide to CAQDAS methods and procedures, while Paulus et al. (2013) provide a comprehensive guide that discusses the use of data analysis software along with additional digital tools. Coding and retrieval are two specific functions performed by CAQDAS and these are discussed in Chapter 6. If you are interested in finding out more about software that is available for qualitative analyses of audio see Chapter 2, of video see Chapter 55 and of images see Chapter 16.

Questions for reflection

Epistemology, theoretical perspective and methodology

* Have you made clear methodological decisions before you make CAQDAS decisions? How can you ensure that CAQDAS will support your qualitative analyses, rather than oppose them? Do specific packages support certain methodologies (grounded theory, ethnography or qualitative content analysis, for example)? What research do you need to undertake to ensure that the right CAQDAS is chosen? Information that will help with these decisions can be obtained from the Restore Repository, National Centre for Research Methods (www.restore.ac.uk/lboro/research/soft ware/caqdas_primer.php) and from Kuş Saillard (2011) and Silver and Lewins (2014).

* Do you have a clear analytical strategy and do you understand the analytical process through which you need to go to answer your research question? Silver and Lewins (2014: 15) offer advice on this, pointing out that there are various stages of analysis in a research project including the

literature review, transcribing, coding, linking data and concepts and writing up. CAQDAS can only be used in certain stages of the analysis.

- Do you have a clear understanding of advantages and disadvantages (or pros and cons) of using CAQDAS? Your attitudes towards this will depend on your theoretical perspective and methodological standpoint, and on personal experience. For example, possible advantages can include flexibility, speed of analysis, the ability to handle large amounts of complex data, improved validity (and the opportunity to illustrate and justify the analysis), improved accuracy and freedom from monotonous and repetitive tasks. Possible disadvantages can include incorrect use of software, lack of training, over-reliance on output, lack of in-depth analysis and judgement, incompatibility (format and software), too much concentration on coding and retrieval, partial retrieval or error in retrieval. See Jackson et al. (2018) for an enlightening and entertaining discussion on researcher criticism of QDAS and see Woods et al. (2016a) for an interesting discussion on how researchers use QDAS software and tools.

- Do you have a clear understanding of what CAQDAS can and cannot do? For example, Rodik and Primorac (2015) illustrate how perceptions and expectations of software functions and capabilities can be influenced by previous experiences of statistical software use (researchers expected qualitative software to analyse data in a similar way to statistical software packages, for example).

Ethics, morals and legal issues

- Do you have a clear understanding of software licences? This can include commercial licences that provide unrestricted use for commercial purposes, along with a full support package; discounted educational licences for exclusive use in academic teaching and research, with a full support package; and open source licences that grant users certain rights to use, copy, modify and, in certain cases, redistribute source code or content. Comprehensive information about open source licences can be obtained from the Open Source Initiative (https://opensource.org/licenses).

- How do you intend to address issues surrounding qualitative data sharing and archiving? Mannheimer et al. (2018) discuss these issues in relation to data repositories and academic libraries, highlighting three challenges: obtaining informed consent for data sharing and scholarly use, ensuring data are legally and ethically shared and ensuring that shared data cannot be de-identified.

- How will you address issues of encryption and controlled access? Useful information on sharing, encrypting and controlling access to data can be obtained from the UK Data Service (www.ukdataservice.ac.uk).

- What is the impact of CAQDAS on researcher reflexivity? Woods et al. (2016b) will help you to address this question.

Practicalities

- Silver and Rivers (2016: 593) found that 'successful use of CAQDAS technology amongst postgraduate students is related to methodological awareness, adeptness in the techniques of analysis and technological proficiency'. How do you intend to develop these skills, understanding and awareness?

- What software is available? Is it available without cost from your institution or is it open source? If not, can you afford the software? What support is offered? What features and functions are available? Do they meet your needs? Is the software easy to use? What training will be required?

- What tasks do you wish to undertake using CAQDAS? This can include:
  - handling and storing data;
  - searching, querying and visualising data;
  - comparing themes;
  - uncovering connections and interrelationships;
  - spotting patterns and trends;
  - locating particular words or phrases;
  - inserting keywords or comments;
  - counting occurrences of words or phrases and attaching numerical codes;
  - retrieving text;
  - testing theory;
  - building theory.

Useful resources

The Computer Assisted Qualitative Data Analysis (CAQDAS) Networking Project at the University of Surrey in the UK (www.surrey.ac.uk/computer-assisted-qualitative-data-analysis) was established in 1994 and 'provides information, advice, training and support for anyone undertaking qualitative or mixed methods analysis using CAQDAS packages'. Information about training courses, collaboration projects and events can be found on the website.

The Restore Depository, National Centre for Research Methods (www.restore.ac.uk/lboro/research/software/caqdas.php) is maintained by 'a multi-disciplinary team of social science researchers who collect, update and restore web resources created by ESRC funded projects'. You can find comparative reviews of the main CAQDAS packages, along with a useful primer on CAQDAS on this website.

Key texts

Belser, A., Agin-Liebes, G., Swift, T., Terrana, S., Devenot, N., Friedman, H., Guss, J., Bossis, A. and Ross, S. (2017) 'Patient Experiences of Psilocybin-Assisted Psychotherapy: An Interpretative Phenomenological Analysis', *Journal of Humanistic Psychology*, 57(4), 354–88, first published April 28, 2017, 10.1177/0022167817706884.

Bringer, J., Johnston, L. and Brackenridge, C. (2006) 'Using Computer-Assisted Qualitative Data Analysis Software to Develop a Grounded Theory Project', *Field Methods*, 18(3), 245–66, first published August 1, 2006, 10.1177/1525822X06287602.

Jackson, K., Paulus, T. and Woolf, N. (2018) 'The Walking Dead Genealogy: Unsubstantiated Criticisms of Qualitative Data Analysis Software (QDAS) and the Failure to Put Them to Rest', *The Qualitative Report*, 23(13), 74–91, retrieved from https://nsuworks.nova.edu/tqr/vol23/iss13/6.

Kuş Saillard, E. (2011) 'Systematic versus Interpretive Analysis with Two CAQDAS Packages: NVivo and MAXQDA', *Forum Qualitative Sozialforschung/Forum: Qualitative Social Research*, 12(1), Art. 34, retrieved from http://nbn-resolving.de/urn:nbn:de:0114-fqs1101345.

Mannheimer, S., Pienta, A., Kirilova, D., Elman, C. and Wutich, A. (2018) 'Qualitative Data Sharing: Data Repositories and Academic Libraries as Key Partners in Addressing Challenges', *American Behavioral Scientist*, first published June 28, 2018, 10.1177/0002764218784991.

Paulus, T., Jackson, K. and Davidson, J. (2017) 'Digital Tools for Qualitative Research: Disruptions and Entanglements', *Qualitative Inquiry*, 23(10), 751–6, first published September 27, 2017, 10.1177/1077800417731080.

Paulus, T., Lester, J. and Dempster, P. (2013) *Digital Tools for Qualitative Research*. London: Sage.

Robins, C. and Eisen, K. (2017) 'Strategies for the Effective Use of NVivo in a Large-Scale Study: Qualitative Analysis and the Repeal of Don't Ask, Don't Tell', *Qualitative Inquiry*, 23, 768–78, 10.1177/1077800417731089.

Rodik, P. and Primorac, J. (2015) 'To Use or Not to Use: Computer-Assisted Qualitative Data Analysis Software Usage among Early-Career Sociologists in Croatia', *Forum Qualitative Sozialforschung/Forum: Qualitative Social Research*, 16(1), Art. 12, http://nbn-resolving.de/urn:nbn:de:0114-fqs1501127.

Silver, C. and Lewins, A. (2014) *Using Software in Qualitative Research: A Step-by-Step Guide.* London: Sage.

Silver, C. and Rivers, C. (2016) 'The CAQDAS Postgraduate Learning Model: An Interplay between Methodological Awareness, Analytic Adeptness and Technological Proficiency', *International Journal of Social Research Methodology*, 19(5), 593–609, first published September 14, 2015, 10.1080/13645579.2015.1061816.

Woods, M., Macklin, R. and Lewis, G. (2016b) 'Researcher Reflexivity: Exploring the Impacts of CAQDAS Use', *International Journal of Social Research Methodology*, 19 (4), 385–403, first published April 2, 2015, 10.1080/13645579.2015.1023964.

Woods, M., Paulus, T., Atkins, D. and Macklin, R. (2016a) 'Advancing Qualitative Research Using Qualitative Data Analysis Software (QDAS)? Reviewing Potential versus Practice in Published Studies Using ATLAS.ti And NVivo, 1994–2013', *Social Science Computer Review*, first published August 27, 2015, 10.1177/0894439315596311.

CHAPTER 10

Data analytics

Overview

Data analytics refers to the process of examining datasets, or raw data, to make inferences and draw conclusions. The term also encompasses the tools, techniques and software that are used to scrape, capture, extract, categorise, analyse, visualise and report data, and the models, predictions and forecasts that can be made. This differentiates data analytics from data analysis, which refers to the processes and methods of analysing data, rather than the tools, techniques and outcomes. As such, data analysis can be seen to be a subset, or one component, of data analytics. There are various fields of data analytics that are used within different occupations, organisations and disciplines. Each of these has its own entry in this book as there are specialist tools, techniques and software used, depending on the subject, field of study and purpose of the research. This chapter provides an overview and introduction to data analytics and should be read together with any of the chapters listed below that are of interest. It is important to note, however, that these categories are not distinct entities: it will become clear that there are numerous overlaps and connections, and that titles are fluid and flexible. Relevant chapters include:

- big data analytics to examine extremely large and complex data sets (Chapter 3);
- business analytics to explore an organisation's data on processes, performance or product (Chapter 4);
- game analytics to explore data related to all forms of gaming (Chapter 20);
- HR analytics to examine and use people-data to improve performance (Chapter 22);

- learning analytics to explore data about learners and their contexts (Chapter 24);

- social media analytics to examine data from social media platforms (Chapter 52);

- web and mobile analytics to explore web data and data from mobile platforms (Chapter 58).

Data analytics is used by researchers across a wide variety of disciplines and fields of study including business and management, retail, education, politics, travel and tourism, health and medicine, geography and the computer sciences. Examples of research projects that have used or assessed data analytics include research into the effects of reviewer expertise on hotels' online reputation, online popularity and financial performance (Xie and Fung So, 2018); a project to develop a strategic and comprehensive approach to evaluation in public and non-profit organisations (Newcomer and Brass, 2016); a field experiment to understand vehicle maintenance mechanisms of a connected car platform, using real-time data analytics (Kim et al., 2018); and a practical project for students to use data analytics to enable a sponsoring organisation to manage workforce attrition proactively (King, 2016).

There are different data analytics methods and techniques and a selection of these is given below (in alphabetical order).

- Descriptive analytics uses a range of measures to describe a phenomenon. It is used on historical or current data to illustrate what has happened, or what is happening, and provides a description, summary or comparison that can be acted on as required. It is a popular method in business analytics (Chapter 4).

- Diagnostic analytics is used to explore why something has happened. It examines existing data and determines what additional data need to be collected to explain or discover what has happened. Visualisation tools (Chapter 13) and computer models (Chapter 7) can be used in the diagnosis.

- Location analytics is used to gain insight from the location or geographic element of business data. It uses visualisation to interpret and analyse geographic data (historical or real-time). More information about data visualisation can be found in Chapter 13 and more information about location awareness and location tracking technology can be found in Chapter 27.

- Network analytics is used to analyse and report network data and statistics to identify patterns and highlight trends. It enables network administrators or operators to troubleshoot problems, make decisions, further automate the network, increase self-managing network capabilities and adjust resources as required.

- Predictive analytics is used to make predictions based on the quantitative analyses of datasets. It uses existing datasets to determine patterns and predict what will happen in the future. It is a popular method in business analytics (Chapter 4), utilises a variety of data mining techniques (Chapter 12) and is aligned closely with predictive modelling (Chapter 45).

- Prescriptive analytics draws together insights gained from descriptive, diagnostic and predictive analytics to help make decisions or work out the best course of action. It includes modelling, simulation and optimisation so that decisions can be made that will achieve the best outcome.

- Real-time analytics accesses data as soon as they enter the system. It enables organisations or researchers to view live data such as page views, website navigation or shopping cart use. It can also be called dynamic analytics or on-demand analytics, although there are subtle differences in the terms (dynamic analytics include analysing, learning and responding and on-demand analytics involves requests for data, rather than a continuous feed of data, for example).

- Security analytics refers to the process of using data collection, aggregation and analysis tools for monitoring, detecting, reporting and preventing cyberattacks and security breaches. It can help to detect anomalies and determine threats.

- Speech and voice analytics explore and analyse recordings or transcripts of what people say in a conversation and how people speak in a conversation (tone, pitch and emotion, for example). They can be used to assess performance, make predictions and inform future development in organisations that use speech and voice (call centres, for example).

- Text analytics uses tools and software to analyse text. Kent (2014) provides a brief overview of the technologies behind analysing text, gives practical examples and looks into language that is used to describe the processes of text analytics.

- Video analytics refers to the process of using software to analyse video automatically for purposes such as motion detection, facial recognition, people counting, licence plate reading and tracking a moving target. A component of video analytics is video analysis (see Chapter 55).

- Visual analytics combine information visualisation (see Chapter 13), computer modelling (see Chapter 7) and data mining (see Chapter 12) to produce interactive and static visualisations that aid the process of analysis. Brooker et al. (2016) provide an interesting example of how this technique can be used to help understand Twitter data.

Useful resources and references are provided below for those who are interested in finding out more about data analytics for their research. It is also important to read more about the various types of data analytics (see relevant chapters listed above) and the different ways of undertaking data analytics (see above). An understanding of data mining (see Chapter 12) and data visualisation (see Chapter 13) is also important.

Questions for reflection

Epistemology, theoretical perspective and methodology

- How raw is raw data? Can raw data be objective? Have data been manipulated? Are there biases present? Is accessed restricted? Do data serve any political or organisational interests? Who benefits from data generation and distribution? The field of critical data studies will help you to address these and other methodological issues: a good starting point is Richterich (2018), who provides an excellent overview, along with interesting material on ethics, social justice and big data.

- Diesner (2015: 1) points out that 'data for analysis and conducting analytics involves a plethora of decisions, some of which are already embedded in previously collected data and built tools'. Although these decisions 'have tremendous impact on research outcomes, they are not often obvious, not considered or not being made explicit'. How will you ensure that all decisions you make are transparent, clear and made explicit? Diesner (2015) provides interesting and useful information on this topic.

- O'Neil (2017) provides detailed accounts of how data analytics, mathematical models and algorithms can reinforce discrimination, despite existing perceptions that data are objective, cannot lie and models are incontestable. How will you address these arguments and challenges when applying data analytics? How can you ensure that methods and models are transparent and clear, rather than opaque or invisible? O'Neil provides an enlightening discussion that will help you to think about these issues in considerable depth.

Ethics, morals and legal issues

- Are you familiar, and able to comply, with all relevant terms and conditions? This could be terms and conditions of the organisation supplying data (see Chapter 3 for a list of organisations that provide open data for research purposes), or the terms and conditions of the platform that generates user-driven data, for example (see Chapter 52 for more information about social media platforms). It can also include terms and conditions or rules and regulations of your funding body, university or research ethics committee, for example.

- Has informed consent been given by all those surrendering data? Are they clear about what they are consenting to and how data are to be used? Are data to be used only for purposes for which consent has been given? Have individuals been given the opportunity to decline the use of their data?

- How will results/insights from data be used? Can action cause harm to individuals? Can individuals see results, question results and seek change when mistakes have been made?

- Who owns data and subsequent results/insights? How do owners and managers protect data and keep them secure? How will you keep your data and analyses protected and secure? Guidance on these ethical issues and those raised above can be obtained from the UK Government's Data Ethics Framework (www.gov.uk/government/publications/data-ethics-framework/data-ethics-framework).

Practicalities

- Do you intend to collect your own data or use existing data for your research? If you are going to collect your own data, what methods do you intend to use? If you are going to use existing sources, are data accessible, freely available, valid, reliable and accurate (and has informed consent been given: see above)?

- Do you have a thorough understanding of data analytics software and tools that are available? Are user-friendly guides, tutorials or short courses available? Some examples of tools and software are given below and in the individual analytics chapters listed above.

- What costs are involved (software and access to data, for example)? Is software freely available through your institution or is it open source or available with a free trial, for example?

Useful resources

There is a vast range of digital tools and software packages available for data analytics, with differing functions, capabilities, purposes and costs. Some examples that are available at time of writing are given below, in alphabetical order. Consult the relevant analytics chapter (listed above) if you are interested in data analytics tools and software for specific purposes.

- Apache Spark (https://spark.apache.org);
- Clearify QQube (https://clearify.com);
- Dataiku (www.dataiku.com);
- Logi Info (www.logianalytics.com/analytics-platform);
- Looker (https://looker.com);
- Mozenda (www.mozenda.com);
- Python (www.data-analysis-in-python.org);
- SAS (www.sas.com);
- Splunk (www.splunk.com);
- Tableau (www.tableau.com).

INFORMS (www.informs.org) is a membership organisation that 'promotes best practices and advances in operations research, management science, and analytics to improve operational processes, decision-making, and outcomes'. You can find publications, conferences, networking communities (including the Analytics Society) and professional development services on the website, and find useful articles in the 'Analytics Magazine' (http://analytics-magazine.org).

Key texts

Brooker, P., Barnett, J. and Cribbin, T. (2016) 'Doing Social Media Analytics', *Big Data & Society*, first published July 8, 2016. doi:10.1177/2053951716658060.

Diesner, J. (2015) 'Small Decisions with Big Impact on Data Analytics', *Big Data & Society*, first published December 1, 2015. doi:10.1177/2053951715617185.

Kent, E. (2014) 'Text Analytics – Techniques, Language and Opportunity', *Business Information Review*, 31(1), 50–53, first published April 11, 2014. doi:10.1177/0266382114529837.

Kim, J., Hwangbo, H. and Kim, S. (2018) 'An Empirical Study on Real-Time Data Analytics for Connected Cars: Sensor-Based Applications for Smart Cars', *International Journal of Distributed Sensor Networks*, first published January 27, 2018. doi:10.1177/1550147718755290.

Data analytics

King, K. (2016) 'Data Analytics in Human Resources: A Case Study and Critical Review', *Human Resource Development Review*, 15(4), 487–95, first published November 20, 2016. doi:10.1177/1534484316675818.

Newcomer, K. and Brass, C. (2016) 'Forging a Strategic and Comprehensive Approach to Evaluation within Public and Nonprofit Organizations: Integrating Measurement and Analytics within Evaluation', *American Journal of Evaluation*, 37(1), 80–99, first published March 13, 2015. doi:10.1177/1098214014567144.

O'Neil, C. (2017) *Weapons of Math Destruction: How Big Data Increases Inequality and Threatens Democracy*. London: Penguin.

Richterich, A. (2018) *The Big Data Agenda: Data Ethics and Critical Data Studies*. London: University of Westminster Press.

Siege, E. (2016) *Predictive Analytics: The Power to Predict Who Will Click, Buy, Lie, or Die*. Hoboken, NJ: Kogan Page, Inc.

Xie, K. and Fung So, K. (2018) 'The Effects of Reviewer Expertise on Future Reputation, Popularity, and Financial Performance of Hotels: Insights from Data-Analytics', *Journal of Hospitality & Tourism Research*, 42(8), 1187–209, first published December 4, 2017. doi:10.1177/1096348017744016.

Data collection and conversion

Overview

Data collection and conversion refers to the process of bringing together or gathering data and converting them from one media form, or format, to another (through export or data translation software, for example). This is done to enable interoperability so that computer systems or software can exchange data and make use of data (data formatting and the interoperability of GIS datasets from different sources, for example: see Chapter 21). Data capture is a term that is sometimes used interchangeably with data collection; however, data capture refers more specifically to the process of a device reading information from one system or media source and transporting it directly to another system or media source (bar code readers or intelligent document recognition, for example). Data collection is a more general term that refers to gathering information in various ways from various sources and delivering it to a system or media source (and can, therefore, include data capture techniques). Similarly, data migration is a term that is sometimes used interchangeably with data conversion. However, data migration refers specifically to the process of transferring data from one system to another, whereas data conversion refers to the process of transforming or translating data from one format to another. Data conversion is used to extract data from one source, translate or transform them, and prepare them for loading onto another system or media source (ensuring that all data are maintained, with embedded information included in the conversion: the target format must be able to support all the features and constructs of the source data).

There are a variety of applications for data collection and conversion, including:

- Informing research and facilitating data analysis: identifying and examining relevant sources, collecting data and converting to the required format. It enables data to be moved effectively between systems for analysis. In geospatial and spatial analysis, for example, this can include conversion from spreadsheets to mapping tools, conversion from image files to georeferenced imagery, geocoding to convert street addresses into spatial locations, conversion from vector to raster or raster to vector, or converting and testing digital data layers (see Chapters 21 and 54). In research that uses mixed methods approaches, it can involve the conversion of text into numerical data and the conversion of numerical data into text (Nzabonimpa, 2018) and in qualitative research it can include converting textual data into numerical values (Saldaña, 2016).

- Improving business performance: gathering, managing and transforming raw business data into useful and actionable information for business intelligence (Chapter 4).

- Standardising and unifying records: converting legacy data, which may be old and obsolete, or bringing together and converting disparate records into one system for specialist access (medical records, for example).

- Depositing research results and datasets in repositories: converting data to the preferred format and ensuring data integrity during conversion.

- Preserving digital information: converting information into the required format and storing over the long-term, ensuring that it can survive future changes in storage media, devices and data formats, for example. The UK Data Service provides comprehensive information and advice about file formats for data preservation (details below).

Research that has used, assessed or critiqued data collection and conversion methods include a study into converting or migrating adverse event databases (Steiner et al., 2001); an assessment and development of recommendations on converting paper records to electronic records in an Indian eye hospital (Missah et al., 2013); research into the possibilities and impossibilities of converting narratives into numerical data and numerical data into narratives (Nzabonimpa, 2018); an exploration of the library system migration experiences at the King Fahd University of Petroleum and Minerals (Khurshid and Al-Baridi, 2010); and an assessment and evaluation of the process of converting legacy data to the Medical Dictionary for Regulatory Activities (Doan, 2002).

If you are interested in finding out more about data collection and conversion for your research it is useful to visit your IT services department or computing services where trained members of staff will be able to offer advice and guidance on the most appropriate systems and software to use and those that are available at your institution. It is useful to visit some of the websites listed in useful resources below as these provide examples of software, services and organisations that are available to offer information, advice and guidance to those interested in data collection and conversion. You might also find it of interest to obtain more information about coding and retrieval methods (see Chapter 6). Saldaña (2016) provides a comprehensive and user-friendly guide for those interested in coding qualitative data and more information about computer-assisted qualitative data analysis software can be found in Chapter 9. If you are a business person, or a researcher working in business and management, you might find it of interest to obtain more information about business analytics (Chapter 4), big data analytics (Chapter 3) and HR analytics (Chapter 22).

Questions for reflection

Epistemology, theoretical perspective and methodology

- Are you interested in short-term data processing or long-term data pre-servation? File formats associated with proprietary data analysis software can change or become obsolete over the long-term, whereas standard, open formats that are more widely available might be less likely to change over the long-term. Guidance on suitable file formats can be obtained from the UK Data Service (details below).

- Is data conversion necessary? Are there more effective, efficient and cheaper alternatives? For example, if you have only a small number of documents, images or cases that need converting, is it easier to re-enter data in the desired format? This may be preferable in cases where data will be lost or where conversion is complex, for example.

- What influence does data collected from unmanaged external sources (sensors, data streams or Internet of things, for example) have on trace-ability? What impact will conversion have on traceability?

- How can you ensure data integrity during collection and conversion? This involves issues of accuracy, consistency and quality control (calibration of instruments, using standardised methods, minimising bias, setting validations rules, consistent and detailed labelling and demonstrating authenticity, for example).

- How can you avoid problems with lost data or inexact data when under-taking data conversion? This can include missing value definitions or decimal numbers, variable labels, reduction in resolution, lost headers and footers or missing text formatting such as bold and italics, for example. Solutions can include using specific software to prevent loss, using approximation in cases where specific formats cannot be converted and reverse engineering to achieve close approximation, for example.

Ethics, morals and legal issues

- How do you intend to keep stored data or data in transit safe from hackers, unauthorised access or misuse? This can involve data encryption and decryption (translating data into a cipher, ciphertext or code so that information cannot be understood by unauthorised people, and translat-ing data back to their original form when required). Encryption is used to protect private or sensitive digital data.

- Have you considered issues of copyright? Do you need to seek copyright permission with regard to data ownership? The UK Data Service (details below) provides detailed information, pointing out that most research outputs such as spreadsheets, databases, publications, reports and com-puter programs are protected by copyright. Data can be reproduced for non-commercial research purposes without infringing copyright under the fair dealing concept, as long as the data source, distributor and copy-right holder are acknowledged. The website also contains useful informa-tion about copyright duration in the UK.

Practicalities

- What are your goals for data conversion (e.g. uniformity, standardisation, analysis, sharing or preservation)? What are the benefits to be gained?

- How long will data collection and conversion take and how much will it cost?

- What sources are available for data collection and are datasets open and accessible (the Linked Open Data cloud, for example: see useful resources)? Additional sources are listed in Chapter 3.

- How can you ensure successful data conversion? Steiner et al. (2001) will help you to address this question.

- Do you have a good understanding of the processes involved in data collection and conversion? This includes:
  - Identifying suitable and relevant data sources, checking for inconsistent or missing data and complying with relevant licenses.
  - Identifying data destination(s), defining data requirements and checking translation and conversion capabilities (conceptual similarities, for example).
  - Loading and reviewing data, checking for accuracy and validating data.

Useful resources

There are a variety of companies and organisations that provide data collection and conversion services, software and/or tools. Some of these are for specific purposes such as geospatial analysis or business intelligence, some are specifically for the conversion of one data format to another and others are ETL (extract, transform and load) tools. The following list provides examples of what is available at time of writing and is ordered alphabetically:

- Centerprise Data Integrator for business users (www.astera.com/centerprise);
- CloverETL Open Source Engine (www.cloveretl.com/products/open-source);
- Datastream Data Conversion Services for business intelligence (www.dscs.com/data);
- Jaspersoft ETL (https://community.jaspersoft.com/project/jaspersoft-etl);
- Prehnit GIS data collection, conversion and quality control services (https://prehnit.hr/consulting/data-collection-conversion-quality-control);
- ResCarta Data Conversion Tool (www.rescarta.org/index.php/sw/40-data-conversion-tool);
- Safe Software Data Conversion for data conversion across a range of environments (www.safe.com/fme/key-capabilities/data-conversion);
- Schneider Data Collection and Conversion for geospatial data (http://schneidercorp.com/services/geospatial/data-collection-conversion);
- Stat/Transfer for data transfer between software packages (https://stattransfer.com);

- Talend ETL tool for data integration and data transformation (https://sourceforge.net/projects/talend-studio).

The UK Data Service (www.ukdataservice.ac.uk) provides 'access to local, regional, national and international social and economic data' and offers 'guidance and training for the development of skills in data use'. Guidance on file formats recommended and accepted by the UK Data Service for data sharing, reuse and preservation can be obtained from this website, along with comprehensive information about preparing and depositing data.

The Linked Open Data cloud diagram can be accessed at https://lod-cloud.net. The image on the website shows datasets that have been published in the Linked Data format. It includes datasets in the fields of geography, government, life sciences, linguistics, media, publications, social networking, cross domain and user-generated. Information about contributing to the diagram, in the required format, is included on the site.

Key texts

Doan, T. (2002) 'Converting Legacy Data to Meddra: Approach and Implications', *Therapeutic Innovation & Regulatory Science*, 36(3), 659–66, first published July 1, 2002, 10.1177/009286150203600320.

Khurshid, Z. and Al-Baridi, S. (2010) 'System Migration from Horizon to Symphony at King Fahd University of Petroleum and Minerals', *IFLA Journal*, 36(3), 251–58, first published October 14, 2010, 10.1177/0340035210378712.

Missah, Y., Dighe, P., Miller, M. and Wall, K. (2013) 'Implementation of Electronic Medical Records: A Case Study of an Eye Hospital', *South Asian Journal of Business and Management Cases*, 2(1), 97–113, first published June 6, 2013, 10.1177/2277977913480682.

Nzabonimpa, J. (2018) 'Quantitizing and Qualitizing (Im-)Possibilities in Mixed Methods Research', *Methodological Innovations*, first published July 27, 2018, 10.1177/2059799118789021.

Saldaña, J. (2016) *The Coding Manual for Qualitative Researchers*, 3rd edition. London: Sage.

Steiner, J., Cauterucci, C., Shon, Y. and Muirhead, A. (2001) 'Planning for a Successful Safety Data Conversion', *Therapeutic Innovation & Regulatory Science*, 35(1), 61–69, first published January 1, 2001, 10.1177/009286150103500107.

Data mining

Overview

Data mining is the process that is used to turn raw data into useful information that will help to uncover hidden patterns, relationships and trends. It is a data-driven technique in which data are examined to determine which variables and their values are important and understand how variables are related to each other. It involves applying algorithms to the extraction of hidden information with the aim of building an effective predictive or descriptive model of data for explanation and/or generalisation. The focus is on data sourcing, pre-processing, data warehousing, data transformation, aggregation and statistical modelling. The availability of big data (extremely large and complex datasets, some of which are free to use, re-use, build on and redistribute, subject to stated conditions and licence: see Chapter 3) and technological advancement in software and tools (see below), has led to rapid growth in the use of data mining as a research method.

There are various sub-categories of data mining, including (in alphabetical order):

- Distributed data mining that involves mining distributed data with the intention of obtaining global knowledge from local data at distributed sites. Zeng et al. (2012) provide a useful overview of distributed data mining.

- Educational Data Mining (EDM) that is used to observe how people learn and how they behave when learning, without disturbing their learning (see Chapter 17). It enables researchers to answer educational research questions, develop theory and better support learners. Lee (2019) provides a good example of how EDM is used in research and Chapter 24 discusses learning analytics, which is closely related to EDM.

- Ethno-mining that enables researchers to study human behaviour and culture through combining ethnography with data mining tools and techniques (see Chapter 18).

- Link mining that applies data mining techniques to linkage data. This can involve tasks such as link-based categorisation, link-based identification and predicting link strength (see Chapter 25 for more information about link analysis, which is closely connected to link mining).

- Opinion mining that tracks opinions expressed by people about a particular topic. This can involve sentiment analysis, which considers whether an opinion expresses a polarity trend toward a particular feeling (Basili et al., 2017).

- Reality mining that uses mobile devices, GPS and sensors to provide an accurate picture of what people do, where they go and with whom they come into contact and communicate. Eagle and Greene (2016) use the term 'reality mining' to refer to the analysis of big data and illustrate the potential of such analyses in their book.

- Social media mining that collects, processes and explores information from social media sites, including posts, comments, likes, images and tweets. Social media analytics (Chapter 52) is closely connected to social media mining.

- Text mining that involves extracting new and unexpected information from a collection of text. Ignatow and Mihalcea (2018) provide a comprehensive introduction to text mining.

- Trajectory data mining that mines trajectory data from moving objects such as people, vehicles, animals and natural phenomena such as hurricanes, tornadoes and ocean currents. Zheng (2015) provides a detailed overview of trajectory data mining.

- Visual data mining that uses information visualisation techniques to explore large datasets (see Chapter 13 for more information about data visualisation and Yu et al., 2009 for information about how visualisation tools can be used in data mining).

- Web mining that uses data mining techniques to extract and explore information from the web. This can include web content mining, web structure mining and web usage mining, for example. Web and mobile analytics (Chapter 58) is closely connected to web mining.

Researchers from a variety of disciplines and fields of study use data mining methods including health and medicine (Chao Huang, 2013; Chaurasia et al., 2018), the social sciences (Kennedy et al., 2017; Oliver et al., 2018; Yu et al., 2009), travel and tourism (Shapoval et al., 2018) and information technology (Zeng et al., 2012; Zheng, 2015). Examples of research projects that have used data mining methods include a study into the behaviour of in-bound tourists to Japan (Shapoval et al., 2018); a study to analyse weight loss treatment for overweight and obese children (Oliver et al., 2018); research into lifestyle and metabolic syndrome of workers in Taiwan (Chao Huang, 2013); research into the prediction of benign and malignant breast cancer (Chaurasia et al., 2018); and a study to look into clients' habits and customer usage patterns in hotels (Garrigos-Simon et al., 2016).

If you are interested in finding out more about data mining Aggarwal (2015), Brown (2014), Leskovec et al. (2014), Hastie et al. (2017) and Witten et al. (2017) all provide comprehensive information and advice. The questions for reflection listed below will help you to think more about data mining tools and techniques, and the list of tools and software will help you to find out more about purpose, functions and capabilities. It is also important that you find out more about the related topics of data analytics (Chapter 10) and data visualisation (Chapter 13). Ethno-mining is a combination of data mining and ethnography and is discussed in Chapter 19 and cluster analysis is a particular technique that can be used in data mining, which is discussed in Chapter 5. You may also find it useful to obtain more information about big data analytics (Chapter 3), data collection and conversion (Chapter 11) and predictive modelling (Chapter 45).

Questions for reflection

Epistemology, theoretical perspective and methodology

- Do you understand the differences between data mining and data analytics, and have you decided which might the best method for your research? Data mining builds or uses algorithms to identify meaningful structure in the data and involves machine learning, artificial intelligence and statistics. Data analytics is the process of ordering and organising raw data to gain useful insights and aid decision-making and can be exploratory, descriptive, predictive, prescriptive and diagnostic, for example (see Chapter 10). Data mining can be used as part of the data analytics process, if appropriate.

Data mining

- Data mining involves knowledge discovery in databases. This area of research has been termed Knowledge Discovery in Distributed and Ubiquitous Environments, KDubiq (Knowledge Discovery in Ubiquitous Environments) and Ubiquitous Knowledge Discovery. Do you understand what is meant by these terms and the relevance to data mining as a research method? May and Saitta (2010) will help you to get to grips with this area of research.

Ethics, morals and legal issues

- Do you have a thorough understanding of issues surrounding privacy, security and the misuse or abuse of data? How can you ensure that your data mining is performed within the bounds of privacy regulations and informed consent?

- How fair are data mining practices? How do consumers/users feel about the practice? Kennedy et al. (2017) provide an enlightening discussion on this topic, based on the perceptions of social media users that were provided in focus group research.

- How can you undertake discrimination-aware data mining and fairness-aware data mining? Berendt and Preibusch (2017: 135) discuss how discrimination can arise through the 'combined effects of human and machine-based reasoning'. They go on to look at how discrimination can occur and consider ways to avoid discrimination and comply with data protection law.

- Are there any legal and technical restrictions on the databases that you wish to use? The 'database right' in the UK and EU, for example, protects collections of data that have required substantial investment to collect together, verify and display content. Also, UK copyright law allows for computational analysis of copyrighted material only if this analysis is for non-commercial purposes.

Practicalities

- Do you feel comfortable with, and understand, your data? It is important to get to grips with your data as it will help you to get the most out of your data. You could, for example, run basic statistical analyses (bivariate, univariate or principle component analysis) to build a better understanding of data before choosing the more complex tools and techniques listed below.

- Do you have a good understanding of data mining tools and techniques? This can include, for example:
  - classification (to assign items to target classes or categories, used to draw conclusions and for prediction);
  - clustering (to group or partition a set of data or group of objects into sub-classes);
  - decision trees (to supply definitive choices and indicate the best path to follow);
  - association rules (to uncover relationships between what may appear to be unrelated data using if/then statements);
  - anomaly detection (to identify items, observations or events that are unusual or do not conform);
  - intrusion detection (to decontaminate databases and improve security);
  - sequential patterns (to identify patterns of ordered events or temporal associations between events);
  - regression (to estimate the relationships among or between variables);
  - neural networks (to recognise and store patterns that may be useful in the future);
  - summarisation (to produce and present relevant and meaningful information in a compact way).

Useful resources

There are a wide variety of tools and software packages available for data mining. Some of these are open source and free to use, whereas others have significant costs involved. Tools and software vary enormously and your choice will depend on your research topic, needs and purpose. The following list provides a snapshot of what is available at time of writing and is listed in alphabetical order:

- Datapreparator (www.datapreparator.com);
- ELKI (https://elki-project.github.io);
- Jubatus (http://jubat.us/en);
- Orange Data Mining (https://orange.biolab.si);

- RapidMiner (https://rapidminer.com);

- SAS Enterprise Miner (www.sas.com/en_gb/software/enterprise-miner.html);

- SAS Visual Data Mining and Machine Learning (www.sas.com/en_gb/software/visual-data-mining-machine-learning.html);

- Shogun (www.shogun-toolbox.org);

- Tanagra (http://eric.univ-lyon2.fr/~ricco/tanagra/en/tanagra.html);

- Weka Data Mining (www.cs.waikato.ac.nz/ml/weka).

Key texts

Aggarwal, C. (2015) *Data Mining: The Textbook*. New York, NY: Springer.

Basili, R., Croce, D. and Castellucci, G. (2017) 'Dynamic Polarity Lexicon Acquisition for Advanced Social Media Analytics', *International Journal of Engineering Business Management*, first published December 20, 2017. doi:10.1177/1847979017744916.

Berendt, B. and Preibusch, S. (2017) 'Toward Accountable Discrimination-Aware Data Mining: The Importance of Keeping the Human in the Loop—And under the Looking Glass', *Big Data*, 5(2), 135–52, first published Jun 1, 2017. doi:10.1089/big.2016.0055.

Brown, M. (2014) *Data Mining for Dummies*. Hoboken, NJ: John Wiley & Sons, Inc.

Chao Huang, Y. (2013) 'The Application of Data Mining to Explore Association Rules between Metabolic Syndrome and Lifestyles', *Health Information Management Journal*, 42(3), 29–36, first published October 1, 2013. doi:10.1177/183335831304200304.

Chaurasia, V., Pal, S. and Tiwari, B. (2018) 'Prediction of Benign and Malignant Breast Cancer Using Data Mining Techniques', *Journal of Algorithms & Computational Technology*, 12(2), 119–26, first published February 20, 2018. doi:10.1177/1748301818756225.

Eagle, N. and Greene, K. (2016) *Reality Mining: Using Big Data to Engineer a Better World*, Reprint edition. Cambridge, MA: MIT Press.

Garrigos-Simon, F., Llorente, R., Morant, M. and Narangajavana, Y. (2016) 'Pervasive Information Gathering and Data Mining for Efficient Business Administration', *Journal of Vacation Marketing*, 22(4), 295–306, first published November 30, 2015. doi:10.1177/1356766715617219.

Hastie, T., Tibshirani, R. and Friedman, J. (2017) *The Elements of Statistical Learning: Data Mining, Inference, and Prediction*, 2nd edition. New York, NY: Springer.

Ignatow, G. and Mihalcea, R. (2018) *An Introduction to Text Mining: Research Design, Data Collection, and Analysis*. Thousand Oaks, CA: Sage Publications, Inc.

Kennedy, H., Elgesem, D. and Miguel, C. (2017) 'On Fairness: User Perspectives on Social Media Data Mining', *Convergence*, 23(3), 270–88, first published June 28, 2015. doi:10.1177/1354856515592507.

Lee, Y. (2019). Using Self-Organizing Map and Clustering to Investigate Problem-Solving Patterns in the Massive Open Online Course: An Exploratory Study. *Journal of Educational Computing Research*, 57(2), 471–90. doi:10.1177/0735633117753364.

Leskovec, J., Rajaraman, A. and Ullman, D. (2014) *Mining of Massive Datasets*, 2nd edition. Cambridge: Cambridge University Press.

May, M. and Saitta, L. (eds.) (2010) *Ubiquitous Knowledge Discovery: Challenges, Techniques, Applications.* Berlin: Springer.

Oliver, E., Vallés-Perez, I., Baños, R., Cebolla, A., Botella, C. and Soria-Olivas, E. (2018) 'Visual Data Mining with Self-Organizing Maps for "Self-Monitoring" Data Analysis', *Sociological Methods & Research*, 47(3), 492–506, first published August 3, 2016. doi:10.1177/0049124116661576.

Shapoval, V., Wang, M., Hara, T. and Shioya, H. (2018) 'Data Mining in Tourism Data Analysis: Inbound Visitors to Japan', *Journal of Travel Research*, 57(3), 310–23, first published April 3, 2017. doi:10.1177/0047287517696960.

Witten, I., Frank, E., Hall, M. and Pal, C. (2017) *Data Mining: Practical Machine Learning Tools and Techniques*, 4th edition. Cambridge, MA: Morgan Kaufmann.

Yu, C., Zhong, Y., Smith, T., Park, I. and Huang, W. (2009) 'Visual Data Mining of Multimedia Data for Social and Behavioral Studies', *Information Visualization*, 8(1), 56–70, first published February 12, 2009. doi:10.1057/ivs.2008.32.

Zeng, L., Li, L., Duan, L., Lu, K., Shi, Z., Wang, M., Wu, W. and Luo, P. (2012) 'Distributed Data Mining: A Survey', *Information Technology and Management*, 13(4), 403–09, first published May 17, 2012. doi:10.1007/s10799-012-0124-y.

Zheng, Y. (2015) 'Trajectory Data Mining: An Overview', *ACM Transactions on Intelligent Systems and Technology*, 6(3), article 29, publication date: May 2015. doi:10.1145/2743025.

CHAPTER 13

Data visualisation

Overview

Data visualisation is a research method that uses visualisation to help researchers better understand data, see patterns, recognise trends and communicate findings. Researchers approaching their work from a variety of epistemological positions and theoretical perspectives use data visualisation as a research method, and it is used in both quantitative and qualitative approaches. It can be both exploratory and explanatory, helping the viewer to understand information quickly, pinpointing emerging trends, identifying relationships and patterns and communicating the story to others. It enables data to become useful and meaningful, providing insight into data and helping researchers to develop hypotheses to explore further. Large datasets can often appear overwhelming: data visualisation methods help to make them more manageable and present data in a visually-engaging way to a much wider audience. Hand-drawn and printed visualisation has been used for hundreds of years, but recent advances in software and visualisation tools, along with the increasing collection and use of big data (Chapter 3), have led to a rapid increase in the use of data visualisation as a digital research method. Information about the history of visualisation can be found on the website of the Centre for Research and Education in Arts and Media (CREAM) at the University of Westminster (http://data-art.net).

There are different ways in which data visualisation can be used in research, depending on the subject and purpose of the research (the following bullet points should not be viewed as distinct entities, but instead can be seen to be fluid and flexible with significant overlap and possible movement between points as research projects develop and progress).

- Data visualisation for solving problems, increasing understanding and advancing knowledge. These visualisations can be used to delve into mathematical, statistical and scientific phenomena and can include digital images from electron microscopes, astrophotography, medical imaging and mathematical diagrams. They can also include computer models and simulations (Chapters 1 and 7).

- Data visualisation for exploration. These visualisations explore patterns, trends, networks and relationships and can include visualisations used in social network analysis (Chapter 53), data analytics (Chapter 10) and social media analytics (Chapter 52), for example.

- Data visualisation for forecasting and predicting. These visualisations are used to predict future outcomes and/or to make forecasts about events that take place at any time. They can be used in predictive analytics (Chapter 10), predictive modelling (Chapter 45) and geospatial analysis (Chapter 21), for example.

- Data visualisation for explanation and description. These visualisations help to explain or describe human behaviour or natural phenomena and can include visualisations used in business intelligence (Chapter 4), mapping (Chapter 60) and education (Chapter 24), for example.

- Data visualisation for communication and for communicating research results. These visualisations seek to provide information, transfer knowledge, create explanations or tell a story. They can include static visualisations such as pie charts, flow charts and graphs, and interactive visualisations used to communicate a particular message or tell a story (Chapter 15), for example.

There are different types of data visualisation: the type(s) that is used depends on research methodology and methods, the purpose and subject of the research, the digital tool or software package that is chosen, preferences of the researcher and the particular stage of a research project. Examples of different types of data visualisation include:

- animated and visual simulations (models that predict or explain);
- data art or information art that blurs the boundaries between art and information;

- interactive visualisations that encourage interaction and connectivity, enabling the user to discover information in data that is not immediately apparent;

- one-, two-, three- and multi-dimensional visualisations such as cartograms, dot distribution maps, bar charts, pie charts and histograms;

- sequential and temporal visualisations, such as connected scatter plots and timelines;

- static and interactive infographics that can be digested quickly and shared with others (these present data in a simple format, are visually-engaging and are seen to be 'link worthy' and/or 'share worthy' and are used, increasingly, in qualitative approaches);

- stratified or ranking visualisations that illustrate structure and groupings, such as dendrograms and tree maps;

- three-dimensional volumetric visualisations, such as fluid flows or topographic scans;

- web, grid or network visualisations that illustrate connections, interconnections and networks such as matrices and hive plots.

If you are interested in using data visualisation in your research project a good starting point is Kirk (2016), who takes time to explore the notion and practice of the visualisation design process, which will enable you to plan, design and move forward with your project. You can also visit Kirk's website (www.visualisingdata.com) to find informative blogs and articles on data visualisation. Another useful source is Healy (2019), which provides a comprehensive introduction to the principles and practice of data visualisation along with practical instruction using R and ggplot2. If you are interested in using Excel charts and graphs, Evergreen (2016) provides user-friendly advice and guidance on using the most appropriate types of visualisation for your research. Ware (2013) explores issues of perception and vision: why we see objects in the way that we do. He provides useful reading and practical guidelines for anyone interested in producing data visualisations. Yau (2013) provides an informative discussion on the graphic side of data analysis and data representation, along with interesting examples. There are a wide variety of digital tools and software packages available for data visualisation, and some of these are listed below. It is useful to visit some of these sites to try out the tools and software so that you can gain a deeper understanding of those that might be useful for your research.

Questions for reflection

Epistemology, theoretical perspective and methodology

- In what way are beliefs about the use, value and meaning of visualisations influenced by epistemological position and theoretical perspective? How does this influence how we produce, understand, respond to or use a visualisation? For example, a researcher approaching their work from an objectivist standpoint, producing data visualisation through diligent and systematic enquiry with strict adherence to scientific method, might view their work as a neutral lens through which to view factual, reliable and valid data. Researchers approaching their work from a constructionist standpoint, however, might think about how the meaning of a visualisation is culturally (and collectively) transmitted and viewed: meaning is constructed in different ways, depending on who produces and who views the visualisation. This leads to further questions:

    o How is understanding and response connected to audience (who is viewing, when, where and how)?

    o How is understanding and response connected to type of visualisation (what has been produced, how, why and in what form)?

    o How do we see what we see? How do we perceive what we see? How do we make judgements about what we see (drawing conclusions, making generalisations and developing models and hypotheses, for example)?

    o How do perceptions of credibility of researchers, data scientists and other 'brokers' affect response to visualisations (Allen, 2018)?

- How do you intend to use data visualisation in your research? For example, you could use it to:

    o explore data, develop hypotheses and construct knowledge;

    o analyse data;

    o aggregate, amalgamate and synthesise data;

    o collaborate and work interactively with others;

    o evaluate data, draw insight and make discoveries;

    o present data and share information.

Ethics, morals and legal issues

- Have the original data been generated with strict adherence to issues such as data corruption, statistical confidentiality, privacy rules, information security, unlawful disclosure, data ownership and sharing of data? Are you aware of these issues and how they might affect your use of data?

- Have you paid close attention to references, attributions, annotations and labelling when using data from different sources?

- Can you ensure that your data visualisation is not misleading, confusing or inaccurate? What procedures and techniques will you develop to demonstrate transparency in the methods you use to create your visualisations? How will you ensure that your visualisations do not lead to false conclusions?

- What bias might be present in data visualisation? This could include:

  o Western-centric, high-level visualisation aimed at a small audience that can be too complex or misleading for certain audiences;

  o biased language that stereotypes or alienates;

  o visualisation that backs up ideology (see Kennedy et al., 2016);

  o narrative bias that chooses a particular story to illustrate;

  o cultural biases of both producer and viewer that can lead to confusion, misunderstanding and alienation.

Practicalities

- Where will you find your data? Are data valid and reliable? Can quality be assured? How, when, why and by whom were data generated?

- Will data need to be converted into a format that you can use (Chapter 11)? Are tools available? Do you need to clean the data (delete duplicates and ensure consistency, for example)? How can you ensure that data are not lost during conversion?

- What processing and analysing tasks do you need to undertake (sorting, filtering and aggregating, for example)?

- What data visualisation tool/software do you intend to use? What tasks can be performed? Are clear instructions available? Is help or training available, if required (in-house or online courses, for example: see below)?

- What processes and/or procedures can corrupt data and subsequent visualisations? Are there particular processes and/or stages when corruption can occur? How can corruption be avoided?

- Do you understand how to use elements such as colour, lines lengths, tone, density and shapes to produce an effective visualisation? Ware (2013) will help you to think about these design issues.

- Once visualisations are produced are they clear and easy to understand? Do they help point to patterns or trends? Again, Ware (2013) will help you to address these questions.

Useful resources

Examples of different types of data visualisation tool and software that are available at time of writing are listed below (in alphabetical order). This list includes tools that create maps, explore networks, produce infographics and visualise words in text. Some are free to access whereas others require registration and/or a subscription.

- Carto (https://carto.com);
- Chartblocks (www.chartblocks.com);
- Datawrapper (www.datawrapper.de);
- Dygraphs (http://dygraphs.com);
- Excel (https://products.office.com/en-GB/excel);
- Gephi (https://gephi.org);
- Paraview (www.paraview.org);
- Pitochart (https://piktochart.com);
- Plotly (https://plot.ly);
- R (www.r-project.org);
- Raw (http://raw.densitydesign.org);
- Tableau public (https://public.tableau.com);
- Visualize Free (http://visualizefree.com);
- Wordle (www.wordle.net).

The Data Visualisation Catalogue (www.datavizcatalogue.com) provides a useful introduction to data visualisation, along with lists of tools that can be used to generate different visualisations.

Researchers at the Oxford Internet Institute, University of Oxford, are in the process of developing 'online wizards to let users easily create and customize interactive visualizations with their own network and geospatial data'. More information about this project, along with visualisation demonstrations, can be obtained from their website (https://blogs.oii.ox.ac.uk/vis/visualization-demos).

'Seeing Data' (http://seeingdata.org) is a research project that 'aims to improve the production of data visualisations and to aid people in developing the skills they need to interact with them'. The website contains informative articles for people who are new to data visualisation, including a section on 'developing visualisation literacy' that is aimed at people who are not experts in the subject.

There are various MOOCs (massive open online courses) available covering data visualisation. Details of relevant courses can be obtained from MOOC platforms:

- Coursera (www.coursera.org);
- edX (www.edx.org);
- FurtureLearn (www.futurelearn.com);
- open2study (www.open2study.com).

Key texts

Allen, W. (2018) 'Visual Brokerage: Communicating Data and Research through Visualisation', *Public Understanding of Science*, 27(8), 906–22, first published February 5, 2018, 10.1177/0963662518756853.

Evergreen, S. (2016) *Effective Data Visualization: The Right Chart for the Right Data*. Thousand Oaks, CA: Sage.

Healy, K. (2019) *Data Visualization A Practical Introduction*. Princeton, NJ: Princeton University Press.

Kennedy, H., Hill, R., Aiello, G. and Allen, W. (2016) 'The Work that Visualisation Conventions Do', *Information, Communication and Society*, 19(6), 715–35, published online March 16, 2016.

Kirk, A. (2016) *Data Visualisation: A Handbook for Data Driven Design*. London: Sage.

Ware, C. (2013) *Information Visualization: Perception for Design*. Waltham, MA: Elsevier.

Yau, N. (2013) *Data Points: Visualization that Means Something*. Indianapolis, IN: Wiley Publishing, Inc.

CHAPTER 14

Digital ethnography

Overview

The term 'digital ethnography' is an umbrella term that encompasses all forms of digital technology-based ethnography including online (Chapter 37), mobile (Chapter 31) and offline ethnography involving digitalisation (Varis, 2014). The term also encompasses other online ethnography such as virtual ethnography, netnography, network ethnography, cyber ethnography, webethnography and webnography, all of which are discussed in Chapter 37. It is important to note, however, that terms used to describe digital ethnography are discretionary, flexible and fluid, with no settled and clear typology for different approaches or research focus. Varis (2014) provides a detailed list, with references, of the various terms that can be used to describe research that adopts digital ethnography as an approach. She also provides a brief and insightful discussion on the history of digital ethnography.

Ethnography is the study of individuals, groups or communities in their own environment over an extended period of time. The focus is to describe and interpret cultural behaviour and phenomena. Although ethnography is a methodology rather than a method (it provides an overall framework to guide research rather than a specific technique or tool for collecting and analysing data), it has been included in this book because there are many 'how to' issues involved in digital ethnography. A related method is ethno-mining, which enables researchers to study human behaviour and culture through combining ethnography with data mining tools and techniques: the two approaches are combined, integrated and interweaved to form a more complete picture, understanding or interpretation of the behaviour or culture under study (Chapter 18).

Digital ethnography

The 'digital' of ethnography includes a vast array of technologies, platforms, communication channels and media channels that are available for collecting and/or analysing data in ethnographic research. This includes:

- mobile devices and apps (Chapters 31 and 50);
- computers and software;
- the internet, the web and the dark web;
- social media and social networks (Chapter 53);
- news sites, newsgroups and wikis;
- chat rooms, discussion groups, forums and study circles;
- blogs, vlogs, videos and podcasts (including personal diaries and logs: Chapter 30);
- micro-blogging;
- email, texting and messaging;
- cameras, webcams and videoconferencing;
- digital objects, images and artefacts (Chapter 48);
- gaming and virtual worlds (Chapters 20 and 56);
- digital sensors, wearables and eye-tracking technology (Chapters 49, 57 and 19);
- datasets (big data, small data and open data) and data mining tools (Chapters 3, 12 and 18);
- data analysis software and data visualisation tools (Chapters 9 and 13);
- mass media, public and individual discourses about the internet or web.

A variety of research methods can be adopted within digital ethnography including online interviews (Chapter 40), online observations (Chapter 41), online focus groups (Chapter 39) and researching digital objects (Chapter 48). Ethnographers can also adopt offline digital methods, such as tools that enable researchers to collect, store and analyse data digitally (Chapters 8 and 9). Researchers can adopt a mixed methods approach within their methodological framework, combining methods such as statistical analysis, discourse analysis, small stories research or offline participant observation, for example (see Chapter 37 for relevant references). Examples of research projects that have used, assessed or critiqued digital ethnography include research into a dark-web drug use community (Barratt and Maddox, 2016); a study into how content users navigate 'imagined audiences' on Twitter (Marwick and boyd, 2011); an assessment of digital ethnography as a formal methodology for criminal and hidden online populations (Ferguson, 2017); a discussion on 'how digital tools can make the ethnographic approach a collaborative analysis of human experience' (Cordelois, 2010); research into responses to internet censorship and tracking in Turkey (Cardullo, 2015); and an assessment of pregnancy apps as complex ethnographic environments (Barassi, 2017).

If you are interested in finding out more about digital ethnography for your research, a good understanding of epistemology, theoretical perspective and/or methodology that can inform and guide digital ethnography is important. Hjorth et al. (2017) provide a good starting point, with the first part of the book providing a topical debate about digital ethnography. Hammersley and Atkinson (2019) discuss ethnography as a reflexive process and include a new chapter on ethnography and the digital world in the fourth edition of their book. Hine (2015) proposes and discusses a multi-modal approach to ethnography for the internet and Pink et al. (2015) provide practical and comprehensive advice on ethnographic research methods (researching experiences, localities, social worlds and events, for example). Information about the practicalities of undertaking digital ethnography in virtual worlds can be found in Boellstorff et al. (2012), along with a brief history of ethnography and a discussion about 'myths' associated with ethnography. If you are interested in finding out more about digital anthropology (the study of human societies to gain a deeper understanding of human social and cultural life, and what it means to be human), Horst and Miller (2012) provide an interesting collection of papers, with some useful case studies and methodological discussion. There are a wide variety of digital tools and software packages available for those interested in digital ethnography, and details of these can be found in Chapter 18 (ethno-mining), Chapter 31 (mobile ethnography) and Chapter 37 (online ethnography).

Questions for reflection

Epistemology, theoretical perspective and methodology

* Have you thought about, and come to terms with, your epistemological position and theoretical perspective and the influence that these have on your choice of digital ethnography as a methodology? For example, reflexive

or critical ethnography (involving an ideological critique and questioning the status quo of power relations) and naturalistic ethnography (founded on positivism and based on the legacy of colonialism) both have different influences on the way that your study is conducted and data analysed. How will you address these issues? Pink et al. (2015) provide detailed information about ethnography, theoretical perspective and disciplinary approach, which will help you to address these questions.

- How do you intend to take account of possible contextual complexities of online communities or communicators? For example, people within a social network can come from very different spheres of a person's life: how do individuals imagine their audience? Marwick and boyd (2011) call this 'context collapse' and discuss this concept in their paper.

- What level of engagement will you adopt? How can you deal with what de Seta (2017) in her guest blog for Anthro{dendum} describes as the false choice between naturalist lurking or active involvement (see below)?

- How do you intend to recognise and acknowledge researcher bias and preconceived notions (the influence of personal or research agendas, personal history and political beliefs, for example)? What quality controls will you develop and use to reduce bias, when sampling, selecting and reporting? Is it possible to use mixed methods or triangulation, to reduce problems that could occur as a result of bias?

Ethics, morals and legal issues

- Are you aware of issues surrounding illegal, unlawful, illicit or dangerous behaviour in the digital world and possible consequences? Are you clear about how the law works in the relevant country (if you are subpoenaed to attend court, for example)? See Barratt and Maddox (2016) for an example of how to conduct research in an online environment where sensitive and illicit activities are discussed.

- What are the potential harms and benefits for participants taking part in the research (conflict, disagreement, exposure or legitimisation, for example)?

- When using material produced by others do you have a clear and full understanding of what is meant by public domain, copyright, fair dealing or fair use and Creative Commons licences? 'Public domain' is a legal term that refers to works on which rights have been forfeited, waived or expired. These rights, however, can be country-specific (the works might be in the public domain in one country, but subject to rights in another). 'Copyright' grants the creator exclusive rights to publish, reproduce and distribute the work for a number of years. It is infringed if another person does this without seeking permission, even if the work is referenced correctly. It is possible to use 'fair dealing' (UK) or 'fair use' (US) as a defence against copyright infringement. Creative Commons licences makes the work available to the public for certain uses while copyright is still maintained (a 'No Rights Reserved' licence, for example).

- How do you intend to address issues of informed consent, anonymity, confidentiality, privacy and authenticity? See Chapter 40 for more information about obtaining informed consent, Chapter 52 for more information about anonymity and Chapter 37 for more information about privacy and authenticity.

- Will your proposed project obtain ethical approval? Research Ethics Committees or Ethics Review Boards are bodies that have been set up around the world to protect the rights, safety, health and well-being of human subjects involved in research. The work of approval and review bodies is to make an assessment of, and ruling on, the suitability of research proposals, investigators, research methods and facilities.

- Have you thought about data protection and security issues? Will you need a data protection and management plan before ethical approval will be granted?

Practicalities

- What data collection methods do you intend to use? This can include, for example, online observation (Chapter 41), face-to-face interviews utilising mobile digital technology (Chapter 8) or online interviewing (Chapter 40). Do you have the required understanding and experience of these methods, or do you need to update your knowledge? Boellstorff et al. (2012) and Hine (2015) will help you to reflect on these questions.

- What data analysis methods do you intend to use? This can include, for example, content analysis, discourse analysis, thematic analysis, computer-assisted analysis or a combination of a variety of methods. Hammersley and Atkinson (2019) will help you to think more about data analysis and writing research.

- Are digital tools and software available, accessible, affordable and easy to use (or can you undertake the required training and become competent in their use)? This can include, for example, ethnographic apps for observation or data analysis (Chapter 31), online ethnographic research tools and templates (Chapter 37) and qualitative data analysis software (Chapter 9).

- How do you intend to access the online group, community or communication channel? Is it public or private? Do you need to obtain permission or obtain entry through some type of gatekeeper?

- Are your observations overt or covert? Will people know that they, or their online communication, are part of a research project? Will you use a real name, pseudonym or avatar, for example (see Chapter 56 for more information about the use of avatars in research)? If you are using your real name, will your current online presence have an influence on your research?

Useful resources

Visit the Digital Ethnography Research Centre, RMIT University, Australia (http://digital-ethnography.com/projects) for examples of digital ethnographic research projects.

The Media Anthropology Network, European Association of Social Anthropologists (www.media-anthropology.net) is a network that has been set up to discuss and collaborate on the anthropology of media. You can find interesting e-seminar reports about various aspects of digital ethnography on this site.

Anthro{dendum} (https://anthrodendum.org) is a group blog devoted to 'doing anthropology in public'. On this site you can find interesting articles about digital ethnography, in particular, Gabriele de Seta's guest blog (2017) 'Three Lies of Digital Ethnography': https://anthrodendum.org/2018/02/07/three-lies-of-digital-ethnography [accessed March 15, 2018].

Key texts

Barassi, V. (2017) 'BabyVeillance? Expecting Parents, Online Surveillance and the Cultural Specificity of Pregnancy Apps', *Social Media + Society*, first published May 19, 2017, 10.1177/2056305117707188.

Barratt, M. and Maddox, A. (2016) 'Active Engagement with Stigmatised Communities through Digital Ethnography', *Qualitative Research*, 16(6), 701–19, first published May 22, 2016, 10.1177/1468794116648766.

Boellstorff, T., Nardi, B., Pearce, C. and Taylor, T. (2012) *Ethnography and Virtual Worlds: A Handbook of Method*. Princeton, NJ: Princeton University Press.

Cardullo, P. (2015) '"Hacking Multitude" and Big Data: Some Insights from the Turkish "Digital Coup"', *Big Data & Society*, first published April 14, 2015, 10.1177/2053951715580599.

Cordelois, A. (2010) 'Using Digital Technology for Collective Ethnographic Observation: An Experiment on "Coming Home"', *Social Science Information*, 49(3), 445–63, first published August 23, 2010, 10.1177/0539018410371266.

Ferguson, R.-H. (2017) 'Offline 'Stranger' and Online Lurker: Methods for an Ethnography of Illicit Transactions on the Darknet', *Qualitative Research*, 17(6), 683–98, first published July 26, 2017, 10.1177/1468794117718894.

Hammersley, M. and Atkinson, P. (2019) *Ethnography: Principles in Practice*, 4th edition. Abingdon: Routledge.

Hine, C. (2015) *Ethnography for the Internet*. London: Bloomsbury Academic.

Hjorth, L., Horst, H., Galloway, A. and Bell, G. (eds.) (2017) *The Routledge Companion to Digital Ethnography*. New York, NY: Routledge.

Horst, H. and Miller, D. (eds.) (2012) *Digital Anthropology*. London: Berg.

Marwick, A. and boyd, d. (2011) 'I Tweet Honestly, I Tweet Passionately: Twitter Users, Context Collapse, and the Imagined Audience', *New Media & Society*, 13(1), 114–33, first published July 7, 2010, 10.1177/1461444810365313.

Pink, S., Horst, H., Postill, J., Hjorth, L., Lewis, T. and Tacchi, J. (2015) *Digital Ethnography: Principles and Practice*. London: Sage.

Underberg, N. and Zorn, E. (2014) *Digital Ethnography: Anthropology, Narrative, and New Media*. Austin, TX: University of Texas Press.

Varis, P. (2014) *Digital Ethnography*. Tilburg University, Netherlands: Tilburg Papers in Culture Studies, 104.

CHAPTER 15

Digital storytelling

Overview

Digital storytelling is a participatory and/or collaborative research method that enables people and communities to communicate, share their lives and illustrate their lived realities, from their perspective. This method can be seen as both a form of art and a way to communicate and record information. Digital stories can be nonlinear, interactive and visually-engaging, using a wide variety of features such as photography, music, text, graphics, narration, title screens, infographics, memes and sound effects. New technologies and accessible production and distribution media (mobile technologies and the internet, for example) have led to rapid increases in people compiling and sharing life stories and personal narratives in a variety of ways (podcasts, YouTube videos, film and via social media platforms). The above is a broad definition of the term. A narrower definition is used by researchers who follow the techniques developed by the Centre for Digital Storytelling, now known as Story-Center (www.storycenter.org). This uses workshops to train participants to construct their own, short digital story, ensuring that issues such as confidentiality and ownership are addressed. Wexler et al. (2014) provide a good example of how this workshop technique was used in their research with Alaskan native young people.

Digital storytelling can also be used as a powerful tool for teaching and learning. If you are interested in finding out more about this technique, Clarke and Adam (2012) discuss the use of digital storytelling as a pedagogical tool in higher education contexts and Sheafer (2017) provides an evaluation of digital storytelling as a teaching method in psychology. Alterio and McDrury (2003) give a comprehensive account of different models of storytelling; Ohler (2013) looks at the art, practice and digitisation of digital storytelling; and Dawson (2019) provides a variety of practical activities that use digital storytelling for teaching study skills in higher education.

There are various types and styles of digital story that can be used in research (and in the classroom), depending on the purpose of the study; the type and level of instruction given to potential storytellers; the preferences, understanding and creativity of the storytellers; the level of researcher and participant interaction and collaboration; and the digital technology available. Examples of styles and types include:

- personal narratives, reflective accounts, autobiographies and life histories;

- anecdotes, scenarios and case studies;

- stories recounting specific or historical events;

- stories that retell the story of another person;

- descriptions or depictions of behaviour and actions;

- stories that provide detailed instruction or information,

- fiction, legend or myth.

Examples of projects that have used, assessed and/or critiqued digital storytelling as a research method include an exploration of the ethical complexities of sponsored digital storytelling (Dush, 2013); an investigation into sentimentality in digital storytelling using war veterans' digital stories (McWilliam and Bickle, 2017); research into the digital storytelling practices between an African American mother and son (Ellison and Wang, 2018); an investigation into the power of digital storytelling as a culturally relevant health promotion tool (Briant et al., 2016); and research to better understand the lives of Alaskan native young people (Wexler et al., 2014).

There are a variety of ways in which digital stories can be analysed, depending on theoretical perspective, methodological framework and digital medium. Methods include discourse analysis, content analysis, comparative analysis, frame analysis, thematic analysis and theoretical analysis (a brief definition of different methods of qualitative data analysis can be found in Chapter 9). It can also include the analysis of video and audio: information about video analysis is provided in Chapter 55 and information about audio analysis is provided in Chapter 2. If you decide to use coding for analysis, more information about coding and retrieval is provided in Chapter 6 and if you decide to use computer-assisted qualitative data analysis software (CAQDAS), more information is provided in Chapter 9.

If you are interested in finding out more about digital storytelling as a method of inquiry, it is useful to start with Lambert and Hessler (2018) as they provide a good overview of digital storytelling, along with an exploration of global applications of the method. The book also provides useful advice and guidance for those who are interested in producing their own digital story, based on the workshop method described above. Dunford and Jenkins (2017) provide some interesting case studies from around the world that help to address questions of concept, theory and practice. Mobile technology can be used for digital storytelling, and more information about the variety of mobile methods that are available for research purposes is given in Chapter 32. The questions for reflection listed below will also help you to think more about how digital storytelling can be used in research (and for your research project).

Questions for reflection

Epistemology, theoretical perspective and methodology

- What factors might influence the telling and/or interpretation of a digital story? Issues to consider include:
  - theoretical perspective of the researcher;
  - methodological framework of the research;
  - connection with the story;
  - type of story;
  - digital technology used;
  - emotional impact of the story;
  - previous experience of the storyteller;
  - levels of storyteller/researcher/sponsoring organisation collaboration and involvement.

- How authentic is a digital story and how authentic is the storyteller's voice? How important is authenticity? How do you determine authenticity? What effect do medium, compiling, editing and analysis have on authenticity?

- How can you address the issue of sentimentality in digital storytelling? McWilliam and Bickle (2017: 78) note that sentimentality is one of the most common criticisms of digital storytelling, but point out that 'senti-mentality may ultimately offer precisely the relatable, accessible, connec-tive potential between "ordinary" people that digital storytelling initially sought to offer, but has yet to fully realise, outside of the strictures of professional media'.

- How will you take account of different storytelling experiences, abilities and understanding? Some people are more confident about telling stories, and are more confident about using digital technology. What impact might this have on your research? Asking participants to work in groups, rather than individually, can help to address some of these issues. How-ever, this will have implications for analysis, interpretation and drawing conclusions.

Ethics, morals and legal issues

- How will you address ethical issues such as gaining the necessary permis-sions, data protection, anonymity, confidentiality, informed consent and the right to discontinue or withdraw at any time (and the destruction of record-ings)? Do these vary, depending on the type of digital story and the level of researcher/participant collaboration and involvement? How will you address these issues in cases where a story is told about another person who is not involved in the research and has not given consent? The hand-book available from Transformative Stories (www.transformativestory.org) provides useful guidance on ethical issues.

- How will you explain what is meant by respect to privacy (participants do not need to disclose information if they do not want to) and by protecting privacy (the researcher will not disclose information without their expli-cit consent)? Privacy issues can relate to individuals and to communities.

- Do you understand ethical implications associated with using images (gaining consent from participants captured in images and understanding the difference between ethically appropriate and ethically inappropriate manipulation of images, for example)?

- If you are working with vulnerable people, or with those who have upsetting or harrowing stories to tell, how can you take into account the physical, social and psychological well-being of storytellers and listeners? How can you demonstrate that care will be taken to avoid harm for those telling the stories and for those who might be affected by the stories?

- Do you have the required professional competence to deal with issues that could arise from stories, especially when they might cover sensitive topics or be told by vulnerable people?

Practicalities

- What digital technology and filming equipment is available? Do participants know how to use tools, software and equipment? Is training required? If so, who will carry out the training and are all participants available, and willing, to attend training sessions?

- What costs are involved for attending training sessions and for recording stories? How can you ensure that participants are not left out-of-pocket (in particular, when they are using their own equipment to record and upload their stories)?

- What analysis methods do you intend to use? This could include, for example, narrative analysis, comparative analysis, content analysis, thematic analysis, frame analysis, discourse analysis, video analysis and audio analysis. Is software available and do you know how to use the software? Are tutorials, support and assistance available, if required?

- Do you intend to share stories online, in local communities or at the policy, institutional or research group level? Do storytellers understand how and in what way their stories might be shared?

Useful resources

There are a wide variety of tools and software available to help research participants (and students) produce digital stories. Some examples are given below, in alphabetical order.

- Camtasia for screen recording and video editing (www.techsmith.com/video-editor.html);

- GIMP free and open source software for image editing (www.gimp.org);

- Magix for creating, customising, editing and optimising videos, photos and music (www.magix.com);

- Moovly for creating, editing, exporting and sharing videos, advertisements and presentations (www.moovly.com);

- Pageflow open source software and publishing platform for multimedia storytelling (https://pageflow.io/en);

- Slidestory for photo and narration sharing (www.slidestory.com);

- PIXLR for mobile photo editing (https://pixlr.com);

- Storybird a language arts tool to inspire creative storytelling and improve reading and writing (https://storybird.com);

- Visage for on-brand graphics and brand storytelling (https://visage.co);

- WeVideo for capturing, creating, viewing and sharing videos and movies (www.wevideo.com).

Transformative Stories (www.transformativestory.org) contains a useful handbook for anyone interested in digital storytelling, covering 'conceptual reflections, practical experiences, and methodological guidance on transformative creative and visual storytelling methods'.

If you are interested in using digital storytelling in the classroom the Higher Education Academy in the UK has a useful section on 'Learning through Storytelling' (www.heacademy.ac.uk/enhancement/starter-tools/learning-through-storytelling). Here you can find information about the history of storytelling, how it can be used and how to get started.

The following websites provide examples of digital storytelling projects:

- www.photovoice.org (participatory photography and visual storytelling to build skills within disadvantaged and marginalised communities);

- www.patientvoices.org.uk (telling and sharing of reflective stories to transform healthcare);

- http://myyorkshire.org (digital stories created with communities by museum, library and archive collections in Yorkshire, UK).

Key texts

Alterio, M. and McDrury, J. (2003) *Learning through Storytelling in Higher Education: Using Reflection and Experience to Improve Learning.* London: Routledge.

Briant, K., Halter, A., Marchello, N., Escareño, M. and Thompson, B. (2016) 'The Power of Digital Storytelling as a Culturally Relevant Health Promotion Tool', *Health Promotion Practice*, 17(6), 793–801, first published July 8, 2016. doi:10.1177/1524839916658023.

Clarke, R. and Adam, A. (2012) 'Digital Storytelling in Australia: Academic Perspectives and Reflections', *Arts and Humanities in Higher Education*, 11(1–2), 157–76, first published June 20, 2011. doi:10.1177/1474022210374223.

Digital storytelling

Dawson, C. (2019) *100 Activities for Teaching Study Skills*. London: Sage.

Dunford, M. and Jenkins, T. (eds.) (2017) *Digital Storytelling: Form and Content*. London: Palgrave Macmillan.

Dush, L. (2013) 'The Ethical Complexities of Sponsored Digital Storytelling', *International Journal of Cultural Studies*, 16(6), 627–40, first published October 4, 2012. doi:10.1177/1367877912459142.

Ellison, T. and Wang, H. (2018) 'Resisting and Redirecting: Agentive Practices within an African American Parent–Child Dyad during Digital Storytelling', *Journal of Literacy Research*, 50(1), 52–73, first published January 17, 2018. doi:10.1177/1086296X17751172.

Lambert, J. and Hessler, B. (2018) *Digital Storytelling: Capturing Lives, Creating Community*, 5th edition. New York, NY: Routledge.

McWilliam, K. and Bickle, S. (2017) 'Digital Storytelling and the "Problem" of Sentimentality', *Media International Australia*, 165(1), 77–89, first published August 28, 2017. doi:10.1177/1329878X17726626.

Ohler, J. (2013) *Digital Storytelling in the Classroom: New Media Pathways to Literacy, Learning, and Creativity*, 2nd edition. Thousand Oaks, CA: Corwin.

Sheafer, V. (2017) 'Using Digital Storytelling to Teach Psychology: A Preliminary Investigation', *Psychology Learning & Teaching*, 16(1), 133–43, first published December 22, 2016. doi:10.1177/1475725716685537.

Wexler, L., Eglinton, K. and Gubrium, A. (2014) 'Using Digital Stories to Understand the Lives of Alaska Native Young People', *Youth & Society*, 46(4), 478–504, first published April 9, 2012. doi:10.1177/0044118X12441613.

CHAPTER 16

Digital visual methods

Overview

Digital visual methods (or digital image research) are used to research computer-based or virtual visual images that are shared and displayed online. These images include video, film, photographs (and selfies), drawings, sculptures, graphic novels, infographics, memes, presentations, screenshots, cartoons and comics, for example. They can be shared and displayed on social networking sites, microblogging sites, personal or organisation websites (including digital archives), blogs and image hosting sites, or through instant messaging, texts and email, for example. Digital visual methods can be used in qualitative research to analyse the ways that human beings communicate through visual images or to help understand and explain human interaction and behaviour, and in quantitative research to communicate results, depict phenomena that cannot be seen or explain relations between properties (digital images from electron microscopes, astrophotography, mathematical diagrams, pie charts, flow charts and graphs, for example).

There are different ways that digital visual images can be used for research purposes. These include:

- Studying phenomena or society through the production of digital images or visualisations by the researcher (the researcher produces images such as mathematical representations, models, diagrams, graphs and charts to help describe or explain results, form hypotheses, develop themes or make predications and forecasts, for example). This type of image production can be referred to as data visualisation: relevant methods, techniques, tools and software are discussed in Chapter 13.

- Studying digital images created by others to help understand and explain phenomena, human interaction, behaviour and social life (this can include infographics, cartoons, comics, video and photography). These images have been created not for the purpose of research, but are used for research once the images become available. A good example of this type of research is provided by Roberts and Koliska (2017) in their study of the use of space in selfies on Chinese Weibo and Twitter.

- Studying the platform structures, agencies and media forms that underpin the circulation of digital images: how these might influence image type and how they change over time and among different people and cultures. This can include investigations into images that are generated by code, networked images, repetition of images, authorship of images and the availability, popularity and restrictions of different hosting services or platforms. It can also include cross-platform analysis that considers images (still or moving) across two or more social media platforms. Pearce et al. (2018) provide more information about this method.

- Studying digital visual images that have been created by research participants for the purpose of research (a child can be asked by the researcher to create a digital image on a particular subject, for example). The researcher provides instructions on what to produce and/or provides technology with which to produce the digital image. See Banks (2007) for more information about this method and other visual methods.

- Encouraging others to produce digital images by giving them control of recording equipment and digital technology. This type of participatory or collaborative method enables people and communities to communicate, share their lives and illustrate their lived realities, from their perspective. Participatory visual methods include participatory photography, digital storytelling (Chapter 15) and collaborative or participatory film and video-making. Elwick (2015) provides a good example of this type of research in her work with 'baby-cams'.

- Using digital images alongside other methods in a mixed methods approach to broaden understanding and gain deeper insight. See Shannon-Baker and Edwards (2018) for an interesting discussion on mixed methods approaches that utilise visual methods.

Researchers from a wide variety of disciplines and fields of study use digital visual methods including media and communication studies, visual culture, software studies, sociology, psychology, computer science, information science, biology, health and medicine, geography and environmental science. Examples of research projects that have used, assessed or critiqued digital visual methods include a study into digital photography and how it is signalling a shift in engagement with the everyday image (Murray, 2008); an assessment of the use of a small digital-camera system worn by infants and a tripod-mounted digital camera for researching infants (Elwick, 2015); research into the impact of mobile phone photos shared on social media on soldiers and civilians in Israel and Palestine (Mann, 2018); an exploration into gender inequities and sexual double standards in teens' digital image exchange (Ringrose et al., 2013); and a discourse analysis of how people make meaning through selfies (Veum and Undrum, 2018).

A good understanding of visual methodologies is important if you are interested in digital visual methods for your research. Pink (2012) and Rose (2012) provide detailed information that will help you to get to grips with both theory and practice. Both books are aimed at researchers from the arts, social sciences and humanities. Gubrium and Harper (2013) provide a comprehensive guide for those interested in participatory visual and digital methods and Heath et al. (2010) provide a useful guide for those who are interested in the use of video (see Chapter 55 for information about video analysis). Chapter 13 provides information for those who are interested in data visualisation for their research (to better understand the data, see patterns, recognise trends and communicate findings, for example). If you are interested in digital storytelling using film and photography, more information is provided in Chapter 15 and if you are interested in finding out more about researching digital objects, see Chapter 48 (representations of digital objects exist within digital space and can include accounts and narratives, songs, music, images, photographs, artworks, video games, virtual worlds and virtual environments).

Questions for reflection

Epistemology, theoretical perspective and methodology

- Are digital images a representation of reality? Can digital images be accurate and true? Can they render an objective account? Elwick (2015) adopts an interpretative approach to her research, discussing how and why digital images cannot be viewed as objective accounts and why they should not be taken too literally.

- What role does digital technology play in enabling creators or producers to manipulate, alter or change images? Can we trust what we are seeing? Does it matter if an image (such as a photograph) has been altered digitally? What are the implications for transparency?

- In what ways are digital images influenced by culture, history, society, politics and availability of, and access to, digital technology?

- How might digital viewing be influenced by others, by the viewer and by viewing platforms or media? How might the way an image is viewed be influenced by how many people view, at what time (on multiple screens simultaneously, for example) and on what device? How might number of views, likes or comments influence how, why and when a digital image is viewed?

- Does method of distribution (image hosting sites, personal email or social networking sites, for example) have an effect on digital images and the way that they are received, displayed and viewed (public and private settings, page visibility or password protection, for example)?

- How can you overcome problems that may arise when visual materials might be interpreted differently by researchers and participants? Glaw et al. (2017) propose elicitation interviewing as a way to address this issue in their paper on autophotography.

Ethics, morals and legal issues

- Cox et al. (2014: 5) point out that 'visual methods require researchers to rethink how they need to respond to key ethical issues, including confidentiality, ownership, informed consent, decisions about how visual data will be displayed and published, and managing collaborative processes'. How do you intend to address these issues in your research?

- How might restrictions on digital images (type, size, licences, laws and conventions) influence image, distribution, sharing and viewing?

- What can digital images tell us about power and prejudice? How can digital images be used to reinforce existing power structures?

- How can digital images help to create social change?

- Do you understand copyright rules and regulations concerning digital images? A brief summary of pertinent issues in the UK relating to images and the internet can be obtained from www.gov.uk/government/publications/

copyright-notice-digital-images-photographs-and-the-internet [accessed November 19, 2018].

Practicalities

- Do you have a good understanding of the different methods that can be used to analyse digital images? This can include:

  - quantitative content analysis that is used to quantify visual images (and text) using reliable and previously defined categories that are exhaustive and mutually exclusive;

  - inductive content analysis that uses an inductive approach to develop categories, identify themes and develop theory from visual images;

  - visual semiotics that considers the way that visual images communicate a message through signs, the relationship between signs (including systems of meaning that these create) and patterns of symbolism;

  - visual rhetoric that considers the message being portrayed in an image and the meaning that is being communicated;

  - psychoanalysis that explores how psychoanalytic concepts are articulated through a particular image (space, gaze, spectator, difference, resistance and identification, for example);

  - discourse analysis that involves an interpretative and deconstructive reading of an image but does not provide definitive answers or reveal the truth (how images are given particular meanings, how meanings are produced and how a particular image works to persuade, for example);

  - iconographic analysis that seeks to establish the meaning of a particular image at a particular time and build an understanding of the content and meaning of symbolic values and representations;

  - visual hermeneutics that seeks to understand rather than offer a definitive explanation of the visual image (the use and understanding of visual images is based on socially established symbolic codes and the meaning and interpretation of images are created by social interaction between the person who created the image, the image itself, the viewer and the method of viewing, for example);

- Do you have a good knowledge of platforms, software and tools that are available for hosting, viewing, producing and analysing digital images?

Useful resources

The Image and Identity Research Collective (IIRC) (http://iirc.mcgill.ca) is 'an informal network of professors, researchers, artists and other professionals interested in using images, artistic forms of representation, and/or innovative interdisciplinary research methodologies in work around issues related to identity and self-study'. There is some useful information about visual methodology and analysing visual data on this site.

The National Archives Image Library (https://images.nationalarchives.gov.uk/assetbank-nationalarchives/action/viewHome) in the UK contains over 75,000 images available to download 'spanning hundreds of years of history from The National Archives' unique collections, from ancient maps to iconic advertising'.

Key texts

Banks, M. (2007) *Using Visual Data in Qualitative Research*. London: Sage.

Cox, S., Drew, S., Guillemin, M., Howell, C., Warr, D. and Waycott, J. (2014) *Guidelines for Ethical Visual Research Methods*. Melbourne Social Equity Institute, University of Melbourne, retrieved from https://artshealthnetwork.ca/ahnc/ethical_visual_research_methods-web.pdf [accessed November 16, 2018].

Elwick, S. (2015) 'Baby-Cam' and Researching with Infants: Viewer, Image and (Not) Knowing', *Contemporary Issues in Early Childhood*, 16(4), 322–38, first published November 27, 2015, 10.1177/1463949115616321.

Glaw, X., Inder, K., Kable, A. and Hazelton, M. (2017) 'Visual Methodologies in Qualitative Research: Autophotography and Photo Elicitation Applied to Mental Health Research', *International Journal of Qualitative Methods*, first published December 19, 2017, 10.1177/1609406917748215.

Gubrium, A. and Harper, K. (2013) *Participatory Visual and Digital Methods*. Walnut Creek, CA: Left Coast Press, Inc.

Heath, C., Hindmarsh, J. and Luff, P. (2010) *Video in Qualitative Research*. London: Sage.

Mann, D. (2018) 'I Am Spartacus': Individualising Visual Media and Warfare', *Media, Culture & Society*, first published March 16, 2018, 10.1177/0163443718764805.

Margolis, E. and Pauwels, L. (eds.) (2011) *The SAGE Handbook of Visual Research Methods*. London: Sage.

Murray, S. (2008) 'Digital Images, Photo-Sharing, and Our Shifting Notions of Everyday Aesthetics', *Journal of Visual Culture*, 7(2), 147–63, first published August 1, 2008, 10.1177/1470412908091935.

Paulus, T., Lester, J. and Dempster, P. (2013) *Digital Tools for Qualitative Research*. London: Sage.

Pearce, W., Özkula, S., Greene, A., Teeling, L., Bansard, J., Omena, J. and Rabello, E. (2018) 'Visual Cross-Platform Analysis: Digital Methods to Research Social Media Images',

Information, Communication & Society, first published June 22, 2018, 10.1080/1369118X.2018.1486871.

Pink, S. (ed.) (2012) *Advances in Visual Methodology*. London: Sage.

Ringrose, J., Harvey, L., Gill, R. and Livingstone, S. (2013) 'Teen Girls, Sexual Double Standards and "Sexting": Gendered Value in Digital Image Exchange', *Feminist Theory*, 14(3), 305–23, first published November 14, 2013, 10.1177/1464700113499853.

Roberts, J. and Koliska, M. (2017) 'Comparing the Use of Space in Selfies on Chinese Weibo and Twitter', *Global Media and China*, 2(2), 153–68, first published June 8, 2017, 10.1177/2059436417709847.

Rose, G. (2012) *Visual Methodologies: An Introduction to Researching with Visual Materials*, 3rdedition. London: Sage.

Shannon-Baker, P. and Edwards, C. (2018) 'The Affordances and Challenges to Incorporating Visual Methods in Mixed Methods Research', *American Behavioral Scientist*, 62 (7), 935–55, first published April 27, 2018, 10.1177/0002764218772671.

Veum, A. and Undrum, L. (2018) 'The Selfie as a Global Discourse', *Discourse & Society*, 29(1), 86–103, first published September 3, 2017, 10.1177/0957926517725979.

CHAPTER 17

Educational data mining

Overview

Educational Data Mining (EDM) is a non-intrusive method that utilises machine learning techniques, learning algorithms and statistics to analyse student learning in educational settings for the purposes of assessment, evaluation, teaching and research (enabling researchers and tutors to answer educational research questions, develop theory and better support learners, for example). EDM can be used to understand the learning activities of students, how they behave when learning, what they know and their level of engagement. It enables researchers and tutors to produce profiles of students, monitor progression, predict behaviour and future performance, identify at-risk students and improve retention rates. Through employing EDM techniques researchers, tutors and administrators are able to develop strategies to personalise learning (including personalised content to mobile users), provide individual support, provide feedback to tutors and students and help with planning and scheduling courses, coursework, courseware and assessments.

EDM is aligned closely with learning analytics (Chapter 24): both are concerned with the collection, analysis and interpretation of educational data and both extract information from educational data to inform decision-making. However, they differ from each other in that EDM reduces systems to components and explores each of these components, along with relationships to others, using techniques and methods such as discovery with models, classification, clustering, Bayesian modelling and relationship mining, for example (see Chapter 12 for information about different data mining techniques). Learning analytics, on the other hand, is concerned with an understanding of whole systems, using techniques and methods such as sense-making models, data visualisation (Chapter 13), social network analysis (Chapter 53) and learner success prediction. Baker and Inventado (2016: 86) believe that the differences between the two techniques are to do with emphasis: 'whether human analysis or automated analysis is central, whether phenomena are considered as systems or in terms of specific constructs and their interrelationships, and whether automated interventions or empowering instructors is the goal'. Liñán and Pérez (2015) provide a detailed discussion of the two methods and their specific differences, while pointing out that differences seem to be less noticeable as EDM and learning analytics evolve over time. Indeed, Baker and Inventado (2016) point out that EDM and learning analytics can be treated as interchangeable. As such, it is useful to read this chapter together with Chapter 24 for further insight and to gain a deeper understanding of the two methods.

Researchers working within a variety of disciplines and fields of study use EDM, including sociology, education, educational psychology, education computing and technology, linguistics and learning sciences. Examples of research projects that have used, assessed or critiqued EDM include a study to identify engineering students' English reading comprehension errors in Taiwan (Tsai et al., 2016); an investigation of longitudinal patterns in online learning and participation (Tang et al., 2018); a review of tools that can be used for EDM (Slater et al., 2017); an assessment of the commercial value and ownership of educational data (Lynch, 2017); a review of EDM and natural language processing in regard to nationwide inquiry-based learning, teaching and assessment (Li et al., 2018); and an examination and evaluation of the effectiveness of two wrapper methods for semi-supervised learning algorithms for predicting students' performance in examinations (Livieris et al., 2018).

If you are interested in EDM for your research Khan et al. (2018) provide a good starting point covering issues such as impact on students, the quality and reliability of data, the accuracy of data-based decisions and ethical implications surrounding the collection, distribution and use of student-generated data. ElAtia et al. (2016) present a series of papers that cover four guiding principles of EDM (prediction, clustering, rule association and outlier detection), the pedagogical applications of EDM and case studies that illustrate how EDM techniques have been applied to advance teaching and learning. Peña-Ayala (2014) provides an interesting collection of papers based on four topics: profile, student modelling, assessment and trends and Romero et al. (2010) provide a useful introduction to those new to EDM, although the book could do with a little updating given rapid advancement in digital tools and software over the last few years. It is useful, also, to find out more about data mining (Chapter 12), learning analytics (Chapter 24) and data analytics (Chapter 10). You may also find it useful to obtain more information about cluster analysis (Chapter 5), machine learning (Chapter 29), predictive modelling (Chapter 45) and social network analysis (Chapter 53).

Questions for reflection

Epistemology, theoretical perspective and methodology

- Is EDM or learning analytics the best approach for your research? Do you have a good understanding of the two techniques and their similarities and differences? More information about learning analytics can be found in Chapter 24 and Liñán and Pérez (2015) provide a comprehensive and enlightening discussion on the two techniques.

- What epistemological framework and theoretical perspective will guide your research? Aristizabal (2016: 131) suggests avoiding a 'merely empiro-positivist approach' and goes on to discuss how an interpretivist approach along with Grounded Theory may provide a suitable framework for EDM.

- Do you intend to adopt a 'discovery with models' approach (building a model to predict future performance, for example)? If so, do you have a good understanding of the theory behind this technique, the potential of this approach and issues concerning effectiveness, validity and reliability? Hershkovitz et al. (2013) will help you to answer these questions and more information about predictive modelling is provided in Chapter 45. You might also find it useful to obtain more information about computer modelling and simulation (Chapter 7).

Ethics, morals and legal issues

- Who owns educational data? Is it institutions, educators or students, for example? Who can access data? Can students obtain data that are held about them? Can individuals specify approved usage of data? Lynch (2017) provides an interesting discussion on these questions and other ethical issues.

- Can too much information be obtained about individuals? Should limits be set on how much educational data are collected?

- Are data to be shared? If so, how and with whom? How will stored data and data in transit be protected and kept secure?

- How can educational data be protected against commercial exploitation?

- How can you avoid bias in EDM? This can include language bias, search bias or overfitting avoidance bias, for example. Combining multiple models may help to address problems with bias.

Practicalities

- Do you understand the procedures and process involved in data mining? This can include cleaning, preparing, organising and creating data (manipulating, restructuring, engineering and labelling data, for example), analysing data (discovering and validating relationships, for example), visualising data and making predictions. Different tools and software can be used for each stage of this process: Slater et al. (2017) provide an informative paper that describes each of these stages and the various tools and software that can be used for each stage.

- Do you have a good idea of data mining techniques? This can include association, clustering, classification, decision trees, regression analysis and sequential pattern mining, for example (see Chapter 12 for definitions).

- What tools do you intend to use for EDM? Slater et al. (2017) will help you to consider your choices carefully and Buenaño-Fernández and Luján-Mora (2017) provide a useful review of three open source EDM tools (RapidMiner, Knime and Weka).

- Is your chosen tool(s) fully integrated with other software products or tools to facilitate data preparation and manipulation, statistical testing and reporting? Does it work with a wide range of databases, spreadsheets and visualisation tools, for example?

- What sources of data do you intend to use? This can include university/ academic transcripts, learning or classroom management systems, student identity cards with Radio Frequency Identification (RFID) technology (Chapter 57), intelligent tutoring systems, eye-tracking data (Chapter 19), module or major choices, admissions data and online activity, for example. See Chapter 24 for other sources of internal and external data that can be used for EDM.

Useful resources

There are a wide variety of digital tools and software packages available for EDM. Some of these are open source and free to use, whereas others have significant costs attached. Examples that are available at time of writing are given below, in alphabetical order (more relevant tools and software are listed in Chapters 12 and 24).

- Apache Flink (https://flink.apache.org/);

- DataShop (https://pslcdatashop.web.cmu.edu);

- IBM SPSS Modeler (www.spss.com.hk/software/modeler);

- Knime (www.knime.com);

- NLTK (www.nltk.org);

- Orange Data Mining (https://orange.biolab.si);

- R packages for data mining (e.g. caret, ggplot, dplyr, lattice) (https://cran.r-project.org);

- RapidMiner (https://rapidminer.com);

- Scikit-learn (http://scikit-learn.org);

- Weka Data Mining (www.cs.waikato.ac.nz/ml/weka).

The International Educational Data Mining Society (http://educationaldata mining.org) supports collaborative and scientific development in educational data mining. You can access the Journal of Educational Data Mining, along with details of conferences and useful resources, on the website.

The IEEE Task Force on Education Mining (http://edm.datasciences.org) provides useful information about EDM, relevant research projects, a useful glossary and links to relevant organisations.

Key texts

Aristizabal, A. (2016) 'An Epistemological Approach for Research in Educational Data Mining', *International Journal of Education and Research*, 4(2), 131–38, February 2016, retrieved from www.ijern.com/journal/2016/February-2016/12.pdf [accessed January 1, 2019].

Baker, S. and Inventado, P. (2016) 'Educational Data Mining and Learning Analytics: Potentials and Possibilities for Online Education', In Veletsianos, G. (ed.) *Emergence and Innovation in Digital Learning*. Edmonton: AU Press, 83–98, 10.15215/aupress/9781771991490.01.

Buenaño-Fernández, D. and Luján-Mora, S. (2017) 'Comparison of Applications for Educational Data Mining', *Engineering Education*, 81–85, 2017 IEEE World Engineering Education Conference (EDUNINE), 10.1109/EDUNINE.2017.7918187.

ElAtia, S., Ipperciel, D. and Zaïane, O. (eds.) (2016) *Handbook of Data Mining and Learning Analytics*. Hoboken, NJ: JohnWiley & Sons, Inc.

Hershkovitz, A., Joazeiro de Baker, R., Gobert, J., Wixon, M. and Sao Pedro, M. (2013) 'Discovery with Models: A Case Study on Carelessness in Computer-Based Science Inquiry', *American Behavioral Scientist*, 57(10), 1480–99, first published March 11, 2013, 10.1177/0002764213479365.

Khan, B., Corbeil, J. and Corbeil, K. (eds.) (2018) *Responsible Analytics and Data Mining in Education: Global Perspectives on Quality, Support, and Decision-Making*. Abingdon: Routledge.

Lee, Y. (2018) 'Using Self-Organizing Map and Clustering to Investigate Problem-Solving Patterns in the Massive Open Online Course: An Exploratory Study', *Journal of Educational Computing Research*, first published January 30, 2018, 10.1177/0735633117753364.

Li, H., Gobert, J., Graesser, A. and Dickler, R. (2018) 'Advanced Educational Technology for Science Inquiry Assessment', *Policy Insights from the Behavioral and Brain Sciences*, 5(2), 171–78, first published August 21, 2018, 10.1177/2372732218790017.

Liñán, L. and Pérez, Á. (2015) 'Educational Data Mining and Learning Analytics: Differences, Similarities, and Time Evolution', *International Journal of Educational Technology in Higher Education*, 12(3), 98–112, first published July, 2015, 10.7238/rusc. v12i3.2515.

Livieris, I., Drakopoulou, K., Tampakas, V., Mikropoulos, T. and Pintelas, P. (2018) 'Predicting Secondary School Students' Performance Utilizing a Semi-Supervised Learning Approach', *Journal of Educational Computing Research*, first published January 18, 2018, 10.1177/0735633117752614.

Lynch, C. (2017) 'Who Prophets from Big Data in Education? New Insights and New Challenges', *Theory and Research in Education*, 15(3), 249–71, first published November 7, 2017, 10.1177/1477878517738448.

Peña-Ayala, A. (ed.) (2014) *Educational Data Mining: Applications and Trends*. Cham: Springer.

Romero, C., Ventura, S., Pechenizkiy, M. and Baker, R. (eds.) (2010) *Handbook of Educational Data Mining*. Boco Raton, FL: CRC Press.

Slater, S., Joksimović, S., Kovanovic, V., Baker, R. and Gasevic, D. (2017) 'Tools for Educational Data Mining: A Review', *Journal of Educational and Behavioral Statistics*, 42(1), 85–106, first published September 24, 2016, 10.3102/1076998616666808.

Tang, H., Xing, W. and Pei, B. (2018) 'Time Really Matters: Understanding the Temporal Dimension of Online Learning Using Educational Data Mining', *Journal of Educational Computing Research*, first published July 4, 2018, 10.1177/0735633118784705.

Tsai, Y., Ouyang, C. and Chang, Y. (2016) 'Identifying Engineering Students' English Sentence Reading Comprehension Errors: Applying a Data Mining Technique', *Journal of Educational Computing Research*, 54(1), 62–84, first published October 6, 2015, 10.1177/0735633115605591.

CHAPTER 18

Ethno-mining

Overview

Ethno-mining (or ethnomining) is a method first discussed in Aipperspach et al. (2006) and developed in Anderson et al. (2009) that enables researchers to study human behaviour and culture through combining ethnography with data mining tools and techniques. Ethnography is the study of individuals, groups or communities in their own environment over a long period of time. The focus is to describe and interpret cultural behaviour and phenomena. The world is observed from the point of view of research participants rather than the ethnographer and stories are told through the eyes of the community as they go about their activities. Data mining is the process that is used to turn raw data into useful information that will help to uncover hidden patterns, relationships and trends. It involves applying algorithms to the extraction of hidden information with the aim of building an effective predictive or descriptive model of data for explanation and/or generalisation (Chapter 12). The two approaches may appear very different, but ethno-mining enables them to be combined, integrated, interweaved and used together to form a more complete picture, understanding or interpretation of the behaviour or culture under study.

Ethno-mining uses data created by people as they carry out their every-day activities, using technology such as smartphones and apps (Chapter 50), the web (Chapter 58), social media (52), digital sensors (Chapter 49), wearables (Chapter 57) and location tracking devices (Chapter 27). Data mining tools and techniques enable researchers to explore this data using methods such as classification, clustering, anomaly detection, regression and summarisation (see Chapter 12 for a description of these and other data mining methods). It is a data-driven technique in which data are examined to determine which variables and their values are important and understand how variables are related to each other. Relationships, patterns and trends can then be explored further with individuals and groups using methods such as interviews, focus groups and participant observation (online or offline), and self-reported behaviour can be checked and cross-referenced with mined data.

Ethno-mining is used by researchers approaching their work from a number of disciplines including the computer sciences (Aipperspach et al., 2006; Anderson et al., 2009), the humanities (Hsu, 2014) and the social sciences (Knox and Nafus, 2018). Examples of research projects that have used ethno-mining techniques include a study of Danish and US Facebook users aged 18–20 years (Bechmann, 2014); a study into the relationship between wireless laptops and space use in the home (Aipperspach et al., 2006:); a study to understand how and why players play an online text-based, fighting, role playing game (Shadoan and Dudek, 2013); research into the experiences of Asian American musicians playing independent rock music (Hsu, 2014); and research into how people congregated and communicated without technology during the Egyptian uprising in 2011 (Shamma, 2013).

Ethno-mining is closely connected to digital ethnography, which encompasses all forms of technology-based ethnography or digital communication (Chapter 14); online ethnography, which uses ethnographic tools, techniques, methods and insight to collect and analyse data from online sources (Chapter 37); and mobile ethnography, which uses mobile technology to undertake ethnographic studies into human activity, behaviour and/or movement (Chapter 31). If you are interested in ethno-mining for your research, it is important to read these chapters as it will give deeper insight into different methodologies and methods of digital ethnography and enable you to work on a suitable framework for your research. An understanding of data mining techniques (Chapter 12), data analytics (Chapter 10) and data visualisation (Chapter 13) will also help you to plan your project. If you feel that you need to understand more about ethnography as a research methodology, Hammersley and Atkinson (2019) provide useful information, and if you want to know more about digital ethnography, Pink et al. (2015) provide a good starting point. A useful collection of papers can be found in Knox and Nafus (2018), which provide interesting reading for those who want to know how ethnographers and data scientists are collaborating to develop new forms of social analysis.

Questions for reflection

Epistemology, theoretical perspective and methodology

- When undertaking ethno-mining, you are mixing methods from different epistemological and theoretical standpoints (positivist assumptions that guide quantitative research and interpretative stances that guide qualitative research, for example). How do you intend to mix, or integrate, these different standpoints? Creswell (2015) provides a brief guide to mixed methods research and will help you to think more about how these approaches can be integrated.

- Anderson et al. (2009: 125) state that ethno-mining is a hybrid method rather than a mixed method. Do you agree with this statement and, if so, why do you think this is the case?

- As we have seen in Chapters 14 and 37, there are various approaches to digital ethnography (virtual ethnography, netnography and network ethnography, for example). Do you have a clear understanding of these approaches, and understand where and how ethno-mining fits with, or stands out from, these approaches? Hsu (2014) touches on these issues and raises many other pertinent methodological issues for those conducting ethnographic studies in the humanities.

- Anderson et al. (2009: 131) point out that 'What ethno-mining does particularly well is it extends the social, spatial and temporal scope of research'. In what way might this be the case and what relevance does this observation have to your research?

- Aipperspach et al. (2006: 1) discuss qualitative and quantitative data (observations and sensor data) and point out that ethno-mining 'provides a means of interpreting that data which produces novel insights by exposing the biases inherent in either type of data alone'. What biases are present in these types of data and how can you recognise, reduce or eliminate such bias on your own work?

Ethics, morals and legal issues

- How will you address issues of anonymity and identity disclosure when adopting ethno-mining methods? Bechmann (2014: 59) for example, points out that the research team will know the identity of a person with whom they come into contact, whereas data that are mined can be presented in the form of aggregated statistics that will help to prevent identity disclosure.

- How will you ensure that participants give informed consent (and understand what is meant by informed consent) and that they know they can withdraw from the study at any time? How will you address issues of informed consent when data are collected through methods about which participants might be unaware (location tracking and sensor data, for example)?

- Hsu (2012) points out that 'certain digital communities are more open to software approaches than others'. She also points out that 'there are borders and boundaries – software- and hardware-dependent – that bind and separate these cyber spaces'. How can such constraints and dilemmas be incorporated into your methodology?

Practicalities

- What devices do you intend to use to collect data? What devices will enable you to minimise disruption to the lives of your participants, while maximising comprehensiveness of data (Anderson et al., 2009: 132)?

- Do you know what digital tools and software packages are available and what is suitable for your research? Examples of relevant tools and software are provided below and in Chapters 12, 31 and 37: it is useful to visit some of the websites listed so that you can get on idea of purpose, functions, capabilities and costs.

- Do you have a good understanding of data mining techniques, such as decision trees, anomaly detection, intrusion detection, clustering, classifying and regression? These techniques are discussed in Chapter 12. There are also a number of free, online courses available for those who need find out more about data mining techniques, such as Data Mining Specializations (www.coursera.org/specializations/data-mining) and Data Mining with WEKA (www.futurelearn.com/courses/data-mining-with-weka).

- Do you have a good understanding of ethnographic methods? Hammersley and Atkinson (2019) provide comprehensive information about ethnographic principles and practices.

- When analysing qualitative and quantitative data, do you intend to address each separately, integrate or interweave your analyses or use one to inform the other, for example? Creswell (2015) will help you to address this question.

Useful resources

There are a wide variety of digital tools and software packages available for ethno-mining. Some of these are data mining, visualisation and analysis tools, whereas others are ethnographic digital tools. Tools and software vary enormously and your choice will depend on your research topic, needs and purpose and on your theoretical framework. Some are open source and free to use, whereas others have significant costs involved. The following list provides a snapshot of what is available at time of writing (in alphabetical order):

- Datapreparator for data pre-processing in data analysis and data mining (www.datapreparator.com);

- Ethos: Ethnographic Observation System for ethnographic research (www.ethosapp.com);

- ExperienceFellow for mobile ethnography (www.experiencefellow.com);

- Indeemo for mobile ethnography and qualitative research (https://indeemo.com);

- Jubatus for online distributed machine learning on data streams (http://jubat.us/en);

- MotivBase ethnography for big data (www.motivbase.com);

- Orange Data Mining for machine learning, data visualisation and data analysis (https://orange.biolab.si);

- RapidMiner software platform for data preparation, data mining and data analytics (https://rapidminer.com);

- ThoughtLight Qualitative Mobile Ethnography App for mobile ethnography (www.civicommrs.com/mobile-insights-app);

- Touchstone Research mobile ethnographic research app for mobile ethnography (https://touchstoneresearch.com/market-research-tools-technologies/mobile-ethnography).

- VisionsLive research software for ethnography (www.visionslive.com/methodologies/ethnography);

- Weka for data mining (www.cs.waikato.ac.nz/ml/weka).

Ethnography Matters (http://ethnographymatters.net) is a website that provides informative articles about a wide variety of digital tools, techniques and methods that can be used in ethnographic studies. The website is 'managed and run by a group of volunteer editors who are passionate about ethnography'. There is a special edition on ethno-mining available on the website: http://ethnographymatters.net/editions/ethnomining.

Key texts

Aipperspach, R., Rattenbury, T., Woodruff, A., Anderson, K., Canny, J. and Aoki, P. (2006) *Ethno-Mining: Integrating Numbers and Words from the Ground Up*, Electrical Engineering and Computer Sciences, University of California at Berkeley, Technical Report No. UCB/EECS-2006-125, October 6, 2006, retrieved from www.eecs.berkeley.edu /Pubs/TechRpts/2006/EECS-2006-125.html [accessed December 20, 2018].

Anderson, K., Nafus, D., Rattenbury, T. and Aipperspach, R. (2009) 'Numbers Have Qualities Too: Experiences with Ethno-Mining', *Ethnographic Praxis in Industry Conference Proceedings*, 2009(1), 123–40, first published December 15, 2009, 10.1111/j.1559-8918.2009.tb00133.x.

Bechmann, A. (2014) 'Managing the Interoperable Self'. In Bechmann, A. and Lomborg, S. (eds.) *The Ubiquitous Internet: User and Industry Perspectives*, 54–73. New York, NY: Routledge.

Creswell, J. (2015) *A Concise Introduction to Mixed Methods Research*. Thousand Oaks, CA: Sage Publications, Inc.

Hammersley, M. and Atkinson, P. (2019) *Ethnography: Principles in Practice*, 4th edition. Abingdon: Routledge.

Hsu, W. (2012) 'On Digital Ethnography: What Do Computers Have to Do with It?' *Ethnography Matters*, first published October 27, 2012, retrieved from http://ethnographymatters.net/blog/2012/10/27/on-digital-ethnography-part-one-what-do-computers-have-to-do-with-ethnography [accessed December 20, 2018].

Hsu, W. (2014) 'Digital Ethnography toward Augmented Empiricism: A New Methodological Framework', *Journal of Digital Humanities*, 3(1), retrieved from http://journalofdigitalhumanities.org/3-1/digital-ethnography-toward-augmented-empiricism-by-wendy-hsu [accessed December 20, 2018] .

Knox, H. and Nafus, D. (eds.) (2018) *Ethnography for a Data-Saturated World*. Manchester: Manchester University Press.

Nafus, D. (2018) 'Working Ethnographically with Sensor Data'. In Knox, H. and Nafus, D. (eds.) *Ethnography for a Data-Saturated World*, 233–51. Manchester: Manchester University Press.

Pink, S., Horst, H., Postill, J., Hjorth, L., Lewis, T. and Tacchi, J. (2015) *Digital Ethnography: Principles and Practice*. London: Sage.

Shadoan, R. and Dudek, A. (2013) 'Plant Wars Player Patterns: Visualization as Scaffolding for Ethnographic Insight', *Ethnography Matters*, first published April 11, 2013, retrieved from http://ethnographymatters.net/blog/2013/04/11/visualizing-plant-wars-player-patterns-to-aid-ethnography [accessed December 12, 2018].

Shamma, A. (2013) 'Tweeting Minarets: A Personal Perspective of Joining Methodologies', *Ethnography Matters*, first published May 6, 2013, retrieved from http://ethnographymatters.net/blog/2013/05/06/tweeting-minarets [accessed December 12, 2018].

CHAPTER 19

Eye-tracking research

Overview

Eye-tracking research uses eye-tracking devices, tools and software to collect, record and analyse eye movement, eye position and gaze data. It can be used to 'unravel attention processes' and 'to investigate other psychological constructs, such as arousal, cognitive load, or perceptual fluency' (Meißner and Oll, 2017: 2). Eye-tracking research is used in both laboratory and field settings as a standalone research method or combined with other methods in mixed or hybrid approaches (see Sarter et al., 2007 for an illustration of how eye-tracking data were used together with behaviour data and mental model assessment). Eye-tracking is used as a research method in a number of disciplines and fields of study including applied linguistics and second language research (Conklin and Pellicer-Sánchez, 2016), medical research (Grundgeiger et al., 2015), medical education (Kok and Jarodzka, 2017), visual analytics (Kurzhals et al., 2016), visitor studies (Eghbal-Azar and Widlok, 2013), marketing and advertising (Ju and Johnson, 2010) and management and organisation studies (Meißner and Oll, 2017). It is also used in transportation research and development, games testing, behavioural studies in virtual reality, product and idea development, virtual design concept evaluation and ergonomics, for example.

Eye-trackers consist of projectors, video cameras and/or sensors, algorithms and software for data presentation. The most common eye-trackers work by beaming infrared or near infrared light via micro-projectors toward the centre of the eye. The reflections (or patterns) in the cornea are tracked by a camera and/or recorded by sensors. Algorithms are then used to calculate eye position, gaze point and movement. Software is used to present data in various forms, such as gaze plots (or scan paths) that visualise the sequence of eye movements and heat maps that illustrate the combined gaze activity of a number of participants.

There are different types of eye-tracking device that can be used by researchers. Examples include:

- Stationary eye-trackers that require participants to be seated in front of a screen or some type of real-world stimuli (an object or person, for example). These are also called screen-based (desktop) eye-trackers, remote eye-trackers (they do not have any physical contact with a person's body) or nonintrusive eye-trackers. Some allow head movement, whereas others prohibit it by using a forehead or chin support. Examples of this type of eye-tracker include:

  o eye-trackers integrated into computer monitors;

  o peripheral eye-tracking devices that can be added to existing screens;

  o standalone eye-trackers that can be mounted on stands or tripods;

  o computers (desktops or tablets) with built-in eye-tracking and eye-operated communication and control systems, used to help people with disabilities to communicate, express themselves and interact with others.

- Mobile eye-trackers that are built into glasses, helmets or other wearable devices. These trackers are also referred to as intrusive devices because they require physical contact with the person. They enable participants to move freely and are used to capture viewing behaviour in real-world environments.

- Virtual reality and augmented reality eye-trackers that enable researchers to gain insight into visual behaviour in simulated or enhanced natural environments.

- Webcam-based eye-trackers that use existing webcams (on desktops or laptops) along with webcam software to collect and analyse data.

Examples of research projects that have used, assessed or critiqued eye-tracking methods include a study into reading behaviour in readers of printed and online newspapers (Leckner, 2012); an investigation into how respondents arrive at socially desirable or undesirable answers in web, face-to-face and paper-and-pencil self-administered modes (Kaminska and Foulsham,

2016); an examination of pilots' automation monitoring strategies and performance on highly automated commercial flight decks (Sarter et al., 2007); an investigation into gaze pattern variations among men when assessing female attractiveness (Melnyk et al., 2014); and research into young women, fashion advertisements and self-reports of social comparison (Ju and Johnson, 2010).

If you are interested in using eye-tracking methods for your research, a good starting point is Holmqvist and Andersson (2017), who provide a comprehensive introduction to the topic. They guide you through the various stages of a research project, including choosing appropriate hardware and software, planning and designing your study and recording and analysing eye-movement data. They also look at the technology and skills required to perform high quality research with eye-trackers. Another good book is Duchowski (2017), which focuses on video-based, corneal-reflection eye trackers and covers issues such as theory and methodology of eye-tracking research, an introduction to the human visual system and choosing and using technology. If you are thinking about using head-mounted eye-tracking devices an up-to-date overview of technology, along with an analysis of 20 essential features of six commercially available head mounted eye-trackers, is provided by Cognolato et al. (2018). A list of eye-trackers that are available at time of writing is provided below. Further information about wearables-based research can be found in Chapter 57.

Questions for reflection

Epistemology, theoretical perspective and methodology

- Are you interested in conducting highly controlled experiments in a laboratory setting (desktop based eye-tracking), conducting research in the real-world to study interaction and behaviour (mobile eye-tracking) or looking at how people act in a simulated environment (virtual eye-tracking)? The methods you choose are guided by epistemology, theoretical perspective and methodological framework and will have implications on generalisability, reproducibility, validity and reliability (experimental control versus ecological validity, for example). Meißner and Oll (2017) provide an enlightening discussion on these issues.

- Kok and Jarodzka (2017: 114) when considering eye-tracking in medical education research, state that 'In order to interpret eye-tracking data properly, theoretical models must always be the basis for designing experiments as well as for analysing and interpreting eye-tracking data. The interpretation of eye-tracking data is further supported by sound experimental design and methodological triangulation'. How will you address these issues when planning your research?

- Do you have a good understanding of technical terms associated with visual perception? For example, Eghbal-Azar and Widlok (2013: 104) discuss 'fixations' (short stops of eye movement), 'saccades' (rapid eye movements between fixations) and 'scan patterns' (a combination of fixations and saccades) and Meißner and Oll (2017) discuss three parts of human visual perception: foveal, parafoveal and peripheral vision.

- Do you have a good understanding of eye-tracking measures (position of fixation, fixation duration, saccade distance or blink rate, for example)? Holmqvist and Andersson (2017) provide a comprehensive list of measures that can be used by researchers.

- Are you interested in gaze sensitivity and gaze response (gaze preference, gaze following and gaze aversion, for example)? If so, Davidson and Clayton (2016) provide interesting information.

- How can you ensure the quality of eye-tracking data? Holmqvist et al. (2012: 45) will help you to think about 'what data quality is and how it can be defined, measured, evaluated, and reported'.

Ethics, morals and legal issues

- Is it possible that eye movement and gaze patterns can reveal the identity of individuals? How can you assure anonymity, in particular, in studies with small samples?

- Eye-tracking research can reveal information about which participants are unaware. What implications does this have for issues of informed consent, privacy and the choice to withhold personal data?

- Is it possible that your eye-tracking research could detect symptoms of neurological or behavioural disorders in individuals? How will you address these issues if they arise? Do you have the professional competence to detect and address these issues?

- Can taking part in eye-tracking research alter decision-making in individuals? If so, what are the implications for your research?

Practicalities

- If using mobile or wearable devices for your research, are they unobtrusive, lightweight and comfortable to wear? Are participants happy to wear the devices? Can participants move around freely? Will wearing the devices affect their behaviour (feeling self-conscious or 'clumsy', for example)?

- What costs are involved (hardware and software)? If your budget is limited, will you need to compromise on some of the quality considerations listed below when considering hardware? Are there good, open source software products available?

- Are devices of the required quality? When choosing devices quality considerations include:

  - accuracy and precision;

  - minimum gaze data loss;

  - reliable and consistent calibration;

  - effective recording and analysis (wireless live view and accurate logging, for example);

  - scalability, portability and interconnection;

  - data export, aggregation, integration and synchronisation.

- What factors can influence accuracy and precision (eye characteristics, calibration procedures and type of experiment, for example)?

- How might blinking affect the performance of eye trackers?

Useful resources

There are a wide variety of eye-trackers available. Examples available at time of writing include (in alphabetical order):

- ASL Mobile Eye-XG Glasses (https://imotions.com/asl-eye-tracking-glasses);

- EyesDecide (www.eyesdecide.com);

- EyeTech (www.eyetechds.com);

- Gazepoint (www.gazept.com);

- GazePointer (https://sourceforge.net/projects/gazepointer);

- Glance (www.mirametrix.com);

- PyGaze (www.pygaze.org);

- Smart Eye (http://smarteye.se);

- Tobii Pro Glasses 2 (www.tobiipro.com/product-listing/tobii-pro-glasses-2).

Examples of eye-tracking data analysis software available at time of writing include the following (in alphabetical order). Some are packaged with specific eye-tracking hardware, whereas others are available for use with existing webcams, for example.

- NYAN 2.0 (www.eyegaze.com/eyegaze-analysis-software);

- OGAMA (www.ogama.net);

- Pupil Capture, Player and Mobile (https://pupil-labs.com/pupil);

- PyGaze Analyser (www.pygaze.org);

- Tobii Pro VR Analytics (www.tobiipro.com/product-listing/vr-analytics).

Key texts

Cognolato, M., Atzori, M. and Müller, H. (2018) 'Head-Mounted Eye Gaze Tracking Devices: An Overview of Modern Devices and Recent Advances', *Journal of Rehabilitation and Assistive Technologies Engineering*, first published June 11, 2018, 10.1177/2055668318773991.

Conklin, K. and Pellicer-Sánchez, A. (2016) 'Using Eye-Tracking in Applied Linguistics and Second Language Research', *Second Language Research*, 32(3), 453–67, first published March 14, 2016, 10.1177/0267658316637401.

Davidson, G. and Clayton, N. (2016) 'New Perspectives in Gaze Sensitivity Research', *Learning and Behavior*, 44(9), 9–17, first published November 18, 2015, 10.3758/s13420-015-0204-z.

Duchowski, A. (2017) *Eye Tracking Methodology: Theory and Practice*, 3rd edition. London: Springer.

Eghbal-Azar, K. and Widlok, T. (2013) 'Potentials and Limitations of Mobile Eye Tracking in Visitor Studies: Evidence from Field Research at Two Museum Exhibitions in Germany', *Social Science Computer Review*, 31(1), 103–18, first published October 31, 2012, 10.1177/0894439312453565.

Grundgeiger, T., Wurmb, T. and Happel, O. (2015) 'Eye Tracking in Anesthesiology: Literature Review, Methodological Issues, and Research Topics', *Proceedings of the Human Factors and Ergonomics Society Annual Meeting*, 59(1), 493–97, first published December 20, 2016, 10.1177/1541931215591106.

Holmqvist, K. and Andersson, R. (2017) *Eye Tracking: A Comprehensive Guide to Methods, Paradigm, and Measures*. Lund, Sweden: Lund Eye-Tracking Research Institute.

Holmqvist, K., Nyström, M. and Mulvey, F. (2012) 'Eye Tracker Data Quality: What It Is and How to Measure It', *Proceedings of the Symposium on Eye Tracking Research and Applications*, 45–52. New York, NY: ACM.

Ju, H. and Johnson, K. (2010) 'Fashion Advertisements and Young Women: Determining Visual Attention Using Eye Tracking', *Clothing and Textiles Research Journal*, 28(3), 159–73, first published April 14, 2010, 10.1177/0887302X09359935.

Kaminska, O. and Foulsham, T. (2016) 'Eye-Tracking Social Desirability Bias', *Bulletin of Sociological Methodology/Bulletin de Méthodologie Sociologique*, 130(1), 73–89, first published March 22, 2016, 10.1177/0759106315627591.

Kok, E. and Jarodzka, H. (2017) 'Before Your Very Eyes: The Value and Limitations of Eye Tracking in Medical Education', *Medical Education*, 51, 114–22, 10.1111/medu.13066.

Kurzhals, K., Fisher, B., Burch, M. and Weiskopf, D. (2016) 'Eye Tracking Evaluation of Visual Analytics', *Information Visualization*, 15(4), 340–58, first published October 27, 2015, 10.1177/1473871615609787.

Leckner, S. (2012) 'Presentation Factors Affecting Reading Behaviour in Readers of Newspaper Media: An Eye-Tracking Perspective', *Visual Communication*, 11(2), 163–84, first published May 21, 2012, 10.1177/1470357211434029.

Meißner, M. and Oll, J. (2017) 'The Promise of Eye-Tracking Methodology in Organizational Research: A Taxonomy, Review, and Future Avenues', *Organizational Research Methods*, first published December 13, 2017, 10.1177/1094428117744882.

Melnyk, J., McCord, D. and Vaske, J. (2014) 'Gaze Pattern Variations among Men When Assessing Female Attractiveness', *Evolutionary Psychology*, first published January 1, 2014, 10.1177/147470491401200113.

Sarter, N., Mumaw, R. and Wickens, C. (2007) 'Pilots' Monitoring Strategies and Performance on Automated Flight Decks: An Empirical Study Combining Behavioral and Eye-Tracking Data', *Human Factors*, 49(3), 347–357, first published June 1, 2007, 10.1518/001872007X196685.

CHAPTER 20

Game analytics

Overview

Game analytics (or gaming analytics) refers to the process of scraping, capturing, extracting and examining data from games, users of games and game companies for research, development and marketing purposes. Through tracking and measuring behaviour and interaction, game developers can make improvements to the game, improve engagement, provide a better user-experience and increase revenue. As game analytics helps to determine what, where, when and how people play games, and behave and interact in games, it can also appeal to academic researchers from a variety of disciplines and fields of study including psychology, sociology, education (in particular, within game-based learning), architecture, visual design, software engineering, marketing, geography and anthropology. Game analytics can be undertaken on all genres and types of computer and mobile game including education, adventure, simulation, role play, real-time strategy, massively multiplayer online, serious games, fitness, puzzle, sports, family, shooting and combat games (see Chapter 56 for information about games that take place in virtual worlds). Examples of research projects that have used game analytics include a project that considers human social interaction in *The Restaurant Game* (Orkin and Roy, 2009); a mixed methods study into engagement and games for learning (Phillips et al., 2014); a study to develop and test an algorithm for creating battle maps in a massively multiplayer online game (Wallner, 2018); and a study that uses game analytics to evaluate puzzle design and level progression in a serious game (Hicks et al., 2016).

Game analytics

There are a variety of digital tools that can be used for game analytics. These include (in alphabetical order):

- A/B testing tools to test different variables within distinct groups of players in the game to understand the positive or negative impact (this can also apply to advertisements in free-to-play games).

- Cohort analysis tools that isolate player groups based on common characteristics (sign-up date, country or usage levels, for example) and compare each group over time to provide insight and discover patterns and trends.

- Data mining tools to understand who is playing a particular game, find out how they play and gain insight into what works and what does not work within a game. More information about data mining tools and techniques is provided in Chapter 12.

- Engagement tools that enable developers to adapt games, depending on who is playing. This can include real-time targeting tools and in-game messaging, for example.

- Event trackers that provide insight into how players are interacting within a game and how well they are achieving the objectives of the game.

- Key performance indicator tools that measure the performance of games, products or companies. They can record progression and achievement within a game and help to ascertain whether users are moving successfully through the game, and they can help to evaluate the success of a company, give a broad understanding of a company's position within the market and show progress towards key goals, for example.

- Optimisation tools that enable developers to improve engagement, retain players, acquire higher value players, improve user experience, increase revenue and keep ahead of the competition.

- Path analysis tools that enable developers and researchers to discover how players move through a game and the routes they take between pre-defined action points.

- Physical and digital touchpoint tools that can identify channels, campaigns, marketing and referrals that find, engage and keep users, and steps that users take to find out about and play certain games (search habits and social media, for example).

- Player modelling that uses computer models of players in a game to study static features of players (gender, age and personality, for example) and dynamic phenomenon (player interaction and in-game choices, for example).

- Segmentation tools that enable developers to understand different user segments and target particular segments (players who are the most competent, the most engaged or spend the most money, for example). It also enables developers to identify under-performing segments and offer gifts, rewards or tips to help them move on.

If you are interested in finding out more about game analytics Seif El-Nasr et al. (2013) provide a good starting point. This is a collection of papers written by experts in industry and research and covers the practical application of game analytics. It could do with a little updating, given rapid technological advances in game analytics, but nevertheless provides useful and pertinent information for researchers who are interested in this method. Loh et al. (2015) provide in-depth coverage of serious games analytics and Baek et al. (2014) provide a collection of papers on serious gaming and social media. ('Serious games' are games that have a more 'serious' purpose and are not just for entertainment: they use learning theories and instructional design principles to maximise learning or training success. 'Serious games analytics' draws, and expands, on game analytics and learning analytics to determine whether or not the game is meeting learning/training objectives: see Chapter 24 for more information about learning analytics.) If you are interested in the wider area of data analytics and entertainment science (movies, games, books and music) a comprehensive guide aimed at 'managers, students of entertainment, and scholars' is provided by Hennig-Thurau and Houston (2018). A brief, personal history of game analytics systems and use, covering flat files, databases, data lakes and serverless systems has been written by Ben Weber and can be obtained from the Towards Data Science website: https://towardsdatascience.com/evolution-of-game-analytics-platforms-4b9efcb4a093 [accessed May 22, 2018]. There are various tools and software packages available if you are interested in game analytics and relevant websites are listed below. It is useful to visit some of these sites so that you can get a better idea of functions, capabilities, purposes, user-friendliness and costs. Also, a deeper understanding of business analytics (Chapter 4), data analytics (Chapter 10), data mining (Chapter 12), data visualisation (Chapter 13), predictive modelling (Chapter 45) and web and mobile analytics (Chapter 58) will help you to think more about your research project.

Questions for reflection

Epistemology, theoretical perspective and methodology

- Do you intend to adopt a theory-driven approach or a process-driven approach to your game analytics? Do you intend to test, statistically, a theory-driven hypothesis using empirical data, or undertake the inductive discovery of patterns and relationships, using specific techniques and algorithms? Elragal and Klischewski (2017) provide a detailed discussion on this topic that will help you to address these questions. Seif El-Nasr et al. (2013) in Chapter 2 of their book, discuss the process for knowledge discovery in data. The term Knowledge Discovery in Databases (KDD) refers to the broad process of finding and discovering knowledge in data, using data mining and data analytics techniques. Gama (2010) and Sanders (2016) provide in-depth information on this topic.

- When considering behavioural data, or user analytics, are these objective data? If so, how complete an understanding of game users do they provide? Do they need to be combined with information about game user perceptions, feelings, emotions, attitudes and motivation, for example? If so, how can you collect and analyse this type of information? Is there room for both quantitative and qualitative approaches and is it possible to combine such approaches?

Ethics, morals and legal issues

- How can you ensure that you, and the gaming company, act ethically and responsibly when collecting and analysing user data? The Digital Analytics Association (www.digitalanalyticsassociation.org/codeofethics) has produced a web analyst code of ethics that covers the topics of privacy, transparency, consumer control, education and accountability. The code can be viewed on their website, along with some useful blogs, news items and dates of events.

- Do you have a clear understanding of what data are 'personal data' when dealing with game analytics and do you understand what you are allowed to do with this data? The General Data Protection Regulation (GDPR) came into force in May 2018 and covers data protection and privacy for all individuals in the EU. Detailed information about the GDPR in the UK can be obtained from the Information Commissioner's Office (https://ico.org.uk/for-organisations/guide-to-the-general-data-protection-regulation-gdpr). You can also find a useful data protection self-assessment tool kit on this site (https://ico.org.uk/for-organisations/guide-to-the-general-data-protection-regulation-gdpr).

- If you intend to use player modelling techniques, how can you ensure that bias is not introduced into the datasets that are used or the models that are developed? These issues can be addressed, in part, by understanding what is meant by bias, recognising and addressing bias, paying attention to modelling assumptions, providing complete, accurate and transparent model descriptions and undertaking a careful evaluation and assessment of outcomes.

- Do you have a good understanding of the ethics of computer games? This will help you to think more deeply about the ethical design of games and the ethical roles and responsibilities of game developers, players, analysts and researchers. A comprehensive guide to this topic is provided by Sicart (2009).

Practicalities

- What source of data do you intend to use? How can you ensure reliability, validity and accuracy of data (the tools that you use will not be effective if there are problems with the data).

- What challenges might affect data acquisition for your research? This can include availability, cost, access, personal preferences, biases, availability of technology and researcher technical ability, for example.

- What tools and software do you intend to use? What functions and capabilities are important to you (design events, virtual currency, monetisation, player progression, player group comparisons, A/B test gameplay or error events, for example)? What support is offered by the software company? What costs are involved?

- Do you have a good understanding of relevant algorithms, their parameters and parameter settings? Can you provide a rationale for algorithm choice and parameter setting? Do you understand how such choices might affect meaning, interpretation and outcomes?

Useful resources

There are various digital tools and software packages that can be used for game analytics. Examples that are available at time of writing include (in alphabetical order):

- Appsee (www.appsee.com);
- CleverTap (https://clevertap.com);
- Countly (https://count.ly);
- DeltaDNA (https://deltadna.com/analytics);
- Firebase (https://firebase.google.com/docs/analytics);
- GameAnalytics (https://gameanalytics.com);
- Leanplum (www.leanplum.com/mobile-analytics);
- Localytics (www.localytics.com);
- Mixpanel (https://mixpanel.com);
- Raygun (https://raygun.com);
- Sonamine (www.sonamine.com/home);
- Unity (https://unity.com/solutions/analytics).

The Gaming Analytics Summit (www.theinnovationenterprise.com/sum mits) takes place annually in the US and covers topics such as data mining, predictive analytics, knowledge discovery and statistics in the gaming industry. The date and agenda for the next event can be obtained from the website.

Key texts

Baek, Y., Ko, R. and Marsh, T. (eds.) (2014) *Trends and Applications of Serious Gaming and Social Media*. Singapore: Springer.

Elragal, A. and Klischewski, R. (2017) 'Theory-Driven or Process-Driven Prediction? Epistemological Challenges of Big Data Analytics', *Journal of Big Data*, 4(19), first published June 23, 2017, 10.1186/s40537-017-0079-2.

Gama, J. (2010) *Knowledge Discovery from Data Streams*, Boca Raton, FL: CRC Press.

Hennig-Thurau, T. and Houston, M. (2018) *Entertainment Science: Data Analytics and Practical Theory for Movies, Games, Books*. Cham: Springer.

Hicks, D., Eagle, M., Rowe, E., Asbell-Clarke, J., Edwards, T. and Barnes, T. (2016) 'Using Game Analytics to Evaluate Puzzle Design and Level Progression in a Serious Game', in *Proceedings of the Sixth International Conference on Learning Analytics & Knowledge*, 440–48. Edinburgh, United Kingdom.

Loh, C., Sheng, Y. and Ifenthaler, D. (eds.) (2015) *Serious Games Analytics: Methodologies for Performance Measurement, Assessment, and Improvement*. Cham: Springer.

Orkin, J. and Roy, D. (2009) 'Automatic Learning and Generation of Social Behavior from Collective Human Gameplay', in *Proceedings of the 8th International Conference on Autonomous Agents and Multiagent Systems (AAMAS)*, 385–92, Budapest, Hungary.

Phillips, R., Horstman, T., Vye, N. and Bransford, J. (2014) 'Engagement and Games for Learning: Expanding Definitions and Methodologies', *Simulation & Gaming*, 45(4–5), 548–68, first published November 4, 2014, 10.1177/1046878114553576.

Sanders, M. (ed.) (2016) *Knowledge Discovery in Databases*. Forest Hills, NY: Willford Press.

Seif El-Nasr, M., Drachen, A. and Canossa, A. (eds.) (2013) *Game Analytics: Maximizing the Value of Player Data*, London: Springer.

Sicart, M. (2009) *The Ethics of Computer Games*, Cambridge, MA: MIT Press.

Wallner, G. (2018) 'Automatic Generation of Battle Maps from Replay Data', *Information Visualization*, 17(3), 239–56, first published July 17, 2017, 10.1177/1473871617713338.

CHAPTER 21

Geospatial analysis

Overview

Geospatial analysis is the process of gathering, exploring, manipulating, analysing and visualising data that contain explicit geographic positioning information or reference to a specific place. This can include Global Positioning System (GPS) data, satellite imagery, photographs, geographic information system (GIS) maps and graphic displays, airplane and drone imagery, remote sensing and historical data, for example. Geospatial analysis is related closely to spatial analysis: indeed, the terms are sometimes used interchangeably (see some of the software packages listed below for examples). However, 'geospatial' relates to data that contains a specific geographic component (locational information) whereas 'spatial' relates to the position, area, shape and size of places, spaces, features, objects or phenomena. More information about spatial analysis and spatial modelling (the analytical process of representing spatial data) can be found in Chapter 54.

The use of geospatial analysis is increasing across a number of disciplines and fields of study including geographical sciences, environmental sciences, archaeology, history, health and medicine, criminal justice and public safety, business, transportation and logistics, urban planning, education and international development. This increasing use is due to a number of factors, including:

- rapid technological advances and evolution in machine learning and artificial intelligence (Chapter 29) and in computer modelling and simulation (Chapter 7);

- the availability and user-friendliness of geospatial analysis tools and software (see below), data mining tools and software (Chapter 12) and visualisation tools (Chapter 13);

- the popularity and increased usage of mobile devices for capturing geospatial data directly from the field (Chapters 27 and 32);

- increasing volumes of geospatial data, linked data and shared data available for analysis;

- the growing movement in community-based, grass-root and crowdsourced geospatial and mapping projects (see Mulder et al., 2016 for an interesting critique of crowdsourced geospatial data in relation to humanitarian response in times of crisis).

Geospatial analysis can be used as a standalone research method or used in mixed methods approaches (see Hites et al., 2013 for an illustration of how geospatial analysis was used together with focus group research and Yoon and Lubienski, 2018 for a discussion on a mixed methods approach relating to school choice). There are a variety of analysis methods and techniques that can be used for geospatial (and spatial) analysis, including:

- network analysis of natural and human-made networks such as roads, rivers and telecommunication lines;

- locational analysis of the spatial arrangement, pattern and order of phenomena;

- surface analysis of physical surfaces to identify conditions, patterns or features (contours, aspects, slopes and hill shade maps, for example);

- distance analysis to consider proximity, efficiency and impact;

- geovisualisation to display information that has a geospatial component to stimulate thinking, generate hypotheses, explore patterns and facilitate communication.

Examples of research projects that have used geospatial analysis include an investigation into condom availability and accessibility in urban Malawi (Shacham et al., 2016); research into the impact of school attendance zone 'gerrymandering' on the racial/ethnic segregation of schools (Richards, 2014); an evaluation of the relationship between UV radiation and multiple sclerosis prevalence (Beretich and Beretich, 2009); an investigation into the influence of geography on community rates of breastfeeding initiation (Grubesic and Durbin, 2016); a study into the extent and pattern of flow of female migrants in Asian countries (Varshney and Lata, 2014); a study to design, develop and implement predictive modelling for identifying sites of archaeological interest (Balla et al., 2013); and research to assess campus safety (Hites et al., 2013).

If you are interested in finding out more about geospatial analysis for your research, a good starting point is de Smith et al. (2018). This book provides a comprehensive discussion on conceptual frameworks and methodological contexts, before going on to consider methods, practical applications and relevant software. The book is freely available on the related website (www.spatialanalysisonline.com), which contains useful blogs, software, resources and publications for anyone interested in the topic. As geospatial analysis can use movement data from mobile devices and other location-based services, you may find Chapters 27 and 33 of interest. You may also find it useful to gather more information about data visualisation (Chapter 13), sensor-based methods (Chapter 49), wearables-based research (Chapter 57) and data mining techniques (Chapter 12). The questions for reflection listed below will help you to think more about using geospatial analysis for your research project.

Questions for reflection

Epistemology, theoretical perspective and methodology

- Do you intend your geospatial analysis to be a quantitative project, or will you incorporate qualitative approaches into your analyses? Yoon and Lubienski (2018: 53) point out that while geospatial analysis 'has traditionally been heavily quantitative in its orientation' there are potential benefits to be gained by incorporating qualitative research. Their paper illustrates why they believe this to be the case.

- In what way are maps and other geographical images socially constructed? In what way might they constitute a political act? How is understanding and response connected to audience (who is viewing, when, where and how)? How is understanding and response connected to the type of map or image (what has been produced, how, why and in what form)?

- How can you overcome problems that may arise due to a misunderstanding of data (or a partial understanding of data), mismeasurement, conjecture or bias (Western-centric, biased language, narrative bias or geographical images that back up ideology, for example)? How can you ensure validity, reliability and reproducibility in your analyses?

Ethics, morals and legal issues

- Do you understand all relevant legislation and regulations? For example, if you intend to use drones for data collection, or use existing drone imagery for your research, can you ensure that the operator complies with the Civilian Aviation Authority (CAA) Air Navigation Order 2016 in the UK or the Federal Aviation Administration's (FAA) Part 107 regulations for Small Unmanned Aerial Systems in the US? Do you know how to fly drones safely? A drone safety code in the UK can be found at http://dronesafe.uk.

- How will you address matters of privacy? This can include tracking individual locations (Chapter 27) or using drones or high resolution satellite imagery, for example. Matters of privacy and other ethical concerns can be addressed by following a suitable Code of Ethics: visit the GIS Certification Institute (GISCI) website for an example (www.gisci.org/Ethics/CodeofEthics.aspx).

- Is it possible that any part of your data collection or analysis activities could cause harm to individuals, organisations, institutions and/or society?

- Is it possible that your geospatial analysis could be used, manipulated or skewed by those in positions of power or by powerful interest groups? Could it be used to marginalise? Mulder et al. (2016) provide a good example of how marginalisation can happen in crowdsourced geospatial projects.

- How can you avoid undue or excessive intrusion into the lives of individuals (location tracking or various types of surveillance that can be used to provide data for geospatial analysis, for example)?

Practicalities

- When obtaining geospatial information from digital libraries, how can you overcome problems with inconsistent metadata (the basic information that is used to archive, describe and integrate data). Inconsistencies can hide or mask relevant geographical resources so that spatially themed queries are unable to retrieve relevant resources. Renteria-Agualimpia et al. (2016: 507) discuss metadata inconsistencies and go on to outline a methodology that is able to detect 'geospatial inconsistencies in metadata collections based on the combination of spatial ranking, reverse geocoding, geographic knowledge organization systems and information-retrieval techniques'. More information about information retrieval methods and techniques can be obtained from Chapter 23.

- Do you have a good understanding of the wide variety of geospatial analysis tools and techniques, including network and locational analysis, distance and directional analysis, geometrical processing, map algebra, grid models, spatial autocorrelation, spatial regression, ecological modelling and digital mapping? If you are interested in finding out more about these tools and techniques, de Smith et al. (2018) provide comprehensive information and advice.

Useful resources

There are a variety of digital tools and software packages available that enable researchers to undertake geospatial analysis and geospatial mapping. Some are aimed at researchers in disciplines such as the social sciences, geography and medicine, whereas others are aimed at business and the search engine optimisation (SEO) sector. Examples available at time of writing are given below, in alphabetical order. For a more comprehensive list of geospatial software, visit www.spatialanalysisonline.com/software.html [accessed June 7, 2018].

- AnyLogic (www.anylogic.com);
- ArcGIS (www.esri.com/arcgis);
- BioMedware software suite (www.biomedware.com/software);
- Didger (www.goldensoftware.com/products/didger);
- GeoVISTA Studio (www.geovistastudio.psu.edu);
- GRASS GIS (https://grass.osgeo.org);
- QGIS (www.qgis.org);
- R (www.r-project.org);
- Sentinel Visualizer (www.fmsasg.com);
- Software of the Center for Spatial Data Science (https://spatial.uchicago.edu/software);
- WhiteboxTools (www.uoguelph.ca/~hydrogeo/WhiteboxTools).

The Open Source Geospatial Foundation (OSGeo) (www.osgeo.org) is 'a not-for-profit organization whose mission is to foster global adoption of open geospatial technology by being an inclusive software foundation devoted to an open philosophy and participatory community driven development'. Details of relevant projects and useful resources, tools and software can be found on the website.

The Geospatial Information & Technology Association (GITA) (www.gita.org) is a member-led, non-profit professional association and advocate for 'anyone using geospatial technology to help operate, maintain, and protect the infrastructure'. Relevant news, publications and events can be found on the website.

The OGC (Open Geospatial Consortium) (www.opengeospatial.org) is 'an international not for profit organization committed to making quality open standards for the global geospatial community'. Relevant blogs, news items and resources can be found on the website.

Key texts

Balla, A., Pavlogeorgatos, G., Tsiafakis, D. and Pavlidis, G. (2013) 'Modelling Archaeological and Geospatial Information for Burial Site Prediction, Identification and Management', *International Journal of Heritage in the Digital Era*, 2(4), 585–609, first published December 1, 2013, 10.1260/2047-4970.2.4.585.

Beretich, B. and Beretich, T. (2009) 'Explaining Multiple Sclerosis Prevalence by Ultraviolet Exposure: A Geospatial Analysis', *Multiple Sclerosis Journal*, 15(8), 891–98, first published August 10, 2009, 10.1177/1352458509105579.

de Smith, M., Goodchild, M. and Longley, P. (2018) *Geospatial Analysis: A Comprehensive Guide to Principles, Techniques and Software Tools*, 6th edition. London: Winchelsea Press.

Grubesic, T. and Durbin, K. (2016) 'Community Rates of Breastfeeding Initiation: A Geospatial Analysis of Kentucky', *Journal of Human Lactation*, 32(4), 601–10, first published August 8, 2016, 10.1177/0890334416651070.

Hites, L., Fifolt, M., Beck, H., Su, W., Kerbawy, S., Wakelee, J. and Nassel, A. (2013) 'A Geospatial Mixed Methods Approach to Assessing Campus Safety', *Evaluation Review*, 37(5), 347–69, first published December 30, 2013, 10.1177/0193841X13509815.

Mulder, F., Ferguson, J., Groenewegen, P., Boersma, K. and Wolbers, J. (2016) 'Questioning Big Data: Crowdsourcing Crisis Data Towards an Inclusive Humanitarian Response', *Big Data & Society*, first published August 10, 2016, 10.1177/2053951716662054.

Renteria-Agualimpia, W., Lopez-Pellicer, F., Lacasta, J., Zarazaga-Soria, F. and Muro-Medrano, P. (2016) 'Improving the Geospatial Consistency of Digital Libraries Metadata', *Journal of Information Science*, 42(4), 507–23, first published August 12, 2015, 10.1177/0165551515597364.

Richards, M. (2014) 'The Gerrymandering of School Attendance Zones and the Segregation of Public Schools: A Geospatial Analysis', *American Educational Research Journal*, 51(6), 1119–57, first published December 1, 2014, 10.3102/0002831214553652.

Shacham, E., Thornton, R., Godlonton, S., Murphy, R. and Gilliland, J. (2016) 'Geospatial Analysis of Condom Availability and Accessibility in Urban Malawi', *International Journal of STD & AIDS*, 27(1), 44–50, first published February 13, 2015, 10.1177/0956462415571373.

Geospatial analysis

Varshney, D. and Lata, S. (2014) 'Female Migration, Education and Occupational Development: A Geospatial Analysis of Asian Countries', *Environment and Urbanization ASIA*, 5(1), 185–215, first published April 16, 2014, 10.1177/0975425314521549.

Yoon, E. and Lubienski, C. (2018) 'Thinking Critically in Space: Toward a Mixed-Methods Geospatial Approach to Education Policy Analysis', *Educational Researcher*, 47(1), 53–61, first published October 23, 2017, 10.3102/0013189X17737284.

CHAPTER 22

HR analytics

Overview

HR analytics refers to the use of data analytics in the field of human resource development to inform more evidence-based decision-making. It is closely related to business analytics (Chapter 4) and is covered by the umbrella term of data analytics (Chapter 10). Data is captured, extracted, categorised and analysed to provide insight, help inform human resource decisions and facilitate policy-making about issues such as hiring, performance, productivity, motivation, reward structures, team building, training and development. There are different types of HR analytics including descriptive, diagnostic, predictive, prescriptive and real-time, and these are discussed in Chapter 10. HR analytics can also be referred to as people analytics, talent analytics, workforce analytics or human capital analytics, although there are subtle differences in the terms (HR analytics, for example, tends to refer to data analytics for and by human resource departments, whereas other terms can be seen to be more wide-ranging and include additional areas or departments that are concerned with people or workforce management). Marler and Boudreau (2017) point out that the adoption of HR analytics, at this present time, is low: there is not a great deal of academic research (and evidence) on the topic. Research that is undertaken tends to be in the fields of business, retail, management and organisation studies (see King, 2016; Khan and Tang, 2016; Rasmussen and Ulrich, 2015, for example).

There are various sources of data for HR analytics including HR systems (employee data, payroll, absence management, benefits, performance reviews, recruitment and workload, for example); business information (operations performance data, sales figures and customer records, for example); and personal information (mobile apps, geo-positioning and wearable technologies, for example). HR analytics involves bringing together or combining these types of data to make interpretations, recognise trends and issues and take proactive steps to improve performance, profitability and/or engagement. HR analytics tends to be undertaken by specialist members of staff working within the organisation, or by outside consultants who are brought in to undertake the work. Within HR analytics, there are a number of subsets, including (in alphabetical order):

- Capability analytics: considers data related to skills, competencies, qualifications, understanding, knowledge and experience of individuals within an organisation. It can be used to inform recruitment and build teams, for example. A related area is competency analytics that seeks to assess and develop workforce competencies.

- Capacity analytics: considers existing work practices, individual workloads, time taken to complete a task, the number of workers required for a task and the number of workers required in an organisation. It is used to help understand present and future capacity so that the organisation can function and grow.

- Culture analytics: considers data related to shared values and beliefs within an organisation. These can be reflected in individual and group patterns of behaviour, competencies, decisions, rules, systems and procedures. It can also include corporate culture within an organisation.

- Employee performance analytics: assesses individual performance to find out who is performing well, who may need additional training and who else may need recruiting. It can be used to maximise performance, reduce costs and identify patterns or traits that lead to poor performance.

- Employee turnover analytics: considers the rate at which employees leave the organisation, looks at why people leave and considers factors that help people to remain within the organisation. It uses data collected from employee satisfaction surveys, employee engagement levels and employee exit interviews, for example.

- Leadership analytics: considers and analyses the performance of leaders with the aim of improving, developing and retaining good and effective leaders. It uses data collected from interviews, surveys and performance evaluation and assessment.

- Recruitment analytics: considers recruitment numbers, trends, decisions, processes and strategies with the aim of optimising hiring processes. It can also include recruitment channel analytics that considers the channels that are used for recruitment to determine which are the most effective and which yield the best individuals for the job.

If you are interested in finding out more about HR analytics an evidence-based review of the method is provided by Marler and Boudreau (2017). Guenole et al. (2017) provide a concise, practical guide whereas more detailed and comprehensive information, including interesting case studies, can be found in Edwards and Edwards (2016) and Bhattacharyya (2017). Other interesting case studies can be found in King (2016) and Rasmussen and Ulrich (2015). Pease et al. (2013) provide useful information aimed at practitioners and Khan and Tang (2016) discuss managerial implications related to potential employee concerns with HR analytics, illustrating how these concerns can have ramifications for organisations wanting to implement change as a result of HR analytics. There are various tools and software packages available for HR analytics: some examples are listed below, including consolidation, prediction, modelling, analysis and visualisation tools. It is useful to visit relevant websites so that you can get an idea of functions, capabilities, purpose and costs, as these vary considerably. You will also find it useful to obtain more information about business analytics (Chapter 4), data analytics (Chapter 10), big data analytics (Chapter 3), data mining (Chapter 12), data visualisation (Chapter 13) and predictive modelling (Chapter 45). Some organisations also use sensor-based methods, mobile devices and wearables to collect, track and analyse employee movement and behaviour. If these are of interest, more information can be found in Chapters 32, 49 and 57.

Questions for reflection

Epistemology, theoretical perspective and methodology

- Is it possible to combine quantitative and qualitative methodologies when adopting HR analytics for a more complete and accurate understanding or interpretation? For example, statistical analyses could show a rise in people leaving a company or a downturn in performance, but might not explain why this is the case. Combining the figures with information from in-depth interviews would provide a more detailed and fuller picture and help to explain trends found in the data.

- When using data from different HR systems and data supplied in different forms, how easy is it to combine and integrate data? More information about data collection and conversion is provided in Chapter 11.

- When developing policy or making decisions based on data-driven evidence, is there any room for individual intuition, understanding, experience or personal knowledge? If there is room, how is it possible to incorporate these within HR analytics? If there is not room, what impact might this have on members of staff who feel that they can contribute, or feel that decisions made as a result of HR analytics are ineffective or wrong, for example?

Ethics, morals and legal issues

- How do employees perceive the use of HR analytics? Do they trust the data, the analysts and the outcomes? How do they feel about issues of privacy? How do they feel about being monitored in real-time? Do they have any suspicions about managerial motives and the impact these might have on how changes are implemented, and the effectiveness of these changes, as a result of HR analytics? Khan and Tang (2016) provide an enlightening discussion on these issues.

- Rasmussen and Ulrich (2015) suggest that HR analytics can be misused to maintain the status quo or drive a certain agenda. How can this be avoided?

- When using organisations' data for research purposes, have employees been made aware that data are to be used for this purpose, and have they given informed consent or signed a disclosure notice, for example? How much control do they have over personal data and how they are shared? This is of particular importance when personal data are collected from mobile apps and wearables, for example.

- If HR analytics are to be used to make decisions that will impact employees' lives, are they fully aware that this is the case (decisions about hiring, firing and promotion, for example)?

- Are all procedures robust and transparent? This is of particular importance when decisions made as a result of analytics have an adverse impact on individuals: the accuracy of calculations and the transparency of procedures can come under scrutiny (and an organisation could face possible legal action). See O'Neil (2017) for examples of instances where individuals are affected by decisions that are wrong yet cannot be challenged.

- Do you understand how to handle personally identifiable information (encryption, security and protection when data are in transit or at rest, for example)? Do you understand what data security measures are mandated by law (privacy laws of the United States or the General Data Protection Regulation in the EU, for example)?

Practicalities

- Rasmussen and Ulrich (2015) point out that analytics should start with business challenges and should lead to decisions that create business success. How does this observation relate to your research?

- How easy is it to access data? Will you be met with resistance (from members of the workforce or managers, for example)? Are data accessible and available for research purposes? Are members of staff happy to share personal data for research purposes? If some members of staff are happy to share personal data, but others are not, what impact will this have on your research?

- What is the quality of data? Are data missing or are errors present? How can you avoid making bad decisions because of inaccurate or ineffective data?

- Do you have the required understanding of descriptive and inferential statistics and of descriptive, predictive or prescriptive analytics and the tools/software that are available (or the ability to acquire the required understanding)? Do you know enough about statistical tests and procedures, such as analysis of variance, correlation and regression analysis? A good introduction to statistics and data analysis is provided by Heumann et al. (2017). If you are a 'non-mathematician' Rowntree (2018) provides a straightforward and user-friendly guide to statistics.

Useful resources

There are various digital tools and software packages available for HR analytics at time of writing. Examples include (in alphabetical order):

- Board (www.board.com/en);

- ClicData (www.clicdata.com);

- crunchr (www.crunchrapps.com);

- Dundas BI (www.dundas.com/dundas-bi);

- Grow BI Dashboard (www.grow.com);

- Oracle HR Analytics (www.oracle.com/us/solutions/business-analytics);

- PeopleStreme (www.peoplestreme.com/product/workforce-analytics);

- Sisense (www.sisense.com);

- Talentsoft Analytics (www.talentsoft.com/hr-software/hr-analytics/software);

- Yellowfin (www.yellowfinbi.com).

Additional digital tools and software packages that can be used for HR analytics are listed in Chapters 3, 4, 10 and 13.

The Chartered Institute of Personnel and Development in the UK has produced a useful guide to HR analytics that covers issues such as strategy, process and data. It can be accessed at www.cipd.co.uk/knowledge/strategy/analytics/factsheet [accessed October 30, 2018]. Registration is required to access the content.

Key texts

Bhattacharyya, D. (2017) *HR Analytics: Understanding Theories and Applications.* London: Sage.

Edwards, M. and Edwards, K. (2016) *Predictive HR Analytics: Mastering the HR Metric.* London: Kogan Page.

Guenole, N., Ferrar, J. and Feinzig, S. (2017) *The Power of People: Learn How Successful Organizations Use Workforce Analytics to Improve Business Performance.* Indianapolis, IN: Pearson Education, Inc.

Heumann, C., Schomaker, M. and Shalabh, (2017) *Introduction to Statistics and Data Analysis: With Exercises, Solutions and Applications in R.* Cham: Springer Nature.

Khan, S. and Tang, J. (2016) 'The Paradox of Human Resource Analytics: Being Mindful of Employees', *Journal of General Management*, 42(2), 57–66, first published May 24, 2017, 10.1177/030630701704200205.

King, K. (2016) 'Data Analytics in Human Resources: A Case Study and Critical Review', *Human Resource Development Review*, 15(4), 487–95, first published November 20, 2016, 10.1177/1534484316675818.

Marler, J. and Boudreau, J. (2017) 'An Evidence-Based Review of HR Analytics', *The International Journal of Human Resource Management*, 28(1), 3–26, published online: November 11, 2016, 10.1080/09585192.2016.1244699.

O'Neil, C. (2017) *Weapons of Math Destruction: How Big Data Increases Inequality and Threatens Democracy*. London: Penguin.

Pease, G., Byerly, B. and Fitz-Enz, J. (2013) *Human Capital Analytics: How to Harness the Potential of Your Organization's Greatest Asset*, Hoboken, NJ: John Wiley & Sons.

Rasmussen, T. and Ulrich, D. (2015) 'Learning from Practice: How HR Analytics Avoids Being a Management Fad', *Organizational Dynamics*, 44(3), 236–42, published online May 27, 2015, 10.1016/j.orgdyn.2015.05.008.

Rowntree, D. (2018) *Statistics without Tears: An Introduction for Non-Mathematicians*. New, updated edition. London: Penguin.

CHAPTER 23

Information retrieval

Overview

Information retrieval (IR) refers to the science of searching for information. For the purpose of this book it refers specifically to searching and recovering relevant information stored digitally in documents, files, databases, digital libraries, digital repositories and digital archives, for example. It also refers to the processes involved in searching and recovering this information such as gathering, storing, processing, indexing, querying, filtering, clustering, classifying, ranking, evaluating and distributing information. Although IR does have a long pre-digital history, it is beyond the scope of this book: if you are interested in finding out about the history of IR research and development, see Sanderson and Croft (2012). IR in the digital environment can be seen to be closely related to data mining. The difference is that IR refers to the process of organising data and building algorithms so that queries can be written to retrieve the required (and relevant) information. Data mining, on the other hand, refers to the process of discovering hidden patterns, relationships and trends within the data (see Chapter 12). IR can be seen as problem-orientated whereas data mining is data-orientated (or data-driven).

There are various subsets within the field of information retrieval in the digital environment. These include (in alphabetical order):

- Blog retrieval: retrieving and ranking blogs according to relevance to the query topic (see Zahedi et al., 2017 for information about improving relevant blog retrieval methods).

- Cross-lingual information retrieval (or cross-language information retrieval): searching document collections in multiple languages and retrieving information in a language other than that expressed in the query.

- Document retrieval: locating meaningful and relevant documents in large collections. This can include records, official papers, legal papers, certificates, deeds, charters, contracts, legal agreements and collections of articles or papers. It can include paper documents that have been scanned or those that have been created digitally.

- Multimedia information retrieval: retrieving information with a combination of content forms (text, images, animation, audio and video, for example).

- Music retrieval: retrieving information from music to categorise, manipulate, analyse and/or create music. See Chapter 2 for more information about music information retrieval.

- Personal information retrieval: retrieving information from personal information archives, personal digital spaces or personal information management systems (on desktops or mobile devices, for example).

- Record retrieval: sometimes used interchangeably with document retrieval, used to describe the process of locating a particular file, document or record.

- Spoken content retrieval: retrieving information from spoken content based on audio rather than text descriptions.

- Text retrieval (or text-based information retrieval): retrieving natural language text data such as academic papers, news articles, web pages, forum posts, emails, product reviews, tweets and blogs.

- Visual information retrieval, which can include:

  o video retrieval: retrieving video sequences from databases through content-based video indexing and retrieval systems, for example.

  o image retrieval: retrieving images from collections of images such as databases and repositories, which can include content-based image retrieval and semantic image retrieval, for example.

See Enser (2008) for a discussion on the evolution of visual information retrieval and Chapter 16 for more information about digital visual methods.

- Web retrieval (or web information retrieval): retrieving information on the Web through the use of search engines, link analysis retrieval, social search engines (or user-generated content searches) and mining, for example. See Singer et al. (2013) for a report on an investigation into the behavioural search characteristics of ordinary Web search engine users.

Examples of research projects that have used, assessed or critiqued information retrieval methods include the introduction and evaluation of an information retrieval approach 'for finding meaningful and geographically resolved historical descriptions in large digital collections of historical documents' (Tulowiecki, 2018); a study into Google's performance in retrieving relevant information from Persian documents (Sadeghi and Vegas, 2017); research on query terms proximity embedding for information retrieval (Qiao et al., 2017); an overview of time-aware information retrieval models (Moulahi et al., 2016); and a study into library users' information retrieval behaviour in an Israeli University library (Greenberg and Bar-Ilan, 2017).

If you are interested in finding out more about information retrieval a good introduction is provided by Manning et al. (2008). The book covers topics such as Boolean retrieval, index construction and compression, clustering, web crawling and link analysis. If you are interested in conducting laboratory experiments in information retrieval, Sakai (2018) provides a good source of information, covering sample sizes, effect sizes and statistical power, along with relevant case studies. If you are approaching your work from a health related environment, Levay and Craven (2018) provide a comprehensive guide, covering theoretical models of information seeking and systematic searching practice, new technologies for retrieving information and considerations for the future. You may also be interested in finding out more about data mining (Chapter 12), researching digital objects (Chapter 48), web and mobile analytics (Chapter 58) and Webometrics (Chapter 59).

Questions for reflection

Epistemology, theoretical perspective and methodology

- What are the differences between data, information and knowledge? Are these differences clear, distinct and recognised? What relevance do such distinctions have for your research? Boisot and Canals (2004) address these questions in relation to information theory and the physics of information.

- What is information? What is retrieval? How might answers to these questions differ, depending on discipline or field of study (computer sciences, philosophy, linguistics, physics, engineering or information science, for example)?

- What theory or theoretical perspective will guide your research? Cole (2011: 1216), for example, proposes a theory of information need for information retrieval, pointing out that 'the central assumption of the theory is that while computer science sees IR as an information- or answer-finding system, focused on the user finding an answer, an information science or user-oriented theory of information need envisages a knowledge formulation/acquisition system'.

Ethics, morals and legal issues

- What factors can influence information seeking and retrieval (socio-cultural, institutional, financial and educational, for example)?

- What types of bias might be present when retrieving information? This can include query sample selection bias (Melucci, 2016), social bias (Otterbacher, 2018) and retrievability bias (Samar et al., 2018).

- Do you have a good understanding of your chosen software licensing agreement (copyright notices for redistributions of source code and redistributions in binary form, and obtaining prior permission for use of names for endorsements or for products derived from the software, for example)?

- What security features are available with your chosen software? This can include authorisation, restricted access, encryption, audit logging, monitoring and compliance with security standards, for example.

Practicalities

- Do you have a good understanding of the types of search that can be performed (structured, unstructured, geolocation or metric, for example)?

- What is the difference in results between IR searches and Structured Query Language (SQL) searches of databases?

- Does your chosen software provide a good range of query types? This can include phrase queries, proximity queries, wildcard queries and range queries.

- What indexing strategies are supported by your chosen software (multi-pass indexing, large-scale single-pass indexing and real-time indexing of document streams, for example)?

- What is the difference between natural language indexing and controlled language indexing? What are the strengths and weaknesses of each?

- Do you have a good understanding of retrieval models? This can include Boolean models, vector space models, Bayesian networks, language retrieval models, link analysis retrieval models, credibility based retrieval models and cross-lingual relevance models, for example.

- What factors enable you to assess the effectiveness of an IR system (this can include relevance, coverage, speed, accuracy, usability, functionality, availability and cost)?

- How can you access information that is not visible to web search engines, such as information residing in corporate databases?

Useful resources

There are a wide variety of digital tools and software packages available for information retrieval. Some of these are open source and free to use, whereas others have significant costs attached. Tools and software vary enormously and your choice will depend on your research topic, needs and purpose. The following list provides a snapshot of what is available at time of writing and is listed in alphabetical order:

- Apache Lucene (http://lucene.apache.org/core);
- Cheshire (http://cheshire.berkeley.edu);
- Elasticsearch (www.elastic.co/products/elasticsearch);
- Hibernate Search (http://hibernate.org/search);
- Lemur (www.lemurproject.org/lemur);
- OpenText™ Search Server (www.opentext.com);
- Solr (http://lucene.apache.org/solr);
- Terrier (http://terrier.org);
- UltraSearch (www.jam-software.com/ultrasearch);
- Windows Search (https://docs.microsoft.com/en-us/windows/desktop/search/windows-search);
- Xapian (https://xapian.org);
- Zettair (www.seg.rmit.edu.au/zettair).

The Information Retrieval Journal (https://link.springer.com/journal/10791) 'provides an international forum for the publication of theory, algorithms, analysis and experiments across the broad area of information retrieval'.

The International Journal of Multimedia Information Retrieval (https://link.springer.com/journal/13735) is an academic archival journal publishing original, peer-reviewed research contributions in areas within the field of multimedia information retrieval.

Key texts

Boisot, M. and Canals, A. (2004) 'Data, Information and Knowledge: Have We Got It Right?', *Journal of Evolutionary Economics*, 14(43), 43–67, published January 2004, 10.1007/s00191-003-0181-9.

Cole, C. (2011) 'A Theory of Information Need for Information Retrieval that Connects Information to Knowledge', *Journal of the Association for Information Science and Technology*, 62(7), 1216–31, first published April 19, 2011, 10.1002/asi.21541.

Enser, P. (2008) 'The Evolution of Visual Information Retrieval', *Journal of Information Science*, 34(4), 531–46, first published June 13, 2008, 10.1177/0165551508091013.

Greenberg, R. and Bar-Ilan, J. (2017) 'Library Metrics – Studying Academic Users' Information Retrieval Behavior: A Case Study of an Israeli University Library', *Journal of Librarianship and Information Science*, 49(4), 454–67, first published April 3, 2016, 10.1177/096100061664003.

Levay, P. and Craven, J. (2018) *Systematic Searching*, London: Facet Publishing.

Manning, C., Raghavan, P. and Schütze, H. (2008) *Introduction to Information Retrieval*. Cambridge: Cambridge University Press.

Melucci, M. (2016) 'Impact of Query Sample Selection Bias on Information Retrieval System Ranking', *2016 IEEE International Conference on Data Science and Advanced Analytics*, published online December 26, 2016, doi:10.1109/DSAA.2016.43.

Moulahi, B., Tamine, L. and Yahia, S. (2016) 'When Time Meets Information Retrieval: Past Proposals, Current Plans and Future Trends', *Journal of Information Science*, 42(6), 725–47, first published September 30, 2015, 10.1177/0165551515607277.

Otterbacher, J. (2018) 'Addressing Social Bias in Information Retrieval'. In Bellot, P. et al. (eds.) *Experimental IR Meets Multilinguality, Multimodality, and Interaction*, 121–127. CLEF 2018. Lecture Notes in Computer Science, 11018, Cham: Springer.

Qiao, Y., Du, Q. and Wan, D. (2017) 'A Study on Query Terms Proximity Embedding for Information Retrieval', *International Journal of Distributed Sensor Networks*, first published February 28, 2017, 10.1177/1550147717694891.

Sadeghi, M. and Vegas, J. (2017) 'How Well Does Google Work with Persian Documents?', *Journal of Information Science*, 43(3), 316–27, first published March 23, 2016, 10.1177/0165551516640437.

Sakai, T. (2018) *Laboratory Experiments in Information Retrieval: Sample Sizes, Effect Sizes, and Statistical Power*, Singapore: Spring Nature Singapore Pte Ltd.

Samar, T., Traub, M., van Ossenbruggen, J., Hardman, L. and de Vries, A. (2018) 'Quantifying Retrieval Bias in Web Archive Search', *International Journal on Digital Libraries*, 19 (1), 57–75, published online April 18, 2017, 10.1007/s00799-017-0215-9.

Information retrieval

Sanderson, M. and Croft, B. (2012) 'The History of Information Retrieval Research', *Proceedings of the IEEE*, 100(Special Centennial Issue), 1444–51, first published April 11, 2012, 10.1109/JPROC.2012.2189916.

Singer, G., Norbisrath, U. and Lewandowski, D. (2013) 'Ordinary Search Engine Users Carrying Out Complex Search Tasks', *Journal of Information Science*, 39(3), 346–58, first published December 17, 2012, 10.1177/0165551512466974.

Tulowiecki, S. (2018) 'Information Retrieval in Physical Geography: A Method to Recover Geographical Information from Digitized Historical Documents', *Progress in Physical Geography: Earth and Environment*, 42(3), 369–90, first published April 29, 2018, 10.1177/0309133318770972.

Zahedi, M., Aleahmad, A., Rahgozar, M., Oroumchian, F. and Bozorgi, A. (2017) 'Time Sensitive Blog Retrieval Using Temporal Properties of Queries', *Journal of Information Science*, 43(1), 103–21, first published December 23, 2015, 10.1177/0165551515618589.

CHAPTER 24

Learning analytics

Overview

Learning analytics refers to the process of capturing, exploring, analysing, visualising and reporting data about learners. This method is used to gain insight into learners and the learning environment, to improve teaching and learning practice and help with retention and support. Learning analytics are also used in business by companies interested in professional development (de Laat and Schreurs, 2013) and in assessing, managing and developing corporate education and training. The method draws on insights, developments and methodologies from a number of methods, techniques and fields of study, including data mining (Chapter 12), educational data mining (Chapter 17), social network analysis (Chapter 53) web and mobile analytics (Chapter 58), sociology, psychology and education. Examples of learning analytics research include a mixed methods project using click-data, user-generated content and student interviews from online higher education courses (Scott and Nichols, 2017), a pilot project to develop experimental approaches to learning analytics with the aim of fostering critical awareness of computational data analysis among teachers and learners (Knox, 2017) and a study into how data might be collected and analysed to improve online learning effectiveness (Martin et al., 2016).

Specific applications of learning analytics include (in alphabetical order):

- Adaptation: modify or adapt learning materials and assessment types.
- Detection: discover patterns and trends and gain deeper insight.
- Direction: manage and guide learners, courses or education/training policy and provide direction for curriculum development.

- Evaluation: gauge programme effectiveness and measure return on educational and technological investment.

- Intervention: take action and provide feedback, guidance and support when required, and provide alerts and notifications for action (for students, tutors, administrators, employees or managers).

- Motivation: set benchmarks and goals, promote self-regulated learning, encourage progression and enable students/employees to view how they are performing and how they can improve.

- Observation: observe how students/employees are learning (individually and collaboratively) and view results of assessments.

- Optimisation: enhance performance (individual and institutional), hone instruction, improve success rates and optimise learner experience.

- Personalisation: provide personalised learning plans, develop personalised learning materials and find suitable learning resources.

- Prediction: identify learners who are at risk of leaving their programme or learners who are at risk of failing their course.

- Visualisation: understand, share, report and act on data.

There are a wide variety of sources of data for learning analytics, and those that are used depend on research focus, purpose and methodology. Sources can be internal (under the control of, and owned by, the educational institution or company) and external (private or public data generated by outside organisations). Examples of data sources that are used for learning analytics are given below. Additional sources are provided in Chapter 17.

Internal sources:

- Data gathered through, and stored within, learning management systems, library systems, student information systems or student records systems. Examples include:

  o student information:

    - demographics;
    - previous academic history;

  o student engagement or activity statistics:

    - attendance levels;

- completion rates;
- enrolments/registrations;
- library usage;
- reading lists (and books that are accessed and withdrawn);
- online activity (navigation habits, reading habits, writing habits and pauses);
- access to IT, WiFi, VLE/intranet and other networking systems;
- learner support/study support unit attendance;

- student performance statistics:
  - assessments;
  - assignment submission;
  - grades;
  - competencies;
  - progression.

- Physical and digital touchpoints:
  - college choices;
  - course choices;
  - module choices;
  - training programme choices;
  - employee performance evaluation, assessment and appraisal interviews.

- Student/employee feedback forms, questionnaires and interviews.
- Data collected from wearables (Chapter 57), sensors (Chapter 49), mobile technologies (Chapter 32), eye-tracking technologies (Chapter 19) and tracking devices that monitor learner behaviour, interaction and/or movement. See the Transformative Learning Technologies Lab, Stanford Graduate School of Education for examples of multimodal learning analytics research projects and papers (https://tltl.stanford.edu/project/multimodal-learning-analytics).

External sources:

- government statistics;

- mobile and web data (see Chapter 58);

- national and international experience, satisfaction and ranking surveys;

- open government data (visit https://opengovernmentdata.org to find out more about open government data).

- research data repositories (visit the Registry of Research Data Repositories to find out what is available: www.re3data.org);

- Linked Data that shares, exposes and integrates global data (see http://linkededucation.org, http://metamorphosis.med.duth.gr and www.gsic.uva.es/seek/useek for examples);

- social media data (see Chapter 52).

If you are interested in finding out more about learning analytics, Sclater (2017) is a good place to start. He provides interesting case studies, illustrates how to undertake institutional projects and assesses the legal and ethical issues associated with learning analytics. Siemens (2013) reviews the historical development of learning analytics, discusses the tools and techniques used by researchers and practitioners and considers future challenges. Slade and Prinsloo (2013) provide a detailed and enlightening discussion on ethical issues and dilemmas associated with learning analytics, along with a brief overview of the purposes and collection of educational data, and the educational purpose of learning analytics. There are a wide variety of tools and software available for learning analytics: some examples are listed below. It is useful to visit relevant websites so that you can get an idea of function, capability, purpose and cost, as these vary considerably. Chapter 17 provides a discussion on the closely related method of educational data mining (EDM), and includes information about the similarities and differences between EDM and learning analytics. An understanding of big data analytics (Chapter 3), data analytics (Chapter 10), data mining (Chapter 12) and data visualisation (Chapter 13) is also important for those who are interested in learning analytics. You may also find it useful to obtain more information about web and mobile analytics (Chapter 58) and social media analytics (Chapter 52).

Questions for reflection

Epistemology, theoretical perspective and methodology

- What is the relationship between epistemology, pedagogy and assessment and how might this relationship relate to learning analytics? Knight et al. (2014) provide answers to this question.

- Should your project be informed by epistemology and theory, or led by technology?

- Slade and Prinsloo (2013: 1510) point out that 'there is an inherent assumption linked to learning analytics that knowledge of a learner's behavior is advantageous for the individual, instructor, and educational provider'. Is this the case? When might there be disadvantages linked to learner analytics?

- What can and cannot be measured by learning analytics? Can learning analytics, for example, measure transformative learning such as changes in the understanding of self, behavioural changes or changes in beliefs? Are there more effective ways to measure such learning?

- How can you make sense of diverse sets of analytical data? Lockyer et al. (2013) explore how learning design might provide the framework for interpreting learning analytics and making sense of data.

Ethics, morals and legal issues

- What are the ethical implications and challenges associated with increased scrutiny of student data? Slade and Prinsloo (2013) discuss these issues in depth, which will help you to reflect on this question. They go on to develop an ethical framework that includes six principles: learning analytics as moral practice; students as agents; student identity and performance are temporal dynamic constructs; student success is a complex and multi-dimensional phenomenon; transparency; and higher education cannot afford not to use data. Jisc provides a useful Code of Practice for Learning Analytics (details below) that covers these and other ethical issues.

- What encryption mechanisms and security policies are in place? How will you (or the software hosting company) protect against data breaches or cyber-attacks (disgruntled students, former employees or outside hackers, for example)? The multitude of connected devices in educational institutions and workplaces can expand the attack surface and enable hackers to gain access in cases where cybersecurity is inadequate, for example.

- How might 'our cultural, political, social, physical, and economic contexts and power relationships shape our responses to the ethical dilemmas and issues in learning analytics' (Slade and Prinsloo, 2013: 1511)?

- Is there a danger that learning analytics can lead to stereotyping (students who are labelled as at risk, for example)? Could this mean that such students are treated unfairly?

Practicalities

- Do you have a good understanding of tools and software packages that are available for learning analytics, and know which are the most appropriate, given the purpose and focus of your research? When choosing tools and software it is important to look for a simple user interface (dashboard or scorecard), data-sharing capabilities, scalable architecture, the ability to process structured and unstructured data, visualisation tools and report-generation tools. The Jisc learning analytics service provides useful information (details below).

- Siemens (2013) suggests that the future development of analytic techniques and tools from learning analytics and EDM will overlap. Do you agree with this suggestion and, if so, do you have the required understanding of relevant tools, techniques and overlaps? More information about EDM can be found in Chapter 17.

- Once your project is complete, how do you intend to disseminate your results, share data, raise awareness and/or act on outcomes and recommendations? Does this involve working with others and, if so, what communication, negotiation and agreement strategies are required? How can you ensure that you are effective in putting your analytics into action?

Useful resources

Examples of digital tools and software packages for learning analytics that are available at time of writing include (in alphabetical order):

- Banner Student Retention Performance (www.ellucian.com/Software/Banner-Student-Retention-Performance);

- Blackboard Analytics (www.blackboard.com/education-analytics/index.html);

- Brightspace Insights (www.d2l.com/en-eu/products/insights);

- KlassData (http://klassdata.com);

- Metrics That Matter (www.cebglobal.com/talent-management/metrics-that-matter.html);

- Moodle Engagement Analytics plug-in (https://moodle.org/plugins/report_engagement);

- SEAtS Learning Analytics (www.seatssoftware.com/learning-analytics);

- The Learning Dashboard (http://acrobatiq.com/products/learning-analytics).

Jisc in the UK has developed 'a learning analytics architecture' and implemented a learning analytics service that are part of the Jisc Data and Analytics service. Videos, toolkits and guides are available on the website (https://docs.analytics.alpha.jisc.ac.uk/docs/learning-analytics/Home) along with details of the Study Goal app. Also, the Jisc learning analytics service provides a suite of software tools and infrastructure for students and staff (www.jisc.ac.uk/learning-analytics) and a detailed Code of Practice (www.jisc.ac.uk/guides/code-of-practice-for-learning-analytics).

The Society for Learning Analytics Research (SoLAR) (https://solaresearch.org) is 'an inter-disciplinary network of leading international researchers who are exploring the role and impact of analytics on teaching, learning, training and development'. You can access the Journal of Learning Analytics and perform computational analyses on the Learning Analytics research literature on the website.

Key texts

de Laat, M. and Schreurs, B. (2013) 'Visualizing Informal Professional Development Networks: Building a Case for Learning Analytics in the Workplace', *American Behavioral Scientist*, 57(10), 1421–38, first published March 11, 2013, 10.1177/0002764213479364.

Knight, S., Buckingham Shum, S. and Littleton, K. (2014) 'Epistemology, Assessment, Pedagogy: Where Learning Meets Analytics in the Middle Space', *Journal of Learning Analytics*, 1(2), 23–47, 10.18608/jla.2014.12.3.

Knox, J. (2017) 'Data Power in Education: Exploring Critical Awareness with the "Learning Analytics Report Card"', *Television & New Media*, 18(8), 734–52, first published January 30, 2017, 10.1177/1527476417690029.

Lockyer, L., Heathcote, E. and Dawson, S. (2013) 'Informing Pedagogical Action: Aligning Learning Analytics with Learning Design', *American Behavioral Scientist*, 57(10), 1439–59, first published March 12, 2013, 10.1177/0002764213479367.

Martin, F., Ndoye, A. and Wilkins, P. (2016) 'Using Learning Analytics to Enhance Student Learning in Online Courses Based on Quality Matters Standards', *Journal of*

Educational Technology Systems, 45(2), 165–87, first published November 15, 2016, 10.1177/0047239516656369.

Sclater, N. (2017) *Learning Analytics Explained*. New York, NY: Routledge.

Sclater, N., Peasgood, A. and Mullan, J. (2016) *Learning Analytics in Higher Education. A Review of UK and International Practice*. Bristol: Jisc, Retrieved from www.jisc.ac.uk/reports/learning-analytics-in-higher-education [accessed May 15, 2018].

Scott, J. and Nichols, T. (2017) 'Learning Analytics as Assemblage: Criticality and Contingency in Online Education', *Research in Education*, 98(1), 83–105, first published August 14, 2017, 10.1177/0034523717723391.

Siemens, G. (2013) 'Learning Analytics: The Emergence of a Discipline', *American Behavioral Scientist*, 57(10), 1380–400, first published August 20, 2013, 10.1177/0002764213498851.

Slade, S. and Prinsloo, P. (2013) 'Learning Analytics: Ethical Issues and Dilemmas', *American Behavioral Scientist*, 57(10), 1510–29, first published March 4, 2013, 10.1177/0002764213479366.

CHAPTER 25

Link analysis

Overview

Link analysis can be defined in three ways, depending on discipline or approach. The first has its roots in social network analysis, which borrows from graph theory (the theoretical study of graphs, their structure and application) and sociometry (the study and measurement of relationships within a group of people). Link analysis, within this approach, is a data analysis technique that is used to evaluate connections or relationships between individuals or objects in a network (social networks, semantic networks or conflict networks, for example). It enables researchers to calculate centrality (a measure of the importance or influence of nodes in a network), which includes degree centrality (the number of direct connections or ties), closeness centrality (the average length of the shortest path) and betweenness centrality (the number of times a node occurs on all shortest paths). More information about social network analysis, including definition of key terms such as nodes, ties, arcs, edges and bridges, can be found in Chapter 53. Link analysis, within this approach, has a long pre-digital history. However, it has been included in this book because this type of link analysis is now carried out using powerful computational tools to identify, process, transform, analyse and visualise linkage data (see below and in Chapter 53 for examples of relevant digital tools and software packages).

Link analysis, within the second approach, has its roots in the discipline of information science, in particular in bibliometrics (the quantitative study of books and other media of communication), scientometrics (the quantitative study of scientific activity and the history of science) informetrics (the quantitative study of information and retrieval) and citation analysis (the quantitative study of citations in documents). Link analysis (or link impact analysis), within this approach, is concerned with the analysis of links, hyperlinks and the graph structure of the Web (see Figuerola and Berrocal, 2013 for an example of a research paper that discusses the analysis of web links). This type of link analysis is a data analysis method that is used in webometrics, which is discussed in Chapter 59. A closely related technique is link mining (the application of data mining techniques to linkage data), which is discussed in Chapter 12. Another closely related area is information retrieval, including web and blog retrieval, which is discussed in Chapter 23.

Link analysis, within the third approach, has its roots in business analytics (the process of examining an organisation's data to measure past and present performance to gain insight that will help with future planning and development). It is used by those working in search engine optimisation (SEO) to improve the search engine rankings of an organisation's website. This involves analysing existing links, building new links and removing harmful links, for example. This approach to link analysis is also used by marketing and development teams to assess their competitors (competitive link analysis, for example) and by webmasters and web designers to troubleshoot and improve security. More information about business analytics can be found in Chapter 4 and more information about HR analytics can be found in Chapter 22.

The three approaches to link analysis described above can be used by researchers from a wide variety of disciplines and fields of study including sociology, politics, the humanities, criminology, business, health and medicine, geography and physics. Examples of research projects from the different approaches include a study into the link analysis of data from on-train recorders and the detection of patterns of behaviour regarding potential safety issues (Strathie and Walker, 2016); an investigation into the potential of web links to act as an indicator of collaboration (Stuart et al., 2007); research into the differences between co-citations in scientific literature and co-links in hypertext (Mutalikdesai and Srinivasa, 2010); and an exploratory study into internal links and search behaviour of Wikipedia users (Wu and Wu, 2011).

If you are interested in using link analysis for your research, Thelwall (2004) provides a good starting point for those approaching their work from the information sciences, while Fouss et al. (2016) provide a useful starting point for computer scientists, engineers, statisticians and physicists. For those approaching their work from a business analytics or SEO perspective, practical, free online courses are useful, such as the Search Engine

Optimization Specialization course offered through the Mooc platform Coursera (www.coursera.org/specializations/seo). It is also useful to view some of the digital tools and software packages that are available for link analysis, and these are listed below. Functions vary, depending on the tool that you choose: some enable you to produce advanced link charts, find common linked entities and produce 3-D and animated link charts, whereas others track, monitor and list links, enable you to respond to alerts and notifications of changes and build high quality links. It is useful to visit some of the sites listed below so that you can get an idea of functions, purpose and costs. A good understanding of data mining techniques (Chapter 12), data visualisation (Chapter 13) and webometrics (Chapter 59), along with reflection on the questions listed below, will help you to further plan your research project.

Questions for reflection

Epistemology, theoretical perspective and methodology

- Do you have a clear understanding of the different approaches to link analysis and associated theoretical perspectives and methodological frameworks (social network theory, network theory and social network analysis, for example)?

- What theoretical perspective, or social theory, will guide your research? Beigi et al. (2016: 539) discuss 'three social science theories, namely Emotional Information, Diffusion of Innovations and Individual Personality, to guide the task of link analysis in signed networks'.

- De Maeyer (2013: 745) suggests that 'there is no guarantee that indication of the author's intentions can be found in the link itself, or in the linked resource'. How will you address this issue in your research? This could include acknowledging the importance of context, noting the difference between web genres and adopting mixed methods approaches that do more than merely count links, for example.

Ethics, morals and legal issues

- What notification ethics are required? For example, the web crawler and link analysis tool aimed at researchers in the social sciences and humanities SocSciBot (http://socscibot.wlv.ac.uk) requests that you enter your email address and asks that you check for complaints about your web crawling (the software emails webmasters of any site you crawl to let them know what you are doing: they can ask you to stop crawling their site and you must agree to do so).

- Do you have a clear understanding of the licence attached to the software or digital tool that you intend to use?

- How do you intend to deal with potential bandwidth overload? SocSciBot deals with this by asking that all users first read Thelwall and Stuart (2006), which provides useful guidance on this and other ethical issues.

- Is link analysis an unobtrusive research method, or is there potential for the researcher to influence what is observed? De Maeyer (2013: 744) points out that 'the mere fact that it is widely known that Google's search engine algorithm draws on links creates a potential bias'. She goes on to suggest that users are becoming more aware of hyperlink structures and that link analysis undertaken by researchers can be spotted by website owners, who may react in some way to the analysis. What are the implications of such observations on your research?

Practicalities

- Do you have a good understanding of which digital tools and software packages are available for the type of link analysis you wish to conduct? Do you have a good idea of costs involved and the amount of technical support offered?

- Some tools may only show a tiny segment of links. What impact will this have on your research?

- Some websites imitate, copy or plagiarise others. How might this affect the validity of inferential statistics? This issue is raised and discussed by Thelwall (2004).

- De Maeyer (2013: 747) poses an interesting question that should be considered when you plan your research: 'when do we stop to purely describe the structure of links as technical objects and when do we start to relevantly exploit links to make sense of a social phenomenon?'

Useful resources

There are a variety of digital tools and software packages available, including web-crawlers and search engine analysis tools, which enable researchers to undertake link analysis. Some are aimed at researchers in disciplines such as the social sciences whereas others are aimed at business and the search engine optimisation (SEO) sector. Examples available at time of writing are given below, in alphabetical order. More useful tools can be found in Chapter 53 (social network analysis) and Chapter 59 (webometrics).

- Hyphe for web corpus curation (http://hyphe.medialab.sciences-po.fr);
- Issue Crawler for network mapping (www.govcom.org/Issuecrawler_in structions.htm);
- Linkody for tracking link building campaigns (www.linkody.com);
- LinkResearchTools for link audit, backlink analysis, competitor analysis and risk monitoring (www.linkresearchtools.com);
- OpenLinkProfiler for analysing website links (www.openlinkprofiler.org);
- Open Site Explorer for link prospecting and link building (https://moz .com/researchtools/ose);
- Sentinel Visualizer for advanced link analysis, data visualisation, social network analysis and geospatial mapping (www.fmsasg.com);
- SocSciBot for web crawling and link analysis in the social sciences and humanities (http://socscibot.wlv.ac.uk);
- Webometric Analyst 2.0 for altmetrics, citation analysis, social web analysis, webometrics and link analysis (http://lexiurl.wlv.ac.uk).

Common Crawl (http://commoncrawl.org) is an open repository of web crawl data that can be accessed and analysed by anyone who is interested in the data, providing 'a corpus for collaborative research, analysis and education'. The Web Data Commons project (http://webdatacommons.org) extracts structured data from Common Crawl and provides the extracted data for public download.

Key texts

Beigi, G., Tang, J. and Liu, H. (2016) 'Signed Link Analysis in Social Media Networks', *Proceedings of the Tenth International AAAI Conference on Web and Social Media* (ICWSM 2016), 539–42, retrieved from www.aaai.org/ocs/index.php/ICWSM/ ICWSM16/paper/download/13097/12783 [accessed November 20, 2018].
De Maeyer, J. (2013) 'Towards A Hyperlinked Society: A Critical Review of Link Studies', *New Media & Society*, 15(5), 737–51, first published December 5, 2012, 10.1177/ 1461444812462851.

Link analysis

Figuerola, C. and Berrocal, J. (2013) 'Web Link-Based Relationships among Top European Universities', *Journal of Information Science*, 39(5), 629–42, first published April 9, 2013, 10.1177/0165551513480579.

Fouss, F., Saerens, M. and Shimbo, M. (2016) *Algorithms and Models for Network Data and Link Analysis*. New York, NY: Cambridge University Press.

Mutalikdesai, M. and Srinivasa, S. (2010) 'Co-Citations as Citation Endorsements and Co-Links as Link Endorsements', *Journal of Information Science*, 36(3), 383–400, first published April 13, 2010, 10.1177/0165551510366078.

Strathie, A. and Walker, G. (2016) 'Can Link Analysis Be Applied to Identify Behavioral Patterns in Train Recorder Data?', *Human Factors*, 58(2), 205–17, first published December 11, 2015, 10.1177/0018720815613183.

Stuart, D., Thelwall, M. and Harries, G. (2007) 'UK Academic Web Links and Collaboration – an Exploratory Study', *Journal of Information Science*, 33(2), 231–46, first published April 1, 2007, 10.1177/0165551506075326.

Thelwall, M. (2004) *Link Analysis: An Information Science Approach*. Amsterdam/ Boston, MA: Elsevier Academic Press.

Thelwall, M. and Stuart, D. (2006) 'Web Crawling Ethics Revisited: Cost, Privacy and Denial of Service', *Journal of the American Society for Information Science and Technology*, 57(13), 1771–79.

Wu, I. and Wu, C. (2011) 'Using Internal Link and Social Network Analysis to Support Searches in Wikipedia: A Model and Its Evaluation', *Journal of Information Science*, 37 (2), 189–207, first published March 4, 2011, 10.1177/0165551511400955.

CHAPTER 26

Live audience response

Overview

Live audience response is a method that enables researchers to collect and record live and spontaneous audience/viewer responses and reactions. It can also be referred to as audience response systems (ARS), personal response systems, group response systems, classroom response systems, participatory audience response or real-time response (RTR). It is an interactive, inclusive and responsive method that can be used for interactive polling, student evaluation and student, employee or audience engagement, for example. It is a method that can also be used in online panel research, using a panel of pre-recruited and profiled volunteers (Chapter 42), and within public or private online research communities to gather attitudes, opinions or feedback over an extended period of time (Chapter 44). There are two types of live audience response: co-located methods in which audience and researcher (or presenter/lecturer) are together in the same place (see Lantz and Stawiski, 2014 for an example of this method) and remote systems in which the audience is at a different location to the researcher (see Hong, 2015 for an example of this method).

Software enables researchers (and presenters or lecturers) to create questions before the event, work out in what format to present visualisations and decide on the method of audience response. Prompts or messages are sent, asking for participants to respond: they can do so using the web (via desktop, laptop or smartphone), through SMS texting via their phones or by using dedicated clickers or keypads. Live responses can be displayed on a presentation screen as the event/presentation progresses from web browsers or embedded into presentation software (PowerPoint, Keynote or Google

Slides, for example). Visualisations change as the audience responds (items are ranked in order or responses with more votes rise to the top of the list, for example). Live responses can also be sent to participants' devices, or hidden completely, depending on the subject and purpose of the research or event. Researchers are able to download reports after the event, analyse findings, produce reports and share results.

There are different methods, techniques and question styles that can be used for live audience response, including:

- multiple choice questions for choosing from pre-defined questions;

- open response or open-ended questions for responding openly and freely (a 'moderator facility' enables the researcher or presenter to check responses before they appear on the presentation screen);

- picture polls for answering questions about a picture (multiple choice, yes/no or true/false questions, for example);

- question and answer sessions for submitting questions or ideas and voting on them (with responses receiving the most votes moving to the top of the list);

- quizzes, games and icebreakers for interaction, learning from each other and getting to know each other;

- rank order for ranking responses that have been provided;

- scales for rating statements on a scale;

- slider bars for dialling or dragging scores along a continuum;

- '100 points' for prioritising different items by distributing 100 points;

- 2 x 2 matrix for rating each item in two dimensions;

- clickable images for clicking anywhere on an image;

- emoji and image word clouds for choosing emoji or images that grow, shrink or move position with each response;

- live word clouds for choosing words that grow, shrink or move position with each response.

Live audience response is a popular method in education, nursing, psychology, politics, media studies, marketing and advertising. Examples of research projects that have used, analysed or critiqued live audience response methods include studies into the use of student response systems in the accounting classroom (Carnaghan et al., 2011), for psychology courses (Dallaire, 2011) and in pharmacology learning (Lymn and Mostyn, 2010); research to assess whether clickers increase retention of lecture material over two days in a controlled environment (Lantz and Stawiski, 2014); a study into audience responses to television news coverage of medical advances (Hong, 2015); an investigation into the impact of audience response systems on student engagement in undergraduate university courses (Graham et al., 2007); and an analysis of undecided swing state voters during the six presidential debates of the 2008 and 2012 US elections (Schill and Kirk, 2014).

If you are interested in finding out more about live audience response methods for your research it is useful to visit some of the software package and digital tool websites listed below, as this gives you an idea of purpose, functions, capabilities and possible costs. Some of these websites contain useful blogs or guidance documents that explain more about live audience response methods, how they are used and how to get started. The questions for reflection listed below will help you to think more about using live audience response as a research method, and the references provided in key texts give examples of research projects that have used, analysed or critiqued live audience response methods. You might also find it useful to find out more about online panel research (Chapter 42) and online research communities (Chapter 44). If you are interested in response systems specifically for teaching, Bruff (2009) provides a good introduction, although the book could do with a little updating, given recent technological and digital advances.

Questions for reflection

Epistemology, theoretical perspective and methodology

- Do you intend to use live audience response as a standalone research method, or combine it with other methods? See Schill and Kirk (2014) for an example of how pre-test and post-test questionnaires were used together with live audience response methods.

- When using live audience response methods, what, exactly, is meant by 'live'? Is 'live' associated with presence, participation, connection, communication, technology, authenticity, reality, timing, immediacy and/or sociality, for example? If associated with presence, how does this fit with remote audience response systems? Van Es (2016) provides an enlightening discussion on the concept of 'liveness' in relation to media studies.

- When using live audience response methods, what, exactly, is meant by 'audience'? Is 'audience' defined by what people do (watching, viewing and listening), where they are, or through participation, connection and relationships, for example? Who is present? Why are they present? Is it an existing 'audience' or one chosen specifically for a research project? If it has been chosen, what sampling procedures have been used (or do you intend to use)? How can you avoid selection bias? How can you ensure that the sample is representative of your research population?

Ethics, morals and legal issues

- Is it possible to guarantee anonymity with the software and tools that you intend to use? For example, some live audience response tools enable researchers to select a 'completely anonymous' option that means that all responses for a particular poll will be anonymous because user data are not collected. However, in live feedback sessions, participants can choose to waive their anonymity, which means that such sessions are not completely anonymous. Researchers can moderate responses and choose to reject them before they appear on the presentation screen, although this function may not be available as standard in all tools. Will your participants be happy with the level of anonymity offered, and how will you discuss this subject prior to the live session?

- When choosing software, is the software company compliant with your local data protection regulations (GDPR in the EU, for example)? Does the software company have a clear and robust privacy policy regarding the collection, use and disclosure of personally identifiable information? Will data be deleted, once your subscription has lapsed?

- Does the software company use third party service providers or affiliates and, if so, do these organisations comply with all relevant legislation? Are clear and transparent written agreements in place?

Practicalities

- What factors need to be taken into account when choosing software (price; intuitive dashboard; ready-to-use questions and surveys; data storage, access and security; free and instant access to new features; functionality, compatibility and scalability, for example)?

- In cases where you are not providing hardware, do all participants have the necessary hardware available for use (desktops, laptops, tablets or smartphones, for example)? What contingency plans will you have in place in cases where participants do not have their own hardware? How can you ensure that all participants are able to use the equipment? What about people who are unable to use the equipment (physical disabilities, problems with reading or phobia of technology, for example)?

- What problems could occur with hardware and software and how can these be overcome (system crashes, loss of power/connection or physical breakage, for example)?

- How can you ensure that all participants respond to all messages/ prompts? Is it possible to ensure that messages, images or videos are not skipped, and that participants are able to give their full attention to the task? Logging and analysing response times can help, for example.

- If you intend to use remote live audience response methods, how can you ensure that all participants are present, available and able to respond when required? What effect might distractions have on their responses, in particular, if they decide to take part when on the move or when in public spaces, for example?

Useful resources

There are a variety of live audience response digital tools and software packages available. Some are aimed at the education sector (lectures, seminars, student response and classroom engagement), some are aimed at the business sector (business meetings, presentations and events) and others provide tools that can be used in a wide variety of sectors for research purposes. Examples available at time of writing are given below, in alphabetical order.

- Clikapad (www.clikapad.com);
- Glisser (www.glisser.com/features/live-polling);
- iClicker (www.iclicker.com);
- ivote-app (www.ivote-app.com);
- Meetoo (www.meetoo.com);
- Mentimeter (www.mentimeter.com);

- ombea (www.ombea.com);

- ParticiPoll (www.participoll.com);

- Poll Everywhere (www.polleverywhere.com);

- presentain (https://presentain.com);

- Slido (www.sli.do);

- Swipe (www.swipe.to);

- The Reactor Suite (www.roymorgan.com/products/reactor/reactor-products);

- Turning Technologies (www.turningtechnologies.com);

- Voxvote (www.voxvote.com).

Key texts

Bruff, D. (2009) *Teaching with Classroom Response Systems: Creating Active Learning Environments.* San Francisco, CA: Jossey-Bass.

Carnaghan, C., Edmonds, T., Lechner, T. and Olds, P. (2011) 'Using Student Response Systems in the Accounting Classroom: Strengths, Strategies and Limitations', *Journal of Accounting Education*, 29(4), 265–83, 10.1016/j.jaccedu.2012.05.002.

Dallaire, D. (2011) 'Effective Use of Personal Response "Clicker" Systems in Psychology Courses', *Teaching of Psychology*, 38(3), 199–204, first published June 14, 2011, 10.1177/0098628311411898.

Firsing, S., Yannessa, J., McGough, F., Delport, J., Po, M. and Brown, K. (2018) 'Millennial Student Preference of Audience Response System Technology', *Pedagogy in Health Promotion*, 4(1), 4–9, first published March 10, 2017, 10.1177/2373379917698163.

Graham, C., Tripp, T., Seawright, L. and Joeckel, G. (2007) 'Empowering or Compelling Reluctant Participators Using Audience Response Systems', *Active Learning in Higher Education*, 8(3), 233–58, first published November 1, 2007, 10.1177/1469787407081885.

Hong, H. (2015) 'Audience Responses to Television News Coverage of Medical Advances: The Mediating Role of Audience Emotions and Identification', *Public Understanding of Science*, 24(6), 697–711, first published August 5, 2014, 10.1177/0963662514544919.

Lantz, M. and Stawiski, A. (2014) 'Effectiveness of Clickers: Effect of Feedback and the Timing of Questions on Learning', *Computers and Human Behavior*, 31, 280–86, 10.1016/j.chb.2013.10.009.

Lymn, J. and Mostyn, A. (2010) 'Audience Response Technology: Engaging and Empowering Non-Medical Prescribing Students in Pharmacology Learning', *BMC Medical Education*, 10(73), 10.1186/1472-6920-10-73.

Schill, D. and Kirk, R. (2014) 'Courting the Swing Voter: "Real Time" Insights into the 2008 and 2012 U.S. Presidential Debates', *American Behavioral Scientist*, 58(4), 536–55, first published October 17, 2013, 10.1177/0002764213506204.

Van Es, K. (2016) *The Future of Live*. London: Polity Press.

Location awareness and location tracking

Overview

Location awareness refers to devices (active or passive) that determine location, and to the process of delivering information about a device's physical location to another application or user. This is done using methods such as GPS tracking, triangulation of radio signals (for mobile phones, for example), Wi-Fi, Bluetooth, ultra-wideband, radio-frequency identification (RFID), infrared, ultrasound and visible light communication (VLC). Location tracking tends to refer to the technologies that physically locate and electronically track and record a person or object, but do not necessarily deliver the information to another user or application. However, the two terms are often used together or interchangeably to refer to hardware and software that is used to record and track location and movement, manipulate data, analyse data, control information and events (give directions or access local services, for example) and send alerts or notifications. This type of hardware and software can include:

- Global Positioning System (GPS) navigation systems in cars and smartphones;
- real-time location systems embedded into mobile phones or navigation systems, based on wireless technologies (fleet or personnel tracking, or network security, for example);
- patient tracking and healthcare monitors and devices (including vulnerable person tracking such as dementia trackers that send an alert when a person leaves their home or a specific area);
- electronic tagging for monitoring and surveillance of offenders on work-release, parole or probation, using GPS and radio frequency technology;

- basic, micro and global trackers (watches, insoles or tags, for example), with functions such as SOS alert, leaving safe zone or geo-fence alert and two-way communication, for children, lone workers, pets, animals or valuable items;

- portable data loggers that can be placed in bags or vehicles to log where the device travels and how quickly it moves, for examination at a later date;

- tourist and heritage guides that use infrared markers, GPS trackers and/or augmented reality technology to guide and inform;

- camera memory cards that tag the location of a picture;

- supply chain management systems (logistics management software and warehouse management systems, for example);

- location-based services on phones, tablets, laptops and desktops (query-based, such as find my phone or find a local service, or push-based, such as providing discounts for local services);

- location-based advertising (or location-based marketing) on phones, tablets, laptops and desktops that target potential customers based on their physical location;

- smart home or smart office technology that reacts when a person enters or leaves the building (switches on lights or heating, or turns off devices, for example);

- software that processes sensor data (accelerometer and gyroscope, for example) in smartphones to determine location and motion (Chapters 49 and 57);

- location analytics software that combines geographic location with business data (thematic mapping and spatial analysis, for example: Chapters 21 and 54).

Location awareness and location tracking are used by researchers from a number of fields and disciplines including the geographical sciences, the social sciences, health and medicine, travel and tourism, engineering, transportation and logistics, environmental protection and waste management, telecommunications, marketing and advertising, business, gaming and robotics. Examples of research projects that have used, assessed or critiqued location awareness and location tracking include an examination of Life360, which is a surveillance app marketed to families that tracks and shares users' location information (Hasinoff, 2017); a study to assess the performance of five portable GPS data loggers and two GPS cell phones for tracking subjects' time location patterns in epidemiological studies (Wu et al., 2010); research into using location tracking and acceleration sensors on smartphones to detect and locate in real-time elderly people who fall (Lee et al., 2018); a study using active location sensors to track electronic waste (Offenhuber et al., 2013); and a study into continuation and change in location-aware collection games (Licoppe, 2017).

If you are interested in finding out more about location awareness and location tracking for your research, de Souza e Silva (2013) provides an enlightening overview of historical, social and spatial approaches to location-aware mobile technologies. It is also useful to find out more about sensor-based methods (Chapter 49), wearables-based research (Chapter 57), geospatial analysis (Chapter 21) and spatial analysis and modelling (Chapter 54). A connected area of study is location analytics, which is mentioned in Chapter 10. The questions for reflection listed below will help you to think more about using location awareness and location tracking for your research, and the list of digital tools and software packages provided below gives further insight into the hardware and software that is available for various types of location awareness and location tracking.

Questions for reflection

Epistemology, theoretical perspective and methodology

- What is meant by the word 'location'? Is it possible to determine an absolute location, or are all locations relative? How is 'location' defined, described and expressed?

- Do location awareness and location tracking devices provide objective, reliable and valid data? What factors might have an influence on objectivity, reliability and validity (flawed methodology, equipment malfunction, calibration problems, hacking, viruses or data analysis mistakes, for example)?

- How might an awareness of other people's location influence social norms in public spaces, and the configuration of social networks (de Souza e Silva, 2013: 118)?

- How will your research approach, design and sample be affected by access to, ownership of and confidence with, relevant mobile or locational technology (by those taking part in the research and by those undertaking the research, for example)? How do you intend to address problems that could arise with participants who are unable or unwilling to agree to the use of relevant technology?

Ethics, morals and legal issues

- Are you aware of relevant rules, regulations and legislation? For example, researchers using vehicle tracking systems in the UK need to comply with the Data Protection Act 2018 that covers issues such as collecting, processing and storing personal data and seeking permission (covert tracking where a driver can be identified but has not given permission is considered to be in breach of the law, for example). You also need to comply with the Human Rights Act 1998, which sets out the right for employees not to be monitored during their private time (there should be a facility to switch off devices for employees and you will need to consider whether this should also be the case for research participants who might not be employees, for example).

- If you intend to use interactive mobile devices and smartphones for your research project, have you received your participants' permission to process captured real-time location data? How do you intend to deal with issues of locational privacy and informed consent? Tsohou and Kosta (2017: 434), for example, present 'a process theory', based on empirical research, that 'reveals how users' valid informed consent for location tracking can be obtained, starting from enhancing reading the privacy policy to stimulating privacy awareness and enabling informed consent'.

- How are perceptions of privacy, sociability and space understood differently in different parts of the world (de Souza e Silva, 2013: 19)?

- If you intend to use location awareness or location tracking technology with vulnerable people, how will you approach issues such as protection of civil liberties, informed consent, avoiding stigma and effective response? These issues, and others, are addressed in a paper produced by the Alzheimer's Society in the UK, which can be accessed at www .alzheimers.org.uk/about-us/policy-and-influencing/what-we-think/safer-walking-technology [accessed August 8, 2018].

- What amount of location data is accessible to the government and what are the implications of such access? This issue is raised and discussed by Karanja et al. (2018).

Practicalities

- When using GPS what factors might influence satellite reception (moving from outdoor to indoor environments and building material and type, for example)?

- What factors need to be taken into account when choosing and using location tracking devices? This can include:
  - type (online, offline or semi-offline, for example);
  - purpose and function;
  - features (on-demand real-time tracking, location update intervals, route history log, geo-fence zones, smartphone notifications and fall alert, for example);
  - cost, warranty and support;
  - power sources and battery life;
  - memory and data storage;
  - reliability (signal loss and spatial accuracy, for example);
  - type of visualisations (satellite, road and hybrid maps and maps to upload, for example);
  - user-friendliness;
  - international use, if required.

Useful resources

There is a wide range of hardware, software and digital tools available for researchers who are interested in location awareness and/or location tracking. Some of these are aimed at the general public, in particular, those who want to track family and friends, whereas others are aimed at researchers (e.g. for data logging or real-time tracking), the health and social care sector (e.g. for monitoring vulnerable individuals) or the business sector (e. g. for tracking employees, remote workers or vehicles, and for location analytics). Examples available at time of writing include (in alphabetical order):

- ArcGIS mapping and analytics software (www.esri.com/en-us/home);
- Brickhouse Security data loggers for tracking people, vehicles or objects (www.brickhousesecurity.com/gps-trackers/gps-data-loggers);
- Emerson™ Location Awareness for real-time monitoring of people (www.emerson.com/en-us/catalog/emerson-location-awareness);
- Galigeo for location intelligence (www.galigeo.com);
- Glympse for location tracking and sharing for business and consumers (https://glympse.com);

- Hibstaff for tracking and managing mobile teams (https://hubstaff.com/features/gps_time_tracking);

- Labor Sync for monitoring and organising employee time and location (www.laborsync.com);

- Leantegra for indoor location intelligence (https://leantegra.com);

- Life360 for locating and connecting family and friends (www.life360.com/family-locator);

- Navisens™ for location and motion monitoring using Smartphones (https://navisens.com);

- Pocketfinder for locating, monitoring, tracking and geo-fencing people and vehicles (https://pocketfinder.com);

- Spyzie for monitoring and controlling digital activity and location (www.spyzie.com);

- Task Monitoring for healthcare tracking, monitoring and security (www.taskltd.com).

Key texts

de Souza e Silva, A. (2013) 'Location-Aware Mobile Technologies: Historical, Social and Spatial Approaches', *Mobile Media & Communication*, 1(1), 116–21, first published January 1, 2013, 10.1177/2050157912459492.

Hasinoff, A. (2017) 'Where are You? Location Tracking and the Promise of Child Safety', *Television & New Media*, 18(6), 496–512, first published December 16, 2016, 10.1177/1527476416680450.

Karanja, A., Engels, D., Zerouali, G. and Francisco, A. (2018) 'Unintended Consequences of Location Information: Privacy Implications of Location Information Used in Advertising and Social Media', *SMU Data Science Review*, 1(3), Article 13, retrieved from https://scholar.smu.edu/datasciencereview/vol1/iss3/13.

Lee, Y., Yeh, H., Kim, K. and Choi, O. (2018) 'A Real-Time Fall Detection System Based on the Acceleration Sensor of Smartphone', *International Journal of Engineering Business Management*, first published January 9, 2018, 10.1177/1847979017750669.

Licoppe, C. (2017) 'From Mogi to Pokémon GO: Continuities and Change in Location-Aware Collection Games', *Mobile Media & Communication*, 5(1), 24–29, first published November 29, 2016, 10.1177/2050157916677862.

Offenhuber, D., Wolf, M. and Ratti, C. (2013) 'Trash Track – Active Location Sensing for Evaluating E-Waste Transportation', *Waste Management & Research*, 31(2), 150–59, first published January 10, 2013, 10.1177/0734242X12469822.

Tsohou, A. and Kosta, E. (2017) 'Enabling Valid Informed Consent for Location Tracking through Privacy Awareness of Users: A Process Theory', *Computer Law & Security Review*, 33(4), 434–57, 10.1016/j.clsr.2017.03.027.

Wu, J., Jiang, C., Liu, Z., Houston, D., Jaimes, G. and McConnell, R. (2010) 'Performances of Different Global Positioning System Devices for Time-Location Tracking in Air Pollution Epidemiological Studies', *Environmental Health Insights*, first published November 23, 2010, 10.4137/EHI.S6246.

Xia, L., Huang, Q. and Wu, D. (2018) 'Decision Tree-Based Contextual Location Prediction from Mobile Device Logs', *Mobile Information Systems*, 2018, article ID, 1852861, 11 pages, 10.1155/2018/1852861.

CHAPTER 28

Log file analysis

Overview

Log file analysis involves collecting, storing, searching and analysing log file data. Log files (or transaction logs) are a list of events that have been recorded, or logged, by a computer or mobile device. There are various types of log file that can be analysed by researchers, including:

- Web server log files that record data about website visits, including a record of the URL/resource that was requested, action taken, time and date, IP of the machine that made the request and user agent or browser type, for example. It can also include visitor patterns, search terms, demographics and visitor paths. Analysis of these log files (sometimes referred to as web log analysis: Jansen et al., 2008) is of interest to researchers from a wide variety of disciplines including the social sciences, psychology, health and medicine, education, business, management and the computer sciences. This type of log file is also analysed by organisations to provide a better understanding of how search engines crawl their website and to detect search engine optimisation (SEO) problems. This helps organisations to know more about the type of information that search engines are finding and enables them to rectify problems and improve the online visibility of their website. The measurement, collection, analysis and reporting of web data is known as web analytics and is discussed in Chapter 58.

- Log files that are generated by operating systems or software utilities (a record of events and processes that can include application, security and system crashes, for example). Analysis of these log files is of interest to researchers working in research and development in the computer sciences (computing engineering, information systems, information technology or software engineering, for example).

- Log files created by software installers (install and uninstall logs, event viewers, instant alerts and troubleshooting logs, for example). Again, analysis of these log files is of interest to researchers and developers working within the computer sciences.

- Mobile device logs that store information such as errors, warnings and messages, and mobile logging software installed for specific research purposes, such as location logging from GPS trajectories (Chapter 27). It can also include key logging software installed for research purposes (as long as informed consent is given). Analysis of these log files is of interest to researchers from a wide variety of disciplines including the social sciences, geography, health and medicine, education business and the computer sciences. More information about mobile methods can be found in Chapter 32.

Log file analysis is a non-reactive method: participants are unaware that their search behaviour is being observed (unless they have agreed to install logging software for specific research purposes). It can be used in the field of Educational Data Mining (EDM), which enables researchers to observe what learners are doing without interrupting their learning processes (Lee, 2018). Bruckman (2006) provides an interesting example of how log file analysis can be used to research behaviour and learning in an online community, and Lee (2018) illustrates how log file analysis can help researchers to understand how students learn in a computer-based learning environment. EDM is discussed in more depth in Chapter 17 and the related field of learning analytics is discussed in Chapter 24. Log file analysis is also used in health research: Sieverink et al. (2014) illustrate how log file analysis can help to improve electronic personal health records and Kelders et al. (2010) use a mixed methods approach, including log file analysis, to evaluate a web-based intervention designed to help people maintain a healthy body weight.

There are various log file analysis tools available. Although functions vary, depending on the chosen software and the purpose of the analysis, in general, researchers are able to collect, store, index, search, correlate, visualize, analyse and report data from log files (see Ribeiro et al., 2018, for an interesting discussion on the use of visualisation tools when analysing game log files and see Chapter 20 for more information about game analytics). If you are interested in using log file analysis for your research project it is useful to find out more about the analysis tools that are available so that you can gain a deeper understanding of functions, capabilities and costs. These include issues such as availability, security, scalability and open source options. Examples are given below. You might also find it useful to gain a deeper understanding of data analytics (Chapter 10), data mining (Chapter 12), web and mobile analytics (Chapter 58), webometrics (Chapter 59) and data visualisation (see Chapter 13). A wide-ranging and thorough discussion of web log analysis, including methodology, method and history, can be obtained from Jansen et al. (2008). This book could do with a little updating, but still provides a useful read for anyone interested in using log file analysis for their research project. The questions for reflection listed below will also help you to move forward with planning your research project.

Questions for reflection

Epistemology, theoretical perspective and methodology

- Do log files contain a true and objective record of events? Is it possible that log files can contain incorrect, misleading or false data? If so, in what way?

- Have you thought about how log file analysis, as a research method, fits within your methodological framework? Jansen et al. (2008), in Chapter 1 of their book, discuss the research methodologies that are applied in web-based research, which will help with your methodological decisions and choices.

- Do you intend to use log file analysis for prediction, clustering, relationship building, discovery with a model or distillation of data for human judgement, for example (Baker, 2010)? Or do you intend to use the method in a mixed methods approach (Kelders et al., 2010)?

- Log files can appear overwhelming due to their size, volume and complexity. There is potential for massive amounts of data from millions of records. How do you intend to make your project manageable? How will you choose software for your analysis and ensure that it can handle the type of analysis required?

Ethics, morals and legal issues

- Do you have a thorough understanding of laws and regulation surrounding personal data recorded in web server logs? IP addresses, usernames and referral information are all defined as personal data under the European Union's General Data Protection Regulation (GDPR), which came into effect in May 2018. This states that you cannot store information without having obtained direct consent for the specific purpose for which you intend to store the information from the people whose information you intend to store. How can you ensure that informed consent is obtained (or has been obtained) when the log files are created? These regulations also state that data must be kept secure, access must be limited and all data should be deleted (including back-ups) when they are no longer needed.

- Can log files be manipulated, interfered with or hacked? When analysing log files, how can you ensure that security has not been breached and that log files have not been altered or deleted by hackers, malware or viruses, for example? How can you check for corruption and error? How can log files themselves be used for detection and reporting purposes? How are these issues addressed when log files are generated and maintained by third parties? Diehl (2016) provides an enlightening and comprehensive discussion on the integrity and security of log files.

- Log files could, potentially, contain records that could be subpoenaed by police officers or courts or could provide evidence of harm to the public (child abuse is suspected or there are life-threatening circumstances involved, for example). How would you address these issues if they were to occur in your research?

Practicalities

- Do you have a good understanding of the advantages and disadvantages of log file analysis? Possible advantages can include simple formatting, automatic recording of web crawlers and no interference from firewalls. Possible disadvantages include the need for regular software updates, problems with access (access via the cache memory of the browser or via a proxy server are not available, for example) and the need for additional storage.

- What costs are involved? Are third party log files freely available or do you intend to develop or purchase your own logging software? Are analysis tools free or are there significant costs involved? If so, who will meet these costs? Are functions, capabilities and user-friendliness linked to costs?

- Do you have a thorough understanding of your chosen analysis tool, or the ability to acquire the necessary knowledge and understanding?
- What log file data do you intend to analyse? For example, web server log files can include the following:
  - IP address and host name;
  - region or country of origin;
  - browser and operating system used;
  - direct access or referral information;
  - search terms;
  - type of search engine used;
  - duration and number of page visits.

If you are interested in additional data that are not contained within log files, such as bounce rates, web analytics may be a more suitable method (see Chapter 58).

- What problems could arise during the analysis process (incorrect procedures, importance bias or lost data, for example)? What cross validation methods do you intend to adopt, in particular, if your goal is prediction or generalisability?

Useful resources

There are a variety of digital tools and software packages available for log file analysis. Some are simple to use, whereas others are more complex due, in part, to wider functionality. A selection of tools available at time of writing is given below, in alphabetical order: it is useful to visit some of these sites so that you can make sure that the right tool is chosen for the right purpose.

- AWStats (https://awstats.sourceforge.io);
- BigQuery from Google (https://cloud.google.com/bigquery/audit-logs);
- Deep Log Analyzer (www.deep-software.com);
- Google Analytics (www.google.com/analytics);
- Logstash (www.elastic.co/products/logstash);
- Logz.io (https://logz.io);

- Screaming Frog Log File Analyser (www.screamingfrog.co.uk/log-file-analyser-1-0);

- Splunk for log management (www.splunk.com/en_us/solutions/solution-areas/log-management.html);

- The Webalizer (www.webalizer.org);

- Visitors (www.hping.org/visitors);

- WebLog Expert (www.weblogexpert.com).

Key texts

Baker, R. (2010) 'Data Mining for Education'. In McGaw, B., Peterson, P. and Baker, E. (eds.) *International Encyclopedia of Education*, 112–18. 3rd edition. Oxford: Elsevier Science.

Bouwman, H., Reuver, M., Heerschap, N. and Verkasalo, H. (2013) 'Opportunities and Problems with Automated Data Collection via Smartphones', *Mobile Media & Communication*, 1(1), 63–68, first published January 1, 2013, 10.1177/2050157912464492.

Bruckman, A. (2006) 'Analysis of Log File Data to Understand Behavior and Learning in an Online Community'. In Weiss, J., Nolan, J., Hunsinger, J. and Trifonas, P. (eds.) *International Handbook of Virtual Learning Environments*, Vols. 1 and 2, 1449–65. Dordrecht: Springer.

Diehl, E. (2016) *Ten Laws for Security*. Cham: Springer.

Jansen, B., Spink, A. and Taksa, I. (eds.) (2008) *Handbook of Research on Web Log Analysis*. Hershey, PA: IGI Global.

Kelders, S., Gemert-Pijnen, J., Werkman, A. and Seydel, E. (2010) 'Evaluation of a Web-Based Lifestyle Coach Designed to Maintain a Healthy Bodyweight', *Journal of Telemedicine and Telecare*, 16(1), 3–7, first published January 19, 2010, 10.1258/jtt.2009.001003.

Lee, Y. (2018) 'Using Self-Organizing Map and Clustering to Investigate Problem-Solving Patterns in the Massive Open Online Course: An Exploratory Study', *Journal of Educational Computing Research*, first published January 30, 2018, 10.1177/0735633117753364.

Ribeiro, P., Biles, M., Lang, C., Silva, C. and Plass, J. (2018) 'Visualizing Log-File Data from a Game Using Timed Word Trees', *Information Visualization*, first published August 2, 2017, 10.1177/1473871617720810.

Sieverink, F., Kelders, S., Braakman-Jansen, L. and Gemert-Pijnen, J. (2014) 'The Added Value of Log File Analyses of the Use of a Personal Health Record for Patients with Type 2 Diabetes Mellitus: Preliminary Results', *Journal of Diabetes Science and Technology*, 8(2), 247–55, first published March 4, 2014, 10.1177/1932296814525696.

CHAPTER 29

Machine learning

Overview

Machine learning refers to the scientific process of facilitating computers to 'learn' (or optimise) by themselves through self-adaption methods and techniques. It is a subfield of artificial intelligence (the theory and development of intelligence in machines), which includes computer vision, robotics and natural language processing, for example. Machine learning is a category of algorithm (a procedure or formula for solving a problem) that allows software applications to learn from previous computations to make repeatable decisions and predictions without the need for explicit, rules-based programming. Algorithms can identify patterns in observed data, build models and make predictions, with the aim of generalising to new settings. Practical applications include fraud detection, credit scoring, email spam and malware filtering, self-driving cars, face recognition technologies and video surveillance in crime detection, for example. Machine learning is also a method used in data analytics (Chapter 10), in particular, in predictive analytics, which is used to make predictions based on historical data and trends in the data. More information about predictive modelling can be found in Chapter 45.

There are two broad categories of machine learning task: supervised learning (both input and desired output are provided and labelled for classification to provide a learning basis for future data processing) and unsupervised learning (information is neither classified nor labelled, allowing the algorithm to act on information without guidance). Data mining that focusses on exploratory data analysis is a type of unsupervised learning (see Chapter 12). Other tasks include semi-supervised learning (an incomplete training signal is given); active learning (the user can be queried interactively for labelling) and reinforcement learning (training data are provided as feedback in a dynamic environment).

Machine learning is used by researchers in a number of disciplines and fields of study, examples of which include hospitality and tourism (Aluri et al., 2018), architecture (Tamke et al., 2018), psychology (Yarkoni and Westfall, 2017), sociology (Fu et al., 2018), health and medicine (Spathis and Vlamos, 2017), renewable energy (Voyant et al., 2017) and cyber security (Buczak and Guven, 2015). Examples of research projects that have used, assessed or critiqued machine learning include a study into how to use machine learning to advance our understanding of personality (Bleidorn and Hopwood, 2018); research into the impact of machine learning on architectural practices (Tamke et al., 2018); a comparison and evaluation of the performances of machine-learning algorithms and manual classification of public health expenditures (Brady et al., 2017); a comparison and assessment of machine learning and traditional methods used to identify what customers value in a loyalty programme (Aluri et al., 2018); an investigation into how machine learning can help psychology become a more predictive science (Yarkoni and Westfall, 2017); research into diagnosing asthma and chronic obstructive pulmonary disease with machine learning (Spathis and Vlamos, 2017); and a study that investigates the use of machine learning to optimise count responses in surveys (Fu et al., 2018).

If you are interested in finding out more about machine learning for your research a good starting point is Alpaydin (2016), who provides comprehensive information that includes a discussion on the evolution of machine learning and a consideration of present day applications for the technology. Another good book for those who are new to the topic is Mueller and Massaron (2016) who provide informative content on the definition and history of machine learning; advice about tools, techniques, software, coding and mathematics; and an interesting discussion on future directions. It is also useful to find out more about data analytics (Chapter 10), big data analytics (Chapter 3), data mining (Chapter 12) and predictive modelling (Chapter 45). The questions for reflection below will help you to think more about using machine learning in your research and useful resources and key texts provide guidance for further reading and research.

Questions for reflection

Epistemology, theoretical perspective and methodology

* Is machine learning universal, neutral and objective? McQuillan (2018) provides an enlightening discussion on issues relating to this question (potential bias in the construction of data, the opaque nature of decision-

making, not knowing how neural networks come to their conclusions and absorbing discrimination into machine learning models, for example). Also, Veale and Binns (2017: 2) point out that existing discrimination in historical data is reproduced when such data are reused (due to the high demand for labelled data in the context of supervised machine learning). They go on to discuss how model choice itself is a political decision.

- How opaque are machine learning algorithms? Burrell (2016) provides an interesting discussion on three forms of opacity in relation to classification and discrimination: 'opacity as intentional corporate or state secrecy', 'opacity as technical illiteracy' and 'opacity as the way algorithms operate at the scale of application'.

- What role, if any, does human judgement have in machine learning?

- Can a machine really learn? Are machines, instead, optimising or improving accuracy of models to generate better and more accurate results?

Ethics, morals and legal issues

- Is there potential for machine learning to cause harm to present or future individuals, groups or populations (social decisions or policy are made on the basis of predictive models, or algorithms are taken more seriously than groups or individuals, for example)? How can you ensure that such problems are avoided in your research, in particular, in cases where machine learning algorithms are used to make socially consequential predictions?

- Is machine learning ethically neutral? McQuillan (2018: 1) states that 'the machine learning assemblage produces a targeting gaze whose algorithms obfuscate the legality of its judgments, and whose iterations threaten to create both specific injustices and broader states of exception'. He goes on to discuss these issues in his paper, and concludes by proposing 'people's councils as a way to contest machinic judgments and reassert openness and discourse'. How might these observations relate to your research?

- How can you ensure that machine learning is fair, accountable and transparent? The Fairness, Accountability, and Transparency in Machine Learning website (www.fatml.org) provides information about their annual event that covers this topic, along with useful papers, events, projects, principles and best practice documents. See Veale and Binns (2017) for an enlightening discussion on fair machine learning in the real world.

Practicalities

- Do you know how to build and operate your machine learning project? Broussard (2018) provides a practical example of how to do this and includes information about potential problems and pitfalls (real-world data collections are often incomplete and those who develop the algorithms are influenced by assumptions, background and biases, for example).

- Do you have a good understanding of specific problems that can occur with machine learning, along with an understanding of how to tackle such problems? This can include:

  - Overfitting: a model learns too well, which has a negative impact on the ability to generalise and can lead to poor model performance. Resampling techniques and validation datasets can be used to limit overfitting.

  - Underfitting: a model has poor performance on training data and poor generalisation to other data. Alternative machine learning algorithms should be used to overcome this problem.

  - Dimensionality: algorithms with more features work in multiple dimensions that make data harder to understand. Dimension reduction or feature extraction/engineering can tackle this problem (but can create groups and alter future outcomes: see Veale and Binns, 2017: 2).

  - Vulnerability: unexpected or unwanted behaviour outside operation specifications is exhibited. Problems can be addressed through careful observation, understanding, reporting and security back-up from the relevant software company.

- Do you have a good understanding of the wide variety of technologies, techniques and methods in machine learning? A brief list includes (in alphabetical order):

  - classification: the division of inputs into two or more classes, with unseen inputs assigned to these classes;

  - clustering: inputs are divided into groups (which are not known beforehand), based on similarities (see Chapter 5 for information on cluster analysis);

  - kernels: tools to analyse relationships or patterns;

  - natural-language processing: the ability to turn text or audio speech into encoded, structured information;

o   neural networks: artificial intelligence processing methods that allow self-learning from experience;

o   regression: for estimating the relationship among inputs;

o   support vector machines: for analysing data using classification, regression and other tasks.

Useful resources

There are a wide variety of digital tools and machine learning software packages available, aimed at researchers, the health sector and business. Some are open source whereas others have significant costs attached. Examples that are available at time of writing are given below, in alphabetical order.

- Accord.NET Framework (http://accord-framework.net);
- ai-one (www.ai-one.com);
- Deeplearning4j (https://deeplearning4j.org);
- Google Cloud AI tools (https://cloud.google.com/products/ai);
- H2O.ai (www.h2o.ai);
- LIBSVM (www.csie.ntu.edu.tw/~cjlin/libsvm);
- Mahout (https://mahout.apache.org);
- mlpack (www.mlpack.org);
- PredictionIO (https://predictionio.apache.org);
- PrediCX (https://warwickanalytics.com/predicx);
- PyTorch (https://pytorch.org);
- TensorFlow (www.tensorflow.org);
- Torch (http://torch.ch).

The Journal of Machine Learning Research (www.jmlr.org) is a peer-reviewed, open access journal that provides scholarly articles in all areas of machine learning. The website contains useful information on machine learning open source software, including machine learning algorithms, toolboxes, coding and programming languages.

Machine Learning (https://link.springer.com/journal/10994) is a peer-reviewed scientific journal that 'publishes articles reporting substantive results on a wide range of learning methods applied to a variety of learning problems'. Open access articles can be found on the website.

Key texts

Alpaydin, E. (2016) *Machine Learning: The New AI*. Cambridge, MA: MIT Press.

Aluri, A., Price, B. and McIntyre, N. (2018) 'Using Machine Learning to Cocreate Value through Dynamic Customer Engagement in a Brand Loyalty Program', *Journal of Hospitality & Tourism Research*, first published January 30, 2018, 10.1177/1096348017753521.

Bleidorn, W. and Hopwood, C. (2018) 'Using Machine Learning to Advance Personality Assessment and Theory', *Personality and Social Psychology Review*, first published May 23, 2018, 10.1177/1088868318772990.

Brady, E., Leider, J., Resnick, B., Alfonso, Y. and Bishai, D. (2017) 'Machine-Learning Algorithms to Code Public Health Spending Accounts', *Public Health Reports*, 132(3), 350–56, first published March 31, 2017, 10.1177/0033354917700356.

Broussard, M. (2018) *Artificial Unintelligence: How Computers Misunderstand the World*. Cambridge, MA: MIT Press.

Buczak, A. and Guven, E. (2015) 'A Survey of Data Mining and Machine Learning Methods for Cyber Security Intrusion Detection', *IEEE Communications Surveys & Tutorials*, 18 (2), second quarter 2016, 1153–76, first published October 26, 2015, 10.1109/COMST.2015.2494502.

Burrell, J. (2016) 'How the Machine 'Thinks': Understanding Opacity in Machine Learning Algorithms', *Big Data & Society*, first published January 6, 2016, 10.1177/2053951715622512.

Fu, Q., Guo, X. and Land, K. (2018) 'Optimizing Count Responses in Surveys: A Machine-Learning Approach', *Sociological Methods & Research*, first published January 30, 2018, 10.1177/0049124117747302.

McQuillan, D. (2018) 'People's Councils for Ethical Machine Learning', *Social Media + Society*, first published May 2, 2018, 10.1177/2056305118768303.

Mueller, J. and Massaron, L. (2016) *Machine Learning For Dummies*. Hoboken, NJ: John Wiley & Sons, Inc.

Spathis, D. and Vlamos, P. (2017) 'Diagnosing Asthma and Chronic Obstructive Pulmonary Disease with Machine Learning', *Health Informatics Journal*, first published August 18, 2017, 10.1177/1460458217723169.

Tamke, M., Nicholas, P. and Zwierzycki, M. (2018) 'Machine Learning for Architectural Design: Practices and Infrastructure', *International Journal of Architectural Computing*, 16(2), 123–43, first published June 13, 2018, 10.1177/1478077118778580.

Veale, M. and Binns, R. (2017) 'Fairer Machine Learning in the Real World: Mitigating Discrimination without Collecting Sensitive Data', *Big Data & Society*, first published November 20, 2017, 10.1177/2053951717743530.

Voyant, C., Notton, G., Kalogirou, S., Nivet, M., Paoli, C., Motte, F. and Fouilloy, A. (2017) 'Machine Learning Methods for Solar Radiation Forecasting: A Review', *Renewable Energy*, 105, May 2017, 569–82, 10.1016/j.renene.2016.12.095.

Yarkoni, T. and Westfall, J. (2017) 'Choosing Prediction over Explanation in Psychology: Lessons from Machine Learning', *Perspectives on Psychological Science*, 12(6), 1100–22, first published August 25, 2017, 10.1177/1745691617693393.

CHAPTER 30

Mobile diaries

Overview

Mobile diaries use mobile technology to enable participants to record everyday behaviour, actions and thoughts. This method has its roots in the diary method, or diary studies, which use qualitative methods to collect and analyse self-reported behaviour. Mobile diaries can be text-based (Jones and Woolley, 2015), audio (Bellar, 2017; Worth, 2009), video (Nash and Moore, 2018), visual (Mattila et al., 2010) or a combination of some or all of the above (Hoplamazian et al., 2018; Plowman and Stevenson, 2012). They are used to assess actions and reactions; monitor activity and behaviour; capture emotions, feelings, sentiments, reflections and inner thoughts; and provide information on movement and location. Mobile diaries bring the researcher closer to the participants' world and enable participants to explore, narrate and describe their world and experiences in their own words while screen recordings record and store real-time behaviour and reactions. They enable researchers to observe real life contexts and bring a research project to life, while encouraging co-interpretation of material by participant and researcher. When using mobile diaries researchers can combine self-reported information with location tracking (Chapter 27), event tracking (Chapter 28), sensor-based data (Chapters 49 and 57) and interviews (Bellar, 2017) to check, verify, cross-reference and/or further explore reflective accounts.

Mobile diaries enable researchers to receive self-reported data (reports, blogs, texts, audio, video or images, for example), respond with new questions or probes (for clarification, elaboration, action or redirection) and test gamification approaches (Chapter 47). They can be solicited or unsolicited. Solicited diaries ask participants to address specific questions or topics, provide prompts or push notifications in the form of images, text or objects, or can be left open to enable participants to make entries according to preference and motivation, usually within a given time or date, and in relation to the research project. Prompts, questions and images can be sent via text or email, or through apps and websites. Unsolicited diaries are provided by participants who have decided to keep a diary for purposes other than the research project, but who are happy to provide their diary when it becomes evident that it is related to the research project (perhaps after completing an interview or filling in a questionnaire, for example). There are various personal diary apps that are used for general diary or journal keeping, and examples are given below.

There are a variety of methodological and theoretical approaches that can frame and guide mobile diaries studies, including an ecocultural approach (Plowman and Stevenson, 2012), ethnography (Nash and Moore, 2018), niche theory (Hoplamazian et al., 2018) and a critical realist stance (Crozier and Cassell, 2016). Examples of research projects that have used mobile diaries as a research method include research into the development of a new, mobile-assisted working model designed to enhance the quality of daily family life (Rönkä et al., 2015); a mixed-methods study examining leadership for women in Science, Technology, Engineering, Mathematics and Medicine fields from five countries in the Global North (Nash and Moore, 2018); research into how young consumers represent and give sense to the relationships they maintain with branded objects when they are internationally mobile (Biraghi, 2017); a study to explore the everyday lives of young children at home (Plowman and Stevenson, 2012); and a study exploring how Evangelical Christians choose and use religious mobile apps, and how app engagement informs their religious identities (Bellar, 2017).

There are a variety of ways in which mobile diaries can be analysed, depending on theoretical perspective, methodological framework and digital medium. If your research is to involve the use of video diaries, information about video analysis is provided in Chapter 55, and if your research involves the use of audio diaries, information about audio analysis is provided in Chapter 2. If you decide to use coding for analysis, more information about coding and retrieval is provided in Chapter 6 and if you decide to use computer-assisted qualitative data analysis software (CAQDAS), more information is provided in Chapter 9. In cases where mobile diaries use visual images, more information about visual analysis methods can be found in Chapter 16 (inductive content analysis, visual semiotics, visual rhetoric and visual hermeneutics, for example).

It is important that you have a good understanding of methodological and practical issues associated with the diary method if you are interested in using mobile diaries in your research. Bartlett and Milligan (2016) and Hyers (2018) provide comprehensive and insightful information on this topic and provide a good starting point for your background research. It is also important that you understand about the variety of mobile methods that are available and these are discussed in Chapter 32. If you are interested in the storytelling aspect of mobile diaries, more information can be found in Chapter 15. There are different digital tools and software packages available for mobile diary research and some are given below: it is useful to visit some of these sites so that you can get an idea of functions, purpose, capabilities and costs.

Questions for reflection

Epistemology, theoretical perspective and methodology

- Are visual, audio and textual diaries a representation of reality? Can pictures, sound and text be accurate and true? Do they record phenomena or social behaviour in an objective way (indeed, is it possible to be objective)?

- How might the use, production and reaction to mobile diaries be influenced by culture, history and politics (of the producer, viewer and reader)?

- What bias may be present in the production and analysis of mobile diaries? How should issues of bias be incorporated into your study?

- Whose point of view will be represented in the mobile diary and how can you be sure that you know which point of view is represented? Plowman and Stevenson (2012) illustrate how choosing to represent the adults' point of view in their research on children's everyday lives was a theoretically motivated decision.

- What level of presence do you intend to adopt, as researcher? Although you are not present when entries are made into mobile diaries, you do have a presence in the form of prompts, instructions and participant perception of the research process, topic and purpose, for example. Nash and Moore (2018) provide insight into these issues in relation to video diaries.

Ethics, morals and legal issues

- How can you ensure the privacy and anonymity of those producing diaries, in particular, in cases where only a few diaries are analysed? How might privacy and anonymity issues affect the way that your mobile diary study is analysed, reported and shared?

- How can mobile diaries be kept secure (encryption, locks, passwords and PINS, for example)?

- Do participants have a private space in which to make entries in their mobile diaries? Is it important that they have private space? Nash and Moore (2018) illustrate how privacy and lack of time impacted on video diaries in their research.

- What impact might involvement in a diary study have on participants? Might this be a negative or positive impact? Crozier and Cassell (2016: 410) in their study of transient working patterns and stress in UK temporary workers, for example, point out that 'participants experienced the audio diary process as cathartic in that it helped them to reflect upon and face the challenges in their employment situation'.

- How can you avoid placing too many demands on participants? Might workload, or perceived workload, lead to participants withdrawing from the study?

- Are participants expected to meet costs associated with mobile diary production? If so, can they afford to meet these costs? Will costs be prohibitive and how will this influence your research? Nash and Moore (2018) illustrate how submission of video diaries was delayed due to prohibitive costs.

Practicalities

- Biraghi (2017: 453) notes that the 'digital native generation' involved in their study 'are eager to self-produce content that discloses self-identity and publicly shares life projects and activities'. Do you think this is the case with your intended participants? Is it possible that you might encounter participants who would be less willing, or able, to engage in this type of self-disclosure?

- What type of mobile diary is most suitable for your research and for your research participants (audio, video, text-based, visual or a combination)? Do you have a good understanding of the strengths and weaknesses of each method?

- How will you ensure that participants own, or have access to, the required mobile technology? Will participants have the required understanding of, and confidence with, the technology to be able to undertake the requested tasks? Will training be necessary? If so, where will training sessions take place, who will run the training sessions and is money available to pay for participants' travel and expenses?

- How can you take account of, or try to overcome, possible problems with hardware, software or connections (mobile phone theft, data loss, lack of memory or viruses, for example)? What back-up systems and file sharing systems will you use? How will you ensure that participants are made aware of compatibility issues?

- Bellar (2017: 117) notes that 'participants could delete and restart diary entries, which may have resulted in altered information'. What impact might this observation have on your research, in particular, if you are hoping to obtain spontaneous thoughts and reactions?

Useful resources

There are a variety of digital tools and software packages available for mobile diary research. Some of this software is general diary software marketed at individuals who are interested in keeping a mobile diary or journal, whereas other software is aimed at researchers who are interested in conducting a mobile diary project, and includes data analysis tools. A snapshot of what is available at time of writing includes (in alphabetical order):

- Contextmapp Mobile Diary Study (https://contextmapp.com);
- Dabble.Me (https://dabble.me);
- Day One (http://dayoneapp.com);
- dscout diary (https://dscout.com/diary);
- EthOS Ethnographic Research Tool Diary Studies (www.ethosapp.com);
- Glimpses (https://getglimpses.com);
- Grid Diary (http://griddiaryapp.com/en);
- Indeemo Mobile Diary Studies (https://indeemo.com/mobile-diary-study);
- Journey HQ (www.journeyhq.com);
- Penzu (https://penzu.com).

Key texts

Bartlett, R. and Milligan, C. (2016) *What Is Diary Method?*. London: Bloomsbury.

Bellar, W. (2017) 'Private Practice: Using Digital Diaries and Interviews to Understand Evangelical Christians' Choice and Use of Religious Mobile Applications', *New Media & Society*, 19(1), 111–25, first published June 13, 2016, 10.1177/1461444816649922.

Biraghi, S. (2017) 'Internationally Mobile Students and Their Brands: Insights from Diaries', *International Journal of Market Research*, 9(4), 449–69, first published July 1, 2017, 10.2501/IJMR-2017-010.

Crozier, S. and Cassell, C. (2016) 'Methodological Considerations in the Use of Audio Diaries in Work Psychology: Adding to the Qualitative Toolkit', *Journal of Occupational and Organizational Psychology*, 89(2), 396–419, first published July 31, 2015, 10.1111/joop.12132.

Hoplamazian, G., Dimmick, J., Ramirez, A. and Feaster, J. (2018) 'Capturing Mobility: The Time–Space Diary as a Method for Assessing Media Use Niches', *Mobile Media & Communication*, 6(1), 127–45, first published October 8, 2017, 10.1177/2050157917731484.

Hyers, L. (2018) *Diary Methods: Understanding Qualitative Research*. Oxford: Oxford University Press.

Jones, A. and Woolley, J. (2015) 'The Email-Diary: A Promising Research Tool for the 21st Century?', *Qualitative Research*, 15(6), 705–21, first published December 31, 2014, 10.1177/1468794114561347.

Mattila, E., Lappalainen, R., Pärkkä, J., Salminen, J. and Korhonen, I. (2010) 'Use of a Mobile Phone Diary for Observing Weight Management and Related Behaviours', *Journal of Telemedicine and Telecare*, 16(5), 260–64, first published May 18, 2010, 10.1258/jtt.2009.091103.

Nash, M. and Moore, R. (2018) 'Exploring Methodological Challenges of Using Participant-Produced Digital Video Diaries in Antarctica', *Sociological Research Online*, first published April 11, 2018, 10.1177/1360780418769677.

Plowman, L. and Stevenson, O. (2012) 'Using Mobile Phone Diaries to Explore Children's Everyday Lives', *Childhood*, 19(4), 539–53, first published April 5, 2012, 10.1177/0907568212440014.

Rönkä, A., Malinen, K., Jokinen, K. and Häkkinen, S. (2015) 'A Mobile-Assisted Working Model for Supporting Daily Family Life: A Pilot Study', *The Family Journal*, 23(2), 180–89, first published March 9, 2015, 10.1177/1066480714565333.

Sonck, N. and Fernee, H. (2013) *Using Smartphones in Survey Research: A Multifunctional Tool. Implementation of a Time Use App: A Feasibility Study*. The Hague: The Netherlands Institute for Social Research.

Worth, N. (2009) 'Making Use of Audio Diaries in Research with Young People: Examining Narrative, Participation and Audience', *Sociological Research Online*, 14(4), 1–11, first published December 11, 2017, 10.5153/sro.1967.

CHAPTER 31

Mobile ethnography

Overview

The term 'mobile ethnography', for the purpose of this book, is used to describe the use of mobile technology (laptops, smartphones, smartwatches or tablets, for example) to undertake ethnographic studies into human activity, behaviour and/or movement. It includes collecting, interpreting and/or analysing data using mobile devices, with participants often acting as co-researchers or co-creators (collecting data, offering interpretations and discussing meanings, for example). The term 'mobile ethnography' is also used to describe ethnographic studies of mobile phenomena, patterns of movement, mobile life or mobile society. Studies of this type that utilise methods such as participant observation rather than mobile technology are beyond the scope of this book, but you can consult Novoa (2015) for an interesting and insightful discussion on the history, definition and examples of this type of mobile or mobility ethnography when used in human geography, and Gottschalk and Salvaggio (2015) for a mobile ethnography of the Las Vegas Strip.

Ethnography has its roots in anthropology and is the study of people and culture: it is field-based, multi-method and multi-faceted. Ethnography is best understood as a methodology, which is the overall framework that guides a research project (Hjorth et al., 2017; Horst and Miller, 2012; Pink et al., 2015). However, mobile ethnography has been included in this book because there are many 'how to' issues involved in the use of mobile technology for this type of study. Mobile ethnography is covered by the umbrella term of 'mobile methods', which use mobile technology to collect, record, interpret and/or analyse quantitative or qualitative data (Chapter 32), and by the umbrella term of 'digital ethnography', which encompasses all forms of digital technology-based ethnography (Chapter 14). It is also connected closely to online ethnography that uses ethnographic tools, techniques, methods and insight to collect and analyse data from online sources (Chapter 37).

Mobile technology helps to provide insight into participant behaviour, enabling the ethnographer to view the world from their perspective. It is real-time and responsive, facilitating the effective and efficient collection of impressions, feedback and locations. It also enables the co-creation of knowledge. Mobile ethnographic apps and software can be used in standalone ethnographic studies, combined with more traditional qualitative methods to gain deeper insight or used as one part of a larger ethnographic study, for example. They can also be used in mixed methods approaches (statistical analyses of location or log data or discourse analyses of conversations and speech, for example) or in auto-ethnographic research (facilitating reflection and writing on mobile devices, for example). Mobile ethnography can be used in a variety of disciplines and fields of study including health and medicine, education, psychology, sociology, marketing and retail. Muskat et al. (2018) provide a good overview of how mobile ethnography is applied in tourism, health and retail, with detailed references for specific types of project. Richardson and Hjorth (2017: 1653) provide an example of research that uses a 'range of data collection methods aimed at capturing the embodiment of mobile media' and Christensen et al. (2011: 227) provide an example of 'mixed methods research into children's everyday mobility'.

There are various apps and software packages available for researchers who are interested in mobile ethnography and some examples are given below (more information about smartphone apps for ethnography and other types of research can be found in Chapter 50). Although features, facilities and functions vary, depending on the app or software package, it is possible to carry out some or all of the tasks listed below:

- capture live, real-time behaviour and actions;
- facilitate live conversations;
- capture audio recordings and live video feedback;
- pose questions (in interviews, focus groups or as a response to particular behaviour or action);
- present stimuli (links, images, videos, audio or texts);

- socialise, share, generate and develop new ideas (with participants and other researchers);
- receive video diaries, screen recordings, photos and notes from participants;
- track location;
- set and utilise 'geo-fences' that provide alerts to specific tasks or questions when a participant enters a particular area;
- alert participants to new tasks (and receive immediate response about new tasks);
- receive automated video transcription;
- receive immediate results and monitor ongoing progress;
- react to results and probe for more information;
- undertake data analysis (automated theming and clustering, for example);
- conduct research in a variety of languages;
- conduct multiple projects simultaneously.

If you are interested in finding out more about mobile ethnography Pink et al. (2015) provide a comprehensive introduction to digital and pre-digital ethnography, covering researching experiences, practices, things, relationships, social worlds, localities and events. Hjorth et al. (2017) present a series of papers on digital ethnography, with Part IV looking specifically at mobile ethnography. Horst and Miller (2012) provide a collection of papers covering digital anthropology, which is useful for those who are interested in the anthropological roots of ethnography. If you are interested in finding out how mobile phone technologies can be combined with GPS in ethnographic research Christensen et al. (2011) provide an enlightening discussion. Additional resources and key texts can be found in Chapters 14 and 37. If you are interested in smartphone ethnographic apps, consult Chapter 50.

Questions for reflection

Epistemology, theoretical perspective and methodology

- Have you considered your theoretical stance and methodological position and the impact these have on methods? For example, mobile ethnography provides an opportunity to co-create knowledge with participants (interpretative, reflective stance), rather than view the objectivity of knowledge (objective, realist stance). Your stance will have an effect on the way you collect, interpret and report data. Mobile diaries, for example, enable researchers to observe real life contexts and bring a research project to life, while encouraging co-interpretation of material by participant and researcher (Chapter 30).

- If you intend to use an ethnographic app or software package as a standalone method in your ethnographic study, have you considered what is meant by 'ethnographic'? Even though apps or software are described as 'ethnographic', the term can be misused or misunderstood (and this can be the case when products are advertised by those who do not have a clear understanding of ethnography and ethnographic methods and techniques). In what way is the app or software ethnographic, and is it the most suitable digital tool to use for your research? Will it help you to answer your research question?

- Mobile ethnography facilitates the collection of deep and extensive non-linear narratives over short and long periods of time. There is potential for vast amounts of data from a wide variety of sources. How will you pull together, store and analyse this data?

- Can mobile self-reporting really be ethnography? Can it replace observational methods? How can you take account of behaviour or actions that are not reported because participants are unaware of them or dismiss them as unimportant, for example?

Ethics, morals and legal issues

- Have you thought deeply about co-construction, social distance and closeness in the relationship between you and participants? How might these issues impact on trust and rapport? Detailed and insightful reflections on these matters can be found in Tagg et al. (2016).

- How can you ensure that you receive full informed consent from all participants? It is important that participants know exactly what they are required to do and what information they will need to provide, before they consent. This is especially so if you intend to gather data from activity logging apps (recording activity such as location, content and browsing, for example) in addition to information uploaded by participants.

- How do you intend to address issues of anonymity, confidentiality, data protection and data security? If you are purchasing software or apps, or using open source software, discuss these issues in-depth with the manufacturer/producer to ensure that they meet your requirements. Think about how you are going to record conversations and the ethical implications associated with participants choosing to disclose personal information in public spaces. More information about legal issues associated with recording mobile phone interviews and information about providing information in public spaces can be found in Chapter 33. More information about issues of respect and disclosure can be found in Chapter 8.

Practicalities

- What costs are involved? Are software, apps, tools and equipment freely available from your institution? If not, are free trials available? What customer support is available if purchases have to be made?

- How might you deal with potential problems such as low participation rates and unmotivated respondents? This will require careful thought about methods and methodology, your project, topic and sample. If you are intending to buy an app or software that has customer support included in the purchase price, will you be able to receive practical help and advice on these issues?

- Some research participants will complete tasks and send data on a regular basis. Others, however, might be too busy, distracted, unwilling or unable to send data on a regular basis. How will you deal with this? Data that are sent, and data that are not sent, can both help to inform your research. On a more practical note you might need to think about how to chase up those who do not send their data (alerts, reminders and incentives, for example).

- How do you intend to take account of people who are unable to use the technology, cannot use the technology or who make mistakes when uploading or sending data, for example? Some of the technical issues can be solved by customer support, depending on which app or software you use, whereas other issues will need to be considered within your overall methodology (when thinking about who is to take part in the research and the technology that is to be used, for example).

Useful resources

There are a variety of digital tools and software packages available for mobile ethnography. Examples that are available at time of writing include (in alphabetical order):

- Ethos: Ethnographic Observation System from Everydaylives Ltd. (www. ethosapp.com);

- ExperienceFellow Mobile ethnography tool (www.experiencefellow.com);

- Indeemo Mobile Ethnography Apps and Qualitative Research (https:// indeemo.com);

- MotivBase Big Data Ethnography (www.motivbase.com);

- Over the Shoulder qualitative research platform (www.overtheshoulder.com);

- ThoughtLight™ Qualitative Mobile Ethnography App (www.civicommrs. com/mobile-insights-app);

- Touchstone Research mobile ethnographic research app (https://touchsto neresearch.com/market-research-tools-technologies/mobile-ethnography).

The Qualitative Report Guide to Qualitative Research Mobile Applications, curated by Ronald J. Chenail (https://tqr.nova.edu/apps) is a comprehensive list of mobile research apps, including those that can be used for ethnographic studies [accessed March 19, 2018].

'Mobile Apps for Ethnographic Research', January 16, 2018 is an interesting blog written by Dick Powis (https://anthrodendum.org/2018/01/16/mobile-apps-for-ethnographic-research-ror2018). This blog discusses the apps that he finds most useful when in the field [accessed March 23, 2018].

A series of articles on mobile and digital ethnography can be found on Ethnography Matters (http://ethnographymatters.net/blog/tag/mobile-apps), which is 'a platform for ethnographers and those using elements of ethno-graphic practice to take part in conversations between academic and applied ethnography in the private and public sector' [accessed December 18, 2018].

Key texts

Christensen, P., Mikkelsen, M., Nielsen, T. and Harder, H. (2011) 'Children, Mobility, and Space: Using GPS and Mobile Phone Technologies in Ethnographic Research', *Journal of Mixed Methods Research*, 5(3), 227–46, first published April 19, 2011, 10.1177/1558689811406121.
Gottschalk, S. and Salvaggio, M. (2015) 'Stuck Inside of Mobile: Ethnography in Non-Places', *Journal of Contemporary Ethnography*, 44(1), 3–33, first published January 8, 2015, 10.1177/0891241614561677.
Hjorth, L., Horst, H., Galloway, A. and Bell, G. (eds.) (2017) *The Routledge Companion to Digital Ethnography*. New York, NY: Routledge.
Horst, H. and Miller, D. (eds.) (2012) *Digital Anthropology*. London: Berg.

Mobile ethnography

Muskat, B., Muskat, M. and Zehrer, A. (2018) 'Qualitative Interpretive Mobile Ethnography', *Anatolia*, 29(1), 98–107, first published November 1, 2017, 10.1080/13032917.2017.1396482.

Novoa, A. (2015) 'Mobile Ethnography: Emergence, Techniques and Its Importance to Geography', *Human Geographies*, 9(1), 97–107, 10.5719/hgeo.2015.91.7.

Pink, S., Horst, H., Postill, J., Hjorth, L., Lewis, T. and Tacchi, J. (2015) *Digital Ethnography: Principles and Practice*. London: Sage.

Pink, S., Sinanan, J., Hjorth, L. and Horst, H. (2016) 'Tactile Digital Ethnography: Researching Mobile Media through the Hand', *Mobile Media & Communication*, 4(2), 237–51, first published December 16, 2015, 10.1177/2050157915619958.

Richardson, I. and Hjorth, L. (2017) 'Mobile Media, Domestic Play and Haptic Ethnography', *New Media & Society*, 19(10), 1653–67, first published July 7, 2017, 10.1177/1461444817717516.

Tagg, C., Lyons, A., Hu, R. and Rock, F. (2016) 'The Ethics of Digital Ethnography in a Team Project', *Applied Linguistics Review*, 8(2–3), 271–92, published online October 29, 2016, 10.1515/applirev-2016-1040.

CHAPTER 32

Mobile methods

Overview

The term 'mobile methods', for the purpose of this book, describes any research method that uses mobile technology to collect, record, interpret and/or analyse quantitative or qualitative data. 'Mobile methods' also refers to research into the mobile (mobility and immobility, dwelling and place-making, potential movement and blocked movement, mobile bodies and mobile contexts, for example: Büscher and Urry, 2009; Cresswell, 2006; Sheller and Urry, 2006). Mobile, mobilities or movement research that is not centred around mobile technology is beyond the scope of this book, but you can find out more about these methods from Cresswell and Merriman (2013) who provide a collection of insightful papers on the topic; from Büscher and Urry (2009) who consider the mobilities paradigm and implications for the relationship between the empirical, theory, critique and engagement; and from Cresswell (2006) who provides a discussion of historical episodes of mobility and regulation.

Mobile methods, as defined in this book, enable researchers to connect with participants across geographical boundaries, instantaneously and when they are on the move, providing the opportunity for collaboration, co-construction and co-creation of knowledge. Research can be interactive and co-operative, working with participants rather than on participants. A fixed time and place for partici-pation in the research project is not required, enabling participants to go about their daily lives while connecting with researchers. Their mobile interaction, such as phone conversations, messaging and texting, and their mobile activity, such as location, movement, browsing and viewing, can all inform research.

Mobile methods

Mobile methods projects can be quantitative, collecting and logging files for analysis, for example (Chapter 28) or qualitative, through in-depth mobile phone interviews, for example (Chapter 33). They can also adopt mixed methods approaches, where a combination of qualitative and quantitative mobile collection and analysis methods are used (Van Damme et al., 2015). Some mobile methods research is highly visible, interactive and participatory, whereas other methods are less visible, working in the background to collect and analyse data. These can be described as active and passive methods (Poynter, 2015; Poynter et al., 2014): for active methods participants use their mobile device to take part in the research (completing a questionnaire or mobile phone interview, for example: see Chapters 33 and 34) and for passive methods, researchers gather information automatically (from log files, sensors or mobile phone-based sensing software, for example: see Chapters 28, 49 and 57).

There are a variety of mobile methods, which include (in alphabetical order):

- computer-assisted personal and telephone interviewing that use mobile devices to collect and record information (Chapter 8);

- ethnographic apps and software that enable researchers to undertake ethnographic studies into human activity, behaviour and/or movement (Chapter 31);

- eye-tracking research that uses mobile eye-trackers built into glasses, helmets or other wearable devices, enabling participants to move freely while capturing viewing behaviour in the real-world environment (Chapter 19);

- geographic positioning technology (e.g. GPS, remote sensing and satellite imagery) for capturing geospatial, flow and movement data directly from the field (Chapter 21);

- location awareness and location tracking that uses mobile devices to record, track and analyse location and movement data (Chapter 27);

- log file analysis for collecting, storing, searching and analysing log file data, which can include key logging for mobiles and mobile device logs, for example (Chapter 28);

- mobile diaries that enable participants to record everyday behaviour, actions and thoughts (Chapter 30);

- mobile phone interviews that require voice or touch button interaction with the researcher or with interactive voice response technology (IVR) (Chapter 33);

- mobile phone surveys that use methods such as SMS and text messages, questionnaires or interviews to collect data about thoughts, opinions, attitudes, behaviour and feelings (Chapter 34);

- mobile sensors and mobile phone-based sensing software to detect, measure and respond to activity, location, proximity or touch, for example (Chapter 49);

- smartphone app-based research to collect, record, store, analyse and visualise data (Chapter 50);

- smartphone questionnaires that use human voice, automated voice, text or video to administer questionnaires (Chapter 51);

- wearables-based research that uses computational devices and/or digital sensors that are worn or carried on the body to track and digitise human activity and/or behaviour (Chapter 57).

Mobile methods can be seen to be an umbrella term that covers all the above methods. If you are interested in finding out more about any of these methods, consult the relevant chapter, where you will find further information about each method, along with useful resources and key texts. Examples of specific projects that use mobile methods include the use of keystroke logging 'to qualitatively investigate the synchronous processes of discursive interaction through mobile devices' (Schneier and Kudenov, 2018); a study of segregation in Estonia, based on mobile phone usage (Järv et al., 2015); a mixed methods approach into mobile news consumption including questionnaires, mobile device logs, personal diaries and face-to-face interviews (Van Damme et al., 2015); an analysis of the potential for smartphone sensor and log data to collect behavioural data in psychological science (Harari et al., 2016); and research to test the feasibility and acceptability of a mobile phone-based peer support intervention for women with diabetes (Rotheram-Borus et al., 2012).

If you are interested in using mobile methods for your research, a detailed methodological discussion about the use of mobile methods, along with practical insight into mobile methods projects from a number of countries, is provided by Toninelli et al. (2015). Information about the history, development and use of various types of mobile method is provided by Poynter et al. (2014). Both of these books are aimed at those studying and working within market research: however, the information provided is relevant for those approaching mobile methods from other disciplines and fields of study. References that are relevant to specific mobile methods, along with useful digital tools and software packages, can be found in the chapters listed above.

Questions for reflection

Epistemology, theoretical perspective and methodology

- What is meant by the word 'mobile'? What are 'mobile methods' in relation to your research? Does the term relate to the free flow of people, information and services, or does it relate to specific technology, tools and devices, for example?

- Do you have a good understanding of the variety of methods that are covered by the umbrella term of mobile methods? How will you choose the most appropriate method(s)? How is this choice guided and informed by epistemology, theoretical perspective and methodological framework?

- How does rapid mobile technological advancement shape social interaction? How is social interaction shaped by familiarity with, and experience of, mobile technology (from the point of view of the researcher and the participant)?

- How is use of mobile technology shaped by demographics, nationality, culture, society, politics and income? How is use shaped by individuals: fear, distrust or dislike of mobile technology, for example? How will you take account of these issues when developing your theoretical and methodological framework and working out your sample?

Ethics, morals and legal issues

- How do you intend to address issues of informed consent, anonymity, confidentiality, data protection and data security? How will you address these issues when participants, who may be co-creators or co-constructors, may not, necessarily, have the same ethical concerns or sensitivity toward data (in particular when they are used to sharing mobile data with friends, family and organisations, for example)? If participants have given informed consent, is it ethical to use all data, or will you encounter situations where data should not, or cannot, be used (where a friend has borrowed a mobile device, for example)? See Tagg et al. (2016) for in-depth discussion and reflection on these issues.

- Mobile technology provides the opportunity for the collection, harvesting or mining of large amounts of data. What implications does this have for issues such as security, management, sharing, curation and preservation? Do you need to develop a Data Management Plan or a Data Management and Sharing Plan and, if so, what needs to be included? Further information and advice about securing, managing, curating and preserving research data can be obtained from the Digital Curation Centre in the UK (www.dcc.ac.uk).

- How will you ensure that mobile traces do not cause harm or increase vulnerability? See Taylor (2016) for an enlightening discussion on these issues.

Practicalities

- Do you have the required financial resources available and access to the relevant mobile technology? Do you have the necessary training and expertise, or know how to access further training?

- Have you considered all practicalities involving optimisation, device heterogeneity and compatibility? Can participants use mobile devices of their choice for your research? What about people who have not upgraded to the latest technology? This is of particular importance when using smartphones in research, for example (see Chapters 50 and 51).

- How can you encourage participation and increase motivation and engagement with your research? This can include:
  - ensuring that participants are interested, willing and able to take part (are comfortable with the mobile technology and understand the purpose and benefit of your research, for example);
  - providing detailed instruction on the use of mobile technology and devices;
  - ensuring optimisation and compatibility of devices;
  - keeping instructions, procedures and questionnaires simple (avoiding the need for long passwords or the input of identification numbers, for example);
  - providing feedback, encouragement and advice/technological support when required;
  - making sure that participants are not left out of pocket if they use their own devices.

Useful resources

Up-to-date, pertinent articles on mobile technology and research can be accessed on the Harvard Business School Working Knowledge website (https://hbswk.hbs.edu/Pages/browse.aspx?HBSTopic=Mobile%20Technol ogy). You can find articles about research ethics, recent technological devel- opments and mobile use on this site [accessed March 23, 2018].

The Software Sustainability Institute (www.software.ac.uk) 'cultivates better, more sustainable, research software to enable world-class research'. It works with researchers, developers and funders to identify key issues and best practice in scientific software. The website contains an interesting and useful blog that covers up-to-date and relevant issues about mobile apps and software [accessed March 23, 2018].

There are a wide variety of digital tools and software packages available for mobile methods research. These are provided in the relevant mobile method chapter listed above.

Key texts

Büscher, M. and Urry, J. (2009) 'Mobile Methods and the Empirical', *European Journal of Social Theory*, 12(1), 99–116, first published February 1, 2009, 10.1177/ 1368431008099642.

Cresswell, T. (2006) *On the Move: Mobility in the Modern Western World*. Abingdon: Routledge.

Cresswell, T. and Merriman, P. (eds.) (2013) *Geographies of Mobilities: Practices, Spaces, Subjects*. Abingdon: Routledge.

Harari, G., Lane, N., Wang, R., Crosier, B., Campbell, A. and Gosling, S. (2016) 'Using Smartphones to Collect Behavioral Data in Psychological Science: Opportunities, Prac- tical Considerations, and Challenges', *Perspectives on Psychological Science*, 11(6), 838–54, first published November 28, 2016, 10.1177/1745691616650285.

Järv, O., Müürisepp, K., Ahas, R., Derudder, B. and Witlox, F. (2015) 'Ethnic Differences in Activity Spaces as A Characteristic of Segregation: A Study Based on Mobile Phone Usage in Tallinn, Estonia', *Urban Studies*, 52(14), 2680–98, first published Septem- ber 22, 2014, 10.1177/0042098014550459.

Merriman, P. (2013) 'Rethinking Mobile Methods', *Mobilities*, 9(2), 167–87, first published May 7, 2013, 10.1080/17450101.2013.784540.

Poynter, R. (2015) 'The Utilization of Mobile Technology and Approaches in Commercial Market Research'. In Toninelli, D., Pinter, R. and de Pedraza, P. (eds.) *Mobile Research Methods: Opportunities and Challenges of Mobile Research Methodologies*. London: Ubiquity Press.

Poynter, R., Williams, N. and York, S. (2014) *The Handbook of Mobile Market Research – Tools and Techniques for Market Researchers*. Chichester: Wiley.

Rotheram-Borus, M., Tomlinson, M., Gwegwe, M., Comulada, W., Kaufman, N. and Keim, M. (2012) 'Diabetes Buddies: Peer Support through a Mobile Phone Buddy System', *The Diabetes Educator*, 38(3), 357–65, first published April 30, 2012, 10.1177/0145721712444617.

Schneier, J. and Kudenov, P. (2018) 'Texting in Motion: Keystroke Logging and Observing Synchronous Mobile Discourse', *Mobile Media & Communication*, 6(3), 309–330, first published December 8, 2017, 10.1177/2050157917738806.

Sheller, M. and Urry, J. (2006) 'The New Mobilities Paradigm', *Environment and Planning A: Economy and Space*, 38(2), 207–26, first published February 1, 2006, 10.1068/a37268.

Tagg, C., Lyons, A., Hu, R. and Rock, F. (2016). 'The Ethics of Digital Ethnography in a Team Project', *Applied Linguistics Review*, 8(2–3), published online October 29, 2016, 10.1515/applirev-2016-1040.

Taylor, L. (2016) 'No Place to Hide? The Ethics and Analytics of Tracking Mobility Using Mobile Phone Data', *Environment and Planning D: Society and Space*, 34(2), 319–36, first published October 6, 2015, 10.1177/0263775815608851.

Toninelli, D., Pinter, R. and de Pedraza, P. (eds.) (2015) *Mobile Research Methods: Opportunities and Challenges of Mobile Research Methodologies*. London: Ubiquity Press.

Van Damme, K., Courtois, C., Verbrugge, K. and De Marez, L. (2015) 'What's APPening to News? A Mixed-Method Audience-Centred Study on Mobile News Consumption', *Mobile Media & Communication*, 3(2), 196–213, first published January 13, 2015, 10.1177/2050157914557691.

CHAPTER 33

Mobile phone interviews

Overview

Mobile phone interviews (or cell phone interviews) are interviews that require respondents to answers questions using a mobile phone, which is personally owned by the respondent or given to them by the researcher. This chapter discusses mobile phone interviews that require voice or touch button interaction with the researcher or with interactive voice response technology (IVR), whereas Chapter 51 discusses questionnaires that are administered via smartphones, using different modes of communication such as voice, text, images or video. Mobile phone interviews can be employed as a research method in both quantitative and qualitative research projects, including mobile phone surveys (see Chapter 34), ethnographic studies (see Chapters 14 and 31) and mixed methods studies (Christensen et al., 2011; Kaufmann, 2018). They enable researchers to contact a large number of participants, across geographical boundaries, at a cost that can be much cheaper and quicker than more traditional face-to-face interviews or pen and paper methods. They also provide the opportunity to reach hard-to-access populations (see Chapter 34).

There a number of ways in which mobile phones can be used for interviewing, including mobile phone-assisted personal interviewing (MPAPI); mobile computer-assisted personal interviewing (mCAPI); computer-assisted self-interviewing (CASI) (audio-CASI is a term that relates to questionnaires administered by phone and video-CASI is a term that relates to questionnaires that appear on screen); computer-assisted personal interviewing (CAPI); and computer-assisted telephone interviewing (CATI). All these terms are discussed in detail in Chapter 8. Mobile phones can also be used to collect log data that can be integrated into qualitative interviews (on mobile phones or face-to-face) for elicitation purposes. These interviews ask participants to reflect on, explain and contextualise action and behaviour based on detailed log data from their mobile phone (Kaufmann, 2018). More information about key logging for mobiles and mobile device logs is provided in Chapter 28.

There are three types of interview that can be conducted by mobile phone, depending on theoretical perspective, methodology and research question:

- Structured mobile phone interviews are used for large-scale mobile phone surveys where multiple interviewers or fieldworkers are used and generalisation is the goal. They are used in mobile phone surveys to ask the same set of standardised questions to all respondents in the same order (see Chapter 34). The questions are grouped into pre-determined categories that will help to answer the research question, or confirm/disconfirm the hypothesis.

- Semi-structured mobile phone interviews enable researchers to ask a standard set of questions, but also allow for additional questions and probing for detail, if required. Data can be quantified and compared and contrasted with data from other mobile phone interviews. It is possible that new themes or ideas can emerge from the interviews.

- Unstructured mobile phone interviews provide the freedom for participants to give information in a way that they wish, with the researcher helping to keep the narrative moving forward. Mobile phones can be used as a prelim to life story interviews and oral history interviews, for example: these begin as mobile phone interviews and then continue as face-to-face interviews due to the amount of time involved, and the amount of data gathered, in the interview.

Mobile phone interviews are used in a variety of disciplines and fields of study including health and medicine (Livingston et al., 2013), public opinion research (Lynn and Kaminska, 2013), sociology (Hinton, 2013) and media and communication (Kaufmann, 2018). Examples of research projects that have used, assessed or critiqued mobile phone interviews include an assessment of response rates and sample biases in mobile phone surveys (L'Engle et al., 2018); an assessment of the impact of mobile phone interviews on survey measurement error (Lynn and Kaminska, 2013); a study into the practical and ethical issues associated with conducting in-depth interviews via mobile phones (Hinton, 2013); and a study into alcohol and drug use using mobile and landline interviews (Livingston et al., 2013).

If you are interested in mobile phone interviews as a research method it is important that you get to grips with a variety of interviewing techniques and methods, and understand how these fit with methodology and theoretical perspective (see questions for reflection, below). Gillham (2005) provides a practical, comprehensive guide to interviewing, which includes a discussion on types of interview, ethics, developing questions and analysis techniques. A more in-depth epistemological, theoretical and methodological discussion can be found in Brinkmann and Kvale (2015). Neither of these books specifically covers mobile phone interviews, but the information provides important background reading for those who are interested in conducting any type of interview for their research. More information about using interviews in mobile phone surveys can be found in Chapter 34; more information about smartphone app-based research is provided in Chapter 50; and more information about smartphone questionnaires is provided in Chapter 51.

Questions for reflection

Epistemology, theoretical perspective and methodology

- Have you thought carefully about whether the type of mobile phone interview you wish to conduct is suitable, given epistemology, theoretical perspective and methodological standpoint? We have seen above that there are different ways to conduct mobile phone interviews: some are used to gather descriptive data whereas others seek to obtain a wider, holistic view of lived experienced. Different types of mobile phone interview create different types of data and different forms of knowledge. Also, transcripts can be analysed in very different ways. An understanding of the connections between epistemology, theoretical perspective and methodology will help you to make decisions about what type of mobile phone interview is most appropriate for your research. For example:

  o If structured interviews are to be used the assumption is that the respondent has experiential knowledge that can be transmitted to the interviewer using a mobile phone. Interview data can be analysed and, if correct procedures have been followed, generalisations can be made to the target population.

  o If semi-structured interviews are to be used the assumption is that experiential knowledge can be transmitted from the participant to the researcher, and that there may be additional experiences/themes that have not been pre-determined by the researcher.

○ If unstructured interviews are to be used the emphasis is on finding meanings and acquiring a deep understanding of people's life experiences. This type of research can take a long time and, as such, may be broken down into several mobile phone interviews or followed with face-to-face interviews. This type of interview may require considerable personal disclosure from the interviewee, and relationships can be built between interviewer and participant over a period of time. Generalisation to the target population is not the goal.

• What sampling frame, methods and procedures do you intend to use? Decisions will be informed by theoretical perspective, methodological framework and possible costs. For example, L'Engle et al. (2018) illustrate how they used an interactive voice response, random-digit dial method for their national mobile phone survey in Ghana, whereas Livingston et al. (2013) used a commercial sample provider for their mobile phone interview research into alcohol and drug use.

Ethics, morals and legal issues

• Are you aware of relevant local, regional and national law concerning unsolicited mobile communication? Are you aware of cultural and social attitudes towards mobile communication? This could include, for example, laws concerning mobile phones and driving, laws and/or attitudes toward acceptable calling times and personal registration on do-not-contact/call lists.

• When conducting in-depth interviews using mobile phones, how can you maintain confidentiality when participants choose to conduct their conversations in public places? How will you deal with these issues if the interview topic is of a sensitive nature, or participants are vulnerable or young, for example? See Hinton (2013) for an insightful discussion into these issues.

• If you intend to record mobile phone interviews (see below), are you aware of, and have you considered, relevant legal issues? For example, in the UK journalists and researchers need to ask permission to record, if they intend to use any of the information from the call. It is possible (although perhaps not ethical) to record calls without seeking permission, but only if the information is for personal use: as soon as it is disclosed to third parties you are in breach of the law. There are some exceptions: rules are complex and involve several pieces of legislation, including:

- Regulation of Investigatory Powers Act 2000 (RIPA);

- Telecommunications (Lawful Business Practice) (Interception of Communications) Regulations 2000 (LBP Regulations);

- Data Protection Act 2018;

- Telecommunications (Data Protection and Privacy) Regulations 1999;

- Human Rights Act 1998.

Practicalities

- How might demographics, mobile phone ownership and attitudes toward mobile phones influence participation rates?

- Will date, day and time of contact have an influence on whether you are able to make contact and whether you are able to complete the interview? Does this differ, depending on demographics? Vicente (2015) provides insight into these issues.

- How will you overcome potential problems with poor reception, or no reception, when you try to contact participants and conduct your interview?

- How do you intend to record information? For example, you could choose to use one of the various recorder apps for Android phones, available from Google Play or a recorder app for iPhones from the App Store. Examples available at time of writing include:

  - Automatic Call Recorder Pro (Android): a trial version is available;

  - TapeACall Pro (iOS, Android): purchase and subscription costs;

  - Call Recorder (Android): free, with ads;

  - Call Recording by NoNotes (iOS): purchase and subscription costs.

When using recording apps ensure that they are compatible and that playback time is not limited (this may be the case with some free apps). Also, some services only enable you to record outbound (or inbound) calls, so ensure that features and facilities are suitable for purpose before purchase. Test all equipment thoroughly and become familiar with its use before you undertake your interviews.

- What issues do you need to think about when developing your interview schedule (this is a list of topics and/or questions that are to be asked or discussed in the interview)? This can include, for example:

  o constructing short, well-worded, non-leading questions, using language that can be understood by participants (avoiding jargon and double-barrelled questions);

  o categorising and grouping relevant questions, while discarding irrelevant questions;

  o ordering questions into a logical sequence, leaving sensitive, personal or challenging questions until the end;

  o asking about experience and behaviour, before asking about opinion and feelings (if relevant);

  o testing and modifying, as appropriate.

- What effect might multi-tasking, distraction and the presence of others have on mobile phone interviews? Lynn and Kaminska (2013) will help you to address this issue.

Useful resources

The European Society of Opinion and Marketing Research (ESOMAR) is 'a membership organization representing the interests of the data, research and insights profession at an international level'. You can access a useful document that provides guidelines about conducting mobile phone surveys and interviews on this website, which covers issues such as costs, safety, contact times, duration and calling protocols:

www.esomar.org/uploads/public/knowledge-and-standards/codes-and-guide lines/ESOMAR_Guideline-for-conducting-Research-via-Mobile-Phone.pdf [accessed January 16, 2019].

Key texts

Brinkmann, S. and Kvale, S. (2015) *InterViews: Learning the Craft of Qualitative Research Interviewing*. London: Sage.

Christensen, P., Mikkelsen, M., Nielsen, T. and Harder, H. (2011) 'Children, Mobility, and Space: Using GPS and Mobile Phone Technologies in Ethnographic Research', *Journal of Mixed Methods Research*, 5(3), 227–46, first published April 19, 2011, 10.1177/ 1558689811406121.

Mobile phone interviews

Gillham, B. (2005) *Research Interviewing: The Range of Techniques.* Maidenhead: Open University Press.

Hinton, D. (2013) 'Private Conversations and Public Audiences: Exploring the Ethical Implications of Using Mobile Telephones to Research Young People's Lives', *Young*, 21 (3), 237–51, first published August 12, 2013, 10.1177/1103308813488813.

Kaufmann, K. (2018) 'The Smartphone as a Snapshot of Its Use: Mobile Media Elicitation in Qualitative Interviews', *Mobile Media & Communication*, first published January 8, 2018, 10.1177/2050157917743782.

L'Engle, K., Sefa, E., Adimazoya, E., Yartey, E., Lenzi, R., Tarpo, C., Heward-Mills, N., Lew, K. and Ampeh, Y. (2018) 'Survey Research with a Random Digit Dial National Mobile Phone Sample in Ghana: Methods and Sample Quality', *PLoS One*, 13(1), e0190902, 10.1371/journal.pone.0190902.

Livingston, M., Dietze, P., Ferris, J., Pennay, D., Hayes, L. and Lenton, S. (2013) 'Surveying Alcohol and Other Drug Use through Telephone Sampling: A Comparison of Landline and Mobile Phone Samples', *BMC Medical Research Methodology*, 13(41), open access, first published March 16, 2013, 10.1186/1471-2288-13-41.

Lynn, P. and Kaminska, O. (2013) 'The Impact of Mobile Phones on Survey Measurement Error', *Public Opinion Quarterly*, 77(2), 586–605, first published Dec 9, 2012, 10.1093/poq/nfs046.

Vicente, P. (2015) 'The Best Times to Call in a Mobile Phone Survey', *International Journal of Market Research*, 57(4), 555–70, first published July 1, 2015, 10.2501/IJMR-2015-047.

CHAPTER 34

Mobile phone surveys

Overview

Mobile phone surveys (or cell phone surveys) are used to collect data about thoughts, opinions, attitudes, behaviour and feelings. They can be used to investigate hypotheses and test theory, to understand and describe a particular phenomenon, or for exploratory purposes, for example. Mobile phone surveys are a methodology (the overall framework that guides a research project) rather than a method (the tools that are used to collect and analyse data). However, mobile phone surveys have been included in this book as there are many practical 'how to' issues involved and, increasingly, the term is used to describe both method and methodology (see some of the digital tool and software package websites listed below for examples of when this occurs). Mobile phones provide the opportunity to undertake survey research that reaches a large amount of people, crosses geographical boundaries and enables connections to be made at any time in any place. They also provide the opportunity to reach hard-to-access populations (Firchow and Mac Ginty, 2017). Mobile phone surveys use samples that are representative of the larger population of interest so that generalisations can be made. Therefore, issues of correlation, causality, reliability and validity are extremely important.

Specific methods that are used in mobile phone surveys include:

- SMS and text message questionnaires and invitations or alerts to open questionnaires (Steeh et al., 2007);

- structured mobile phone interviews where respondents answers questions asked by a researcher or fieldworker (Chapter 33), including:

- computer-assisted telephone interviewing (CATI) where interviews are administered over mobile phone by an interviewer who reads the questionnaire and inputs answers direct into the computer (see Chapter 8);

- mobile phone-assisted personal interviewing (MPAPI) where an interviewer is present to administer a pre-loaded questionnaire, using a mobile phone as part of the interviewing process (Chapter 8);

- interactive voice response technology (IVR) for administering questionnaires, where respondents answer questions via automated self-administered procedures (Gibson et al., 2017);

- questionnaires designed specifically for smartphones, utilising different modes of communication and touchscreens for questionnaire completion (see Chapter 51).

Mobile phone surveys are used in a wide variety of disciplines, including health and medicine (Ali et al., 2017; Carter et al., 2015; Gibson et al., 2017), the social sciences (Firchow and Mac Ginty, 2017; Gilleard et al., 2015; van Heerden et al., 2014) and market research (Vicente, 2015). Examples of research projects that have used, evaluated or critiqued mobile phone surveys include research into the ethical considerations and challenges of mobile phone surveys of non-communicable diseases (Ali et al., 2017); a reflection on the use of mobile phones to reach hard to access populations in conflict affected, low income countries (Firchow and Mac Ginty, 2017); an examination of trends in mobile phone ownership among people over fifty and the implications for survey research (Gilleard et al., 2015); and a study into the feasibility of providing mobile phones to young adults to improve retention rates in a long-term birth cohort study (van Heerden et al., 2014).

If you are interested in carrying out a mobile phone survey, consult Lepkowski et al. (2008) for useful papers on telephone and mobile surveys (although it could do with updating, this book provides some useful theoretical, methodological and methods information that is still pertinent today). Trucano's blog (2014) provides a useful overview of how mobile phones can be used for large-scale data collection and Gilleard et al. (2015) provide a brief yet enlightening overview of mobile phone development and access to mobile phones by age, socio-demographics and socio-economics. The European Society of Opinion and Marketing Research (ESOMAR) provide useful guidelines for conducting mobile phone surveys and more information about this organisation can be found in Chapter 33. There are a wide variety of software packages, tools and apps available for designing, administering and analysing mobile phone questionnaires for survey research (see below). It is important to note, however, that a good knowledge and understanding of survey research and questionnaire design, construction and analysis is important for any researcher considering undertaking a mobile phone survey. It can be easy to neglect these aspects of the research process when software companies offer to take on these tasks on your behalf. Fink (2013), Gillham (2007) and Andres (2012) provide good introductory texts. If you are interested in finding out more about specific research methods that can be used in mobile phone surveys, consult Chapters 8, 33, 50 and 51.

Questions for reflection

Epistemology, theoretical perspective and methodology

• Have you come to terms with how theoretical perspective and methodological standpoint will guide your mobile phone research? For example, closed-ended questionnaires are used to generate statistics in quantitative research. Large numbers are required: they follow a set format and utilise scanning and data capture technology for ease of analysis. For objectivists who hope to describe an objective truth, a stimulus-response model is employed, with standardised questions and answers being the favoured approach for large scale mobile phone surveys. Validity and reliability are extremely important when designing this type of questionnaire and when analysing data (see below). However, if you approach your work from a more subjective standpoint you might assume that both questions and answers have to be understood in terms of the social contexts in which they operate. Your questionnaire or interview schedule and the way it is administered might be very different from the survey approach described above. In this case alternative mobile methods may be more suitable for your research (see Chapters 32, 33 and 50, for example).

• How are you going to produce a population representative mobile phone survey? You could decide to use a random digital dialling approach or a simple random sample of a mobile network provided list, or a list developed from previous surveys, for example. Are these methods representative? Will you use an opt-in requirement (a text asking for participation, for example)? If so, how can you overcome selection bias? How will you take account of phone numbers that do not exist or are inactive? How will you take account of people who do not have mobile phones or those who have hearing difficulties, for example? See Peytchev et al. (2010) for an interesting discussion about coverage bias in landline and cell phone surveys; Firchow and Mac Ginty (2017) for information about increasing the relevance of research to encourage participation; and Gilleard et al. (2015) for information about the declining age divide in mobile usage.

- How can you conduct your survey in a way that will yield unbiased, reliable and valid data? Validity refers to the accuracy of the measurement, asking whether the tests that have been used by the researcher are measuring what they are supposed to measure. Reliability refers to the way that the research instrument is able to yield the same results in repeated trials: are measurements consistent and would other researchers get the same results under the same conditions? Chapter 38 provides more information on validity and reliability.

Ethics, morals and legal issues

- How do you intend to provide introductory information about purpose, intent, motivation, funding organisation, confidentiality, anonymity, potential use of data, methods of data collection and data sharing, for example? How can you ensure that potential respondents have understood the information and that they are able to give voluntary and informed consent?

- Do you understand relevant legal issues concerning the collection, recording and storing of data gathered from mobile phones? Legislation that is relevant to research conducted in the UK is listed in Chapter 33.

- Are you aware of the norms and practices surrounding mobile phone use of your target population? When and how should you contact people (is the method you intend to use for your survey the most appropriate for your target population)? How can you minimise disturbance? How can you avoid undue intrusion? See Ali et al. (2017) for an insightful paper on these issues.

- How can you protect data from hacking and third party ownership? Carter et al. (2015) provide an enlightening discussion on this issue.

Practicalities

- If you are intending to use a number of interviewers/fieldworkers for a large scale mobile phone survey, what training will be required to enable them to become proficient in administering interviews or questionnaires? Will they be able to navigate through the questions and follow correct procedures when entering data or recording information? How can you ensure that adequate time and training is provided?

- How do you intend to test (pilot) your questionnaire? All questionnaires should be piloted on the type of people who will be taking part in the main mobile phone survey. This helps to iron-out ambiguities, point to questions that are unclear or badly-worded and illustrates whether the questionnaire will help to answer the research question. The question-naire should be altered accordingly.

- How do you intend to obtain a high response rate? In mobile phone surveys response rate refers to the number of people who have taken part in the survey divided by the number of people in the sample. It is usually expressed as a percentage the lower the percentage, the more likely some form of bias has been introduced into the research process (certain people with similar traits, characteristics or experiences have been unwilling to respond, for example). This will have an influence on the generalisability of results. Response rates can be increased through the use of text messages and reminders (Steeh et al., 2007), calling at the most appropriate time (Vicente, 2015) and making ques-tionnaires easier to complete (van Heerden et al., 2014).

Further questions for reflection that have relevance to mobile phone surveys can be found in Chapters 8, 33, 50 and 51.

Useful resources

There are a wide variety of digital tools and software packages that can help researchers to design questionnaires and conduct mobile phone surveys. Examples available at time of writing include (in alphabetical order):

- Creative Research Solutions mobile survey software (www.surveysystem. com/mobile-survey-software.htm);

- Interactive SMS questionnaires from Netsize (www.netsize.com/sms-the-natural-flow-for-interactive-questionnaires);

- Nebu Dub InterViewer for survey data collection (www.nebu.com/dub-interviewer-data-collection-software);

- QuestionPro for mobile optimised surveys (www.questionpro.com/mobile/mobile-surveys.html);

- SmartSurvey for SMS surveys (www.smartsurvey.co.uk/sms-surveys);

- SnapSurveys for surveys on mobile devices (www.snapsurveys.com/survey-software/mobile-surveys);

- SurveyGizmo for surveys on mobile devices (www.surveygizmo.com/survey-software-features/mobile-surveys).

Michael Trucano's blog 'Using mobile phones in data collection: Opportunities, issues and challenges' (http://blogs.worldbank.org/edutech/using-mobile-phones-data-collection-opportunities-issues-and-challenges) provides a brief but insightful discussion on how mobile phones can be used for large-scale data collection [posted on April 18, 2014 and accessed March 26, 2018].

Key texts

Ali, J., Labrique, A., Gionfriddo, K., Pariyo, G., Gibson, D., Pratt, B. and Hyder, A. (2017) 'Ethics Considerations in Global Mobile Phone-Based Surveys of Noncommunicable Diseases: A Conceptual Exploration', *Journal of Medical Internet Research*, 19(5), 10.2196/jmir.7326.

Andres, L. (2012) *Designing and Doing Survey Research*. London: Sage.

Carter, A., Liddle, J., Hall, W. and Chenery, H. (2015) 'Mobile Phones in Research and Treatment: Ethical Guidelines and Future Directions', *JMIR mHealth and uHealth*, 3(4), 10.2196/mhealth.4538.

Fink, A. (2013) *How to Conduct Surveys: A Step-by-Step Guide*. Thousand Oaks, CA: Sage.

Firchow, P. and Mac Ginty, R. (2017) 'Including Hard-to-Access Populations Using Mobile Phone Surveys and Participatory Indicators', *Sociological Methods & Research*, first published October 3, 2017, 10.1177/0049124117729702.

Gibson, D., Farrenkopf, B., Pereira, A., Labrique, A. and Pariyo, G. (2017) 'The Development of an Interactive Voice Response Survey for Noncommunicable Disease Risk Factor Estimation: Technical Assessment and Cognitive Testing', *Journal of Medical Internet Research*, 19(5), e112, 10.2196/jmir.7340.

Gilleard, C., Jones, I. and Higgs, P. (2015) 'Connectivity in Later Life: The Declining Age Divide in Mobile Cell Phone Ownership', *Sociological Research Online*, 20(2), 1–13, first published June 1, 2015, 10.5153/sro.3552.

Gillham, B. (2007) *Developing a Questionnaire*, 2nd edition. London: Continuum.

Lepkowski, J., Tucker, C., Brick, J., De Leeuw, E., Japec, L., Lavrakas, P., Link, M. and Sangster, R. (2008) *Advances in Telephone Survey Methodology*. Hoboken, NJ: John Wiley & Sons.

Peytchev, A., Carley-Baxter, L. and Black, M. (2010) 'Coverage Bias in Variances, Associations, and Total Error from Exclusion of the Cell Phone-Only Population in the United States', *Social Science Computer Review*, 28(3), 287–302, first published December 2, 2009, 10.1177/0894439309353027.

Steeh, C., Buskirk, T. and Callegaro, M. (2007) 'Using Text Messages in U.S. Mobile Phone Surveys', *Field Methods*, 19(1), 59–75, first published February 1, 2007, 10.1177/1525822X06292852.

van Heerden, A., Norris, S., Tollman, S., Stein, D. and Richter, L. (2014) 'Field Lessons from the Delivery of Questionnaires to Young Adults Using Mobile Phones', *Social Science Computer Review*, 32(1), 105–12, first published September 24, 2013, 10.1177/0894439313504537.

Vicente, P. (2015) 'The Best Times to Call in a Mobile Phone Survey', *International Journal of Market Research*, 57(4), 555–70, first published July 1, 2015, 10.2501/IJMR-2015-047.

CHAPTER 35

Online analytical processing

Overview

Online analytical processing (OLAP) is a digital method for data discovery that enables researchers to work with very large amounts of data. It uses computer processing to answer queries through the selective extraction and viewing of data stored in databases, facilitating the analysis of multidimensional data from multiple perspectives. It is a method that is used primarily in business and management (Lam et al., 2009), but can also be used in fields of study such as health and medicine (Alkharouf et al., 2005), information sciences (Yin and Gao, 2017) and tourism (Giannopoulos and Boutsinas, 2016). Examples of research projects that have used, assessed or critiqued OLAP include research into developing a Tourism Information System application using OLAP (Giannopoulos and Boutsinas, 2016); the evaluation of an OLAP based intelligent system to help tackle storage location assignment problems in warehouse management (Lam et al., 2009); an evaluation of a complete client–server architecture explicitly designed for mobile OLAP (mOLAP) (Michalarias et al., 2009: see below); the evaluation of a system to detect, explain and resolve bias in decision-support OLAP queries (Salimi et al., 2018); the development of an efficient heuristic algorithm for aggregation in two phases: informational aggregation and structural aggregation (Yin and Gao, 2017); and a study into using OLAP as a data mining tool for gene expression databases (Alkharouf et al., 2005).

OLAP, when used in business, is covered by the umbrella term of business intelligence (the process of gathering, analysing and transforming raw data into accurate and meaningful information to help inform business decision-making and action). Business analytics (Chapter 4) and data mining (Chapter

12) are also covered by the umbrella term of business intelligence: business analytics refers to the process of examining an organisation's data to measure past and present performance to gain insight that will help with future planning and development; data mining involves applying algorithms to the extraction of hidden information with the aim of building an effective predictive or descriptive model of data for explanation and/or generalisation. The difference is that OLAP aggregates information from multiple systems and stores the information in a multidimensional format whereas business analytics and data mining apply statistics, algorithms, ratios and clustering to internal and external business/organisation data. OLAP summarises data and makes forecasts; data mining discovers hidden patterns, relationships and trends; business analytics explores, analyses and provides actionable insight. Despite these differences the terms are sometimes used interchangeably or are considered to be synonymous (when data mining solutions are offered at the database level or when OLAP is used for data mining purposes, for example).

OLAP is an online database query answering system that queries and extracts information that has been inputted by online transaction processing (OLTP). This is an online database modifying system that manages and facilitates transaction-orientated tasks (inserting, deleting and updating, for example) and provides source data for OLAP. There are different categories of OLAP. These include:

- Multidimensional OLAP (MOLAP): facilitates data analysis by using a multidimensional data cube (a method of storing data in a multidimensional form). Data are pre-summarised, optimised and stored so that they can be accessed directly. MOLAP is able to manage, analyse and store large amounts of multidimensional data, facilitate fast data retrieval and process less defined data.

- Relational OLAP (ROLAP): uses data residing in a relational database rather than in a multidimensional database. It can handle large amounts of data, although query time can be long (and processing slow) if the underlying data size is large. It can be accessed by any Structured Query Language (SQL) tool, but is limited by SQL functionalities.

- Hybrid OLAP (HOLAP): combines useful features of MOLAP and ROLAP for greater flexibility when accessing data sources. Data are stored in multidimensional and relational databases and the type of processing required determines which is used. Data handling is flexible: data are stored in a multidimensional database for speculative or theoretical processing and in a relational database for heavy processing, for example.

- Spatial OLAP: combines spatial data sources such as geographic information systems (GIS) with non-spatial data sources. For example, GIS provides the cartographic representation and OLAP provides a multidimensional perspective of data, enabling objects to be viewed and manipulated on maps and facilitating the interactive multidimensional exploration of phenomena.

- Mobile OLAP (mOLAP): users can access OLAP data and applications through mobile devices. Lightweight servers provide a bridge between mobile apps and OLAP cubes, facilitating database connectivity and app security.

- Web OLAP (WOLAP): users can access OLAP data and applications through web browsers or a web interface: this term is used infrequently as most OLAP tools and platforms now provide a web interface.

- Desktop OLAP: users can access sections of data and OLAP applications from their desktop. It enables users to work with the dataset locally, performing multidimensional analyses while disconnected from the source.

If you are interested in using OLAP for your research a useful collection of papers is provided by Wrembel and Koncilia (2007). They are divided into three sections covering modelling and designing, loading and refreshing and efficiency of analytical processing. For information about security and privacy issues see Wang et al. (2010) who provide a detailed and comprehensive discussion on preserving privacy in OLAP. There are a wide variety of platforms, tools and software available if you are interested in OLAP, and some of these are listed below. It is useful to visit some of the sites so that you can get an idea of purpose, functions, capabilities and costs (some are open source and freely available, whereas others have significant costs attached). It is also useful to find out more about business analytics (Chapter 4), data mining (Chapter 12), data analytics (Chapter 10) and geospatial analysis (if you are interested in spatial OLAP: Chapter 21).

Questions for reflection

Epistemology, theoretical perspective and methodology

- Do you have a clear understanding of the similarities and differences between OLAP, business analytics and data mining? Which method is the most appropriate for your research, or is a combination best? OLAP and data mining, for example, can be used to complement each other, with some platforms integrating OLAP and data mining capabilities.

- Do you have a clear understanding of the difference between OLTP and OLAP and understand how the two relate to each other? Which method is most suitable for your research? OLTP, for example, is an online database modifying system that can be used for simple queries, with fast processing and small space requirements (if historical data are archived). OLAP, on the other hand, is an analytical processing system and query answering system that can be used for complex queries and aggregation, but can take longer to process and have much larger space requirements.

- What effect might politics, culture, history and economics have on the development, design and content of, and access to, datasets and databases? What effect might such influences have on OLAP summaries and forecasts?

- How can you determine the relevance of data for OLAP?

- What influence might the time-lag between data origination, processing and analysis have on summaries and forecasts? Can 'real-time' really be real-time?

Ethics, morals and legal issues

- How do you intend to address security issues? This can include security in the relational source system, security in the OLAP server system and security of data in transit from the relational system to the OLAP system, for example. Are all databases used for OLAP protected against internal and external threats such as hacking, theft, cyberattacks and unauthorised access?

- Who owns data? What access control policies are in place for end-users of databases and data warehouses? Might these hinder exploratory OLAP?

- How do you intend to address issues of individual privacy? How can you ensure that personal data remain confidential and that aggregates and derived data do not pose privacy threats (exposing sensitive data, for example)? These issues are addressed in depth by Wang et al. (2010) in their comprehensive guide to privacy issues and OLAP.

- How can you detect, explain, resolve or avoid bias in OLAP queries? Salimi et al. (2018) will help you to address this question.

Practicalities

- Do you have a good understanding of OLAP systems, terms and techniques? This can include:

    o an OLAP cube: a multidimensional database that is used to store data for processing, analysing and reporting purposes;

    o dimension: a dataset of individual, non-overlapping data elements for filtering, grouping and labelling;

    o measure: a numeric value by which the dimension is detailed or aggregated (or an additive numerical value that represents a business metric);

    o fact table: a table that joins dimension tables with measures;

    o star schema: a dimensional model that resembles a star (facts are surrounded with associated dimensions);

    o snowflake schema: a logical arrangement of tables in a multidimensional database that resembles a snowflake;

    o drill down: summarises data to lower levels of a dimension hierarchy;

    o pivot (rotation): rotates the data axis, providing a different perspective to view data;

    o roll-up (consolidation): summarises data along a dimension;

    o sorting: adds, moves or alters the order of dimensions and measures in rows and columns;

    o slice: selects one specific dimension from a given cube and provides a new sub-cube;

    o dice: selects two or more dimensions from a given cube and provides a new sub-cube;

    o data blending: mixes data from different structured and unstructured sources.

- What functions do you require from your chosen OLAP tool? This can include:

    o a full set of cube manipulation features and analytical tools (drill through, pivot, roll-up, time intelligence slicers, 3D data analysis and decomposition analysis, for example);

    o the ability to create customised, interactive charts;

- the ability to switch between visual analysis and grid analysis;

- the ability to share analyses and reports with other users in a secure environment;

- the inclusion of export formats such as Microsoft Excel or PDF.

Useful resources

There are a wide variety of platforms, tools and software available for OLAP with different functions, capabilities and purposes (some concentrate on features for processing large volumes of data whereas others concentrate on high quality visualisations, for example). Examples available at time of writing include (in alphabetical order):

- Apache Kylin (http://kylin.apache.org);

- Axional Mobile OLAP (www.deister.net/en/solutions/mobility/ax-mobile-olap);

- Express Server (www.oracle.com/technetwork/database/database-technologies/express-edition/downloads/index.html);

- icCube (www.iccube.com);

- InetSoft (www.inetsoft.com/info/web_based_olap_server_solution);

- Jedox (www.jedox.com);

- Kyubit Business Intelligence (www.kyubit.com/olap-tools);

- Micro Strategy (www.microstrategy.com);

- Mondrian (https://community.hitachivantara.com/docs/DOC-1009853);

- Oracle Essbase (www.oracle.com/solutions/business-analytics/business-intelligence/essbase/index.html);

- Oracle OLAP (www.oracle.com/technetwork/database/options/olap/index.html);

- PowerOLAP (http://paristech.com/products/powerolap);

- SQL Server Analysis Services (https://docs.microsoft.com/en-us/sql/analysis-services/analysis-services?view=sql-server-2017).

Key texts

Alkharouf, N., Jamison, D. and Matthews, B. (2005) 'Online Analytical Processing (OLAP): A Fast and Effective Data Mining Tool for Gene Expression Databases', *Journal of Biomedicine and Biotechnology*, 2005(2), 181–88, 10.1155/JBB.2005.181.

Giannopoulos, K. and Boutsinas, B. (2016) 'Tourism Satellite Account Support Using Online Analytical Processing', *Journal of Travel Research*, 55(1), 95–112, first published June 19, 2014, 10.1177/0047287514538836.

Lam, C., Chung, S., Lee, C., Ho, G. and Yip, T. (2009) 'Development of an OLAP Based Fuzzy Logic System for Supporting Put Away Decision', *International Journal of Engineering Business Management*, 1(2), 7–12, first published September 1, 2009, 10.5772/6779.

Michalarias, I., Omelchenko, A. and Lenz, H. (2009) 'FCLOS: A Client–Server Architecture for Mobile OLAP', *Data & Knowledge Engineering*, 68(2), 192–220, 10.1016/j.datak.2008.09.003.

Salimi, B., Gehrke, J. and Suciu, D. (2018) 'Bias in OLAP Queries: Detection, Explanation, and Removal', in *Proceedings of the 2018 International Conference on Management of Data*, 1021–35, May 27, 2018, 10.1145/3183713.3196914.

Wang, L., Jajodia, S. and Wijesekera, D. (2010) *Preserving Privacy in On-Line Analytical Processing (OLAP)*. New York, NY: Springer.

Wrembel, R. and Koncilia, C. (eds.) (2007) *Data Warehouses and OLAP: Concepts, Architectures and Solutions*. Hershey, PA: IRM Press.

Yin, D. and Gao, H. (2017) 'A Flexible Aggregation Framework on Large-Scale Heterogeneous Information Networks', *Journal of Information Science*, 43(2), 186–203, first published July 10, 2016, 10.1177/0165551516630237.

CHAPTER 36

Online collaboration tools

Overview

Online collaboration tools are internet- and web-based applications that enable individuals, groups and teams to collaborate in research (and teaching and learning) without the need to meet face-to-face. The term 'virtual collaboration' is often used interchangeably with online collaboration, although virtual collaboration can also encompass non-internet or web-based technology such as audio conferencing. There are various types of collaboration project that utilise online tools including collaboration between sectors such as academia and industry; local, regional or international collaboration between individual researchers, research teams, students or scientific departments; and collaboration between geographically-distributed employers and employees from public, private and voluntary sector organisations. Collaboration projects can be encouraged by social, economic, professional, scientific, political or geographical factors, for example.

There are two types of online collaboration tool: synchronous that enable researchers to share ideas, information and feedback instantaneously (instant messaging and video conferencing, for example) and asynchronous that provide collaboration opportunities that are not instantaneous (email and discussion boards, for example). Examples of online collaboration tools and platforms that can be used for research, teaching and/or learning include:

- general social networking websites and academic networking sites for socialising, networking, sharing, discussing and disseminating information;
- discussion boards, message boards, chat rooms and forums for making contact, sharing ideas and asking/answering questions;

Online collaboration tools

- instant messaging, real-time text messaging, file transfer or video chat for instant feedback, sharing of ideas and dissemination of results;

- website, video, blog, app and music production, collaboration and sharing tools for producing and uploading digital content;

- virtual worlds, virtual environments, virtual maps, virtual workspaces and virtual projects for problem-solving, teamwork, development and research;

- design tools for digital designing, synchronising, sharing and updating;

- data analysis (and analytics) tools and software for collaborative and distributed qualitative and quantitative data analysis;

- presentation tools for creating, collaborating and sharing presentations;

- collaborative search engines (web search engines and enterprise searches within an organisation's intranet) for querying, collaborating, sharing and retrieving information;

- file sharing and management (peer-to-peer file sharing, cloud-based file syncing and sharing services and web-based hyperlinked documents) to share and disseminate documents for feedback, comments and evaluation;

- research, learning and study tools such as citing, referencing and bibliography tools for sharing, managing and recommending scholarly references and note-taking tools for capturing, organising and sharing notes;

- co-operative or collaborative learning environments such as online study groups, online reading groups, globally networked learning environments and Wikis;

- internet conferencing technologies for online meetings and discussion including web conferencing (webinars, webcasts and web meetings), videoconferencing, teleconferencing, virtual reality conferencing and augmented or mixed reality conferencing (to share viewpoints and research spaces, for example);

- management and organisation tools (work management platforms, online calendars and project management tools) for research, development and improving productivity.

Examples of research projects that have used, assessed or critiqued the use of online collaboration tools include an evaluation into the effectiveness of adding the hands and arms of virtual collaborators to a map-based virtual environment (Cheng and Hayes, 2015); an investigation into how gesturing over shared drawings and artefacts can change the way in which virtual team members communicate and perceive themselves as part of a team (Hayes et al., 2013); a comparative review of the features and trade-offs inherent in asynchronous and synchronous communication technology tools that are used to run virtual team projects (Larson et al., 2017); a qualitative study of strategies and competencies used by technical communication and translation students to address challenges when collaborating in global virtual teams (Cleary et al., 2018); and a study into the use of social networking tools in academic libraries (Chu and Du, 2013).

If you are interested in finding out more about online collaboration tools for your research a good starting point is Donelan et al. (2010). This book presents a series of articles grouped into themes such as collaborative technologies, Wikis and instant messaging, online collaboration in action, online communities, virtual worlds and social networking. The book is aimed at those working in the fields of soft computing, information systems, cultural and media studies and communications theory, but also provides useful information for those approaching from other academic disciplines. If you are interested in online or virtual collaboration tools for teaching and learning, comprehensive and practical advice is provided by Flammia et al. (2016). Information is aimed at students and teachers in the fields of business, information technology, communication studies and engineering. If you are interested in the use of Wikis for research collaboration and learning, a useful book is West and West (2009). Although this book could do with some updating, given rapid technological advances, it still contains useful information on Wikis and other collaborative design tools. There are a huge variety of online collaborative tools available, with new technology entering the market at a rapid pace. Some of the more popular tools that are available at time of writing are listed below: it is useful to visit some of these sites so that you can get an idea of functions, purpose and cost to ascertain whether the tools will be useful for your research (or teaching and studying).

Questions for reflection

Epistemology, theoretical perspective and methodology

- Why are you interested in online collaboration? How can online collaboration tools help with your research, teaching or leaning? This can include enabling and facilitating the cross-fertilisation and generation of ideas; the sharing of knowledge, techniques and skills; transferring and building knowledge and skills; and advancing knowledge, for example.

- In what way will online collaboration tools help to extend your network and provide the opportunity to raise your research profile? Will they enable you to present complex ideas/research findings to a more diverse range of people? Will they enable you to raise the visibility of your work? Will they enable you to increase productivity and research output?

- Will your chosen online collaboration tools enable you to establish an effective communication routine that will encourage personal, informal exchange among team members and collaborators? Will it enable collaborators to develop a better understanding of the research problem and help them to develop or use a common vocabulary to communicate results?

Ethics, morals and legal issues

- When collaborating with others online how can you ensure that agreement is reached on issues such as ownership of intellectual property, sharing of potentially sensitive information and accepted author protocols in published material?

- How can you ensure the protection and security of data, in particular, when in transit? How can you ensure that all collaborators have robust data management and sharing plans in place? The UK Data Service (www. ukdataservice.ac.uk/manage-data/collaboration) provides comprehensive advice for collaboration projects, including information about procedures, protocols and file sharing.

- How can you ensure that all parties are trustworthy and upfront about their organisations' strategy and reasons for collaboration?

- Will all parties enjoy the online collaboration, find it mutually beneficial and be challenged by the project?

- How will you foster social, cultural, linguistic and organisational understanding if the project is to cross international, social and organisational boundaries? What organisational differences might have an influence on your collaboration project? This could include the merging of two very different management cultures or the merging of different administrative and technical systems, for example.

Practicalities

- Do all collaborators have the same aims and objectives for the project and do all parties understand the benefits to be gained?

- Who is to take on the management role (if relevant) and does this person have the required experience and skills to run the collaboration project?

- Do all team members have the required knowledge and understanding of online tools and software? Technical understanding is crucial for all those involved in the technical and online communication aspects of the project.

- Are you clear about all costs and is funding available? Can all parties afford the costs involved (and how do these vary if you are involved in international collaboration)? Who will pay for collaboration tools and software? Who will supply equipment? Who owns equipment and tools and what will happen to them once the collaboration project has finished? In general, equipment, tools and software purchased from grant funds for use on the research project (for which the research grant was awarded) belongs to the research organisation (e.g. university, laboratory or museum). In some cases, however, the funding body itself will want to retain ownership of the equipment throughout the period of the grant and possibly beyond.

- Do all members of the team understand their roles and functions within the project? These should be made clear from the outset: everyone needs to understand what is expected from the start so that there are no misunderstandings as the project progresses.

Useful resources

There are a wide variety of online collaboration tools and software packages available. Some are open source and free to use, whereas others have significant costs attached. A selection of the more popular tools available at time of writing is given below, in alphabetical order.

- 99 Chats (www.99chats.com) for setting up four-way chats that can include up to 99 people;

- Blackboard collaborate (www.blackboard.com/online-collaborative-learning/blackboard-collaborate.html) for online learning, web conferences and real time classes;

- Canva (www.canva.com) for creating, collaborating on and sharing slides, posters, flyers and infographics;

- Citeulike (www.citeulike.org) for managing and sharing scholarly references;

- Crocodoc (https://crocodoc.com) for taking, making and sharing notes;

- Dropbox (www.dropbox.com) for storing, sharing and using files;

- Evernote (https://evernote.com) for capturing, organising and sharing notes;

- Google Drive (www.google.com/drive) for file storage, sharing and collaboration;

- Mural (https://mural.co) for visual collaboration and live conversations;

- Prezi (https://prezi.com) for creating presentations and collaborating with groups of up to ten people;

- Scribble (www.scrible.com) for researching, capturing, sharing and bookmarking;

- Sketch (www.sketchapp.com) for digital designing, synchronising, sharing and updating;

- Slack (https://slack.com) for team communication and collaboration;

- Skype (www.skype.com) for free audio and video calls between multiple devices;

- WeVideo (www.wevideo.com) for collaborative video creation;

- WordPress (https://wordpress.org) for creating websites, blogs and apps, and inviting contributions and changes;

- Wunderlist (www.wunderlist.com) to-do lists for students working on the same project;

- Yammer (www.yammer.com) for collaborating across departments, locations and business apps;

- Zoho Projects (www.zoho.eu/projects) project management software for planning, tracking and collaborating.

Key texts

Cheng, X. and Hayes, C. (2015) 'Hand Videos in Virtual Collaboration for Map-Based Planning Activities', *Proceedings of the Human Factors and Ergonomics Society Annual Meeting*, 59(1), 1834–38, first published December 20, 2016, 10.1177/1541931215591396.

Chu, S. and Du, H. (2013) 'Social Networking Tools for Academic Libraries', *Journal of Librarianship and Information Science*, 45(1), 64–75, first published February 17, 2012, 10.1177/0961000611434361.

Cleary, Y., Slattery, D., Flammia, M. and Minacori, P. (2018) 'Developing Strategies for Success in a Cross-Disciplinary Global Virtual Team Project: Collaboration among Student Writers and Translators', *Journal of Technical Writing and Communication*, first published May 27, 2018, 10.1177/0047281618775908.

Donelan, H., Kear, K. and Ramage, M. (eds.) (2010) *Online Communication and Collaboration: A Reader*. Abingdon: Routledge.

Flammia, M., Cleary, Y. and Slattery, D. (2016) *Virtual Teams in Higher Education: A Handbook for Students and Teachers*. Charlotte, NC: Information Age Publishing, Inc.

Hayes, C., Drew, D., Clayson, A. and Cheng, X. (2013) 'In What Ways Do Natural Gestures Change Perceptions of Virtual Teamwork?', *Proceedings of the Human Factors and Ergonomics Society Annual Meeting*, 57(1), 2152–56, first published September 30, 2013, 10.1177/1541931213571479.

Larson, B., Leung, O. and Mullane, K. (2017) 'Tools for Teaching Virtual Teams: A Comparative Resource Review', *Management Teaching Review*, 2(4), 333–47, first published July 14, 2017, 10.1177/2379298117720444.

West, J. and West, M. (2009) *Using Wikis for Online Collaboration: The Power of the Read-Write Web*. San Francisco, CA: Jossey–Bass.

CHAPTER 37

Online ethnography

Overview

The term 'online ethnography' is used to describe the use of ethnographic tools, techniques, methods and insight to collect and analyse data from online sources (blogs, vlogs, discussion forums and boards, chat rooms and social media, for example). The term 'digital ethnography' is sometimes used to describe online ethnography, although this is better understood as an umbrella term that encompasses all forms of digital technology-based ethnography including online, mobile, network and offline ethnography involving digitalisation. More information about digital ethnography is provided in Chapter 14, more information about mobile ethnography can be found in Chapter 31 and more information about online ethnography is provided below.

Ethnography is the investigation into, and systematic recording of, human cultures and human societies. In online ethnography behaviour within the online group or community is observed and recorded, with careful distinction made between scientific perceptions of reality and the perceptions of the online community being investigated (the researcher observes, records and analyses while ensuring that people speak in their own voices). The world is observed from the point of view of research participants rather than the ethnographer and stories are told through the eyes of the online community as they go about their online activities. Causal explanations are avoided. Online ethnographers focus on various aspects of online behaviour and activity. This can include, for example:

- interaction among online users;
- the content produced by online users;

- political, social, cultural and historical meanings that online users ascribe to their actions and interaction;

- the presentation of self and creation and presentation of online identities (including false or misleading identities);

- participation in the formation, interaction and cessation of online forums, networks and groups.

There are different ways to conduct online ethnography, depending on research topic, purpose and focus, and on epistemology and theoretical perspective. Some of these are listed below. It is important to note, however, that these terms are discretionary, flexible and fluid: they have no settled and clear typology and can, on occasions, be used interchangeably. Examples include:

- Virtual ethnography: the investigation of virtual communities, how they are shaped and how they behave. The researcher can choose to become a participant observer through joining the virtual community and observing from within, or can decide to observe from outside the community, if access can be gained. McLelland (2002) and Steinmetz (2012) provide examples of virtual ethnography.

- Network ethnography: the investigation into social networks, using ethnographic tools and techniques. The researcher can choose to become an active participant observer (taking part and asking questions) or a non-active observer (reading and analysing text) in the social network, for example. Berthod et al. (2017: 299) adopt a mixed method approach to network ethnography that 'balances well-established social network analysis with a set of techniques of organizational ethnography'. More information about social network analysis can be obtained from Chapter 53.

- Netnography: the investigation into human online communication. The focus is on reflections and data provided by online communities (textual or multimedia communication, for example). Data are provided naturally and voluntarily by members of the community and can be downloaded by the researcher for analysis. Kozinets (2010, 2015) developed this method of investigation in 1995.

- Cyber ethnography: this term has fallen out of favour in recent years. It used to describe various types of online ethnography including observations from blogs, websites and chat rooms, and the formation of social networks and online groups (Keeley-Browne, 2010). The terms online ethnography or digital ethnography (see Chapter 14) are now more commonly used.

- Webethnography and webnography: these terms are less common, but have been used in market research to describe online ethnographic studies (Prior and Miller, 2012).

Online ethnographic methodologies and methods can be combined, mixed or enmeshed with other approaches, such as with quantitative methods (Snelson, 2015) with small stories research (Georgalou, 2015) and with offline ethnography (Dutta, 2013). Examples of research that has used, assessed or critiqued online ethnography include an exploratory study of vlogging about school on YouTube (Snelson, 2015); an assessment of Facebook 'as a powerful grassroots channel for expressive storytelling within a period of major socio-political upheaval' (Georgalou, 2015); an evaluation of the use of the internet to research gay men in Japan (McLelland, 2002); a methodological assessment of three facets of virtual ethnography: 'space and time, identity and authenticity, and ethics' (Steinmetz, 2012); and an online ethnography of resistance during the Arab Spring (Dutta, 2013).

If you are interested in finding out more about online ethnography for your research, a good starting point is Hine (2015). This book covers both methodology and methods and provides a number of insightful and practical examples of online ethnography and the connections that can be made between online and offline ethnography. If you want to find out more about the practicalities of undertaking online ethnography in virtual worlds, consult Boellstorff et al. (2012) and if you want to find out more about the ethnography of online human communication, including practical methods, consult Kozinets (2010). There are a variety of digital tools and software packages that are available for researchers who are interested in online ethnography and some of these are listed below. Additional tools available for mobile ethnography are listed in Chapter 31 and additional tools available for qualitative data analysis are listed in Chapter 9.

Questions for reflection

Epistemology, theoretical perspective and methodology

- In what way is your research guided by epistemology and theoretical perspective (reflexive or critical ethnography that involves an ideological critique and questions the status quo of power relations and naturalistic ethnography that is founded on positivism and based on the legacy of colonialism, for example)? How might these affect the way that data are collected and analysed?

- How will your methodology frame and guide your research? Georgalou (2015: 4), for example, illustrates how her research was 'situated within the discourse-centered online ethnographic paradigm' that enabled her to take on 'two dimensions: a screen-based and a participant-based one'.

- What level of participation is appropriate? For example, you could choose to become a complete observer, passive participant or active participant within the online community. This will need to be considered within your methodological framework.

- What sampling techniques will you use (judgement sampling, convenience sampling, snowball sampling or theoretical sampling, for example)? See Kurtz et al. (2017) for an enlightening discussion on the systematic sampling of online blogs.

- Online ethnography has the potential to generate huge quantities of online data about the subject under investigation. How are you going to narrow this down into a feasible project?

Ethics, morals and legal issues

- How are you going to address issues of informed consent? This is consent given by a participant to take part in your research, based on an understanding of what is involved. For this consent to be valid it must be given voluntarily and freely by a person who has been informed fully and has the capacity to make the decision. How are you going to ensure that your online participants understand the purpose, benefits and potential risks of taking part in your research and are given time to make their decision? How are you going to address this issue if you intend to become an invisible observer?

- How are you going to address issues of privacy, authenticity and online identity? How might group members present themselves differently in public and private online spaces? Have you considered your own online identity and presence and how this might have an influence on your investigation?

- Have you thought about issues of confidentiality and anonymity? How do you intend to cite information found online (from blogs and social networks, for example) yet respect anonymity and online identities? Kurtz et al. (2017) discuss this issue in relation to the use and citation of public-access blogs.

- If you are going to investigate online communities that have very different social and political beliefs, can you remain neutral and keep your opinions to yourself? Indeed, is this desirable?

- Will your participation within an online group cause problems, anxiety or argument amongst other members? How will you deal with such situations should they arise?

- How will you deal with potential trolls that are set on disrupting or harming the group under investigation (or your study)?

Practicalities

- How do you intend to gain access to the online group or community? Is access freely available, restricted or do you need to be invited to join the group? Is it public or private? Is it a geographically- or socially-bounded community with a clearly defined membership? Are there any group dynamics that might have an influence on access?

- How will you build trust? What will you do if some online participants are suspicious of you? How will you avoid breaching trust?

- How are you going to collect and record data? What equipment do you need? What tools do you intend to use for data collection and analysis? This can include, for example, screen capture software, website archiving tools, data analysis software (Chapter 9), mapping tools (Chapter 21), data visualisation tools (see Chapter 13), content management systems, personal blogs or vlogs, audio and video recordings and digital memos. Are tools available and easy to access?

- How are you going to store data? Will they be safe? Do you understand data protection laws?

- What methods of data analysis are you intending to use? Do you have enough training, understanding and experience with these methods? See Chapter 2 for information about audio analysis, Chapter 55 for information about video analysis and Chapter 9 for information about computer-assisted qualitative data analysis software.

- How will you know when your investigation is complete? How do you intend to withdraw from the online group or community? How will you ensure that you leave on good terms with participants?

Useful resources

There are a variety of digital tools and software packages available for online ethnography. Examples available at time of writing include (in alphabetical order):

- Angelfish Fieldwork qualitative market research ethnography (https://angelfishfieldwork.com/methodologies/ethnography);

- ATLAS.ti software for qualitative and mixed methods data analysis (https://atlasti.com);

- Dedoose cross-platform app for analysing qualitative and mixed methods research (http://dedoose.com);

- Kernwert digital qualitative research software and services (www.kernwert.com);

- MAXQDA software for qualitative and mixed methods research (www.maxqda.com);

- nVivo qualitative data analysis software (www.qsrinternational.com/nvivo/home);

- upBOARD's online ethnographic research tools and templates (https://upboard.io/ethnographic-research-online-tools-templates-web-software);

- VisionsLive research software for ethnography (www.visionslive.com/methodologies/ethnography).

Ethnography Matters (http://ethnographymatters.net) is a website that provides informative articles about a wide variety of digital tools, techniques and methods that can be used in ethnographic studies. The website is 'managed and run by a group of volunteer editors who are passionate about ethnography'.

Key texts

Berthod, O., Grothe-Hammer, M. and Sydow, J. (2017) 'Network Ethnography: A Mixed-Method Approach for the Study of Practices in Interorganizational Settings', *Organizational Research Methods*, 20(2), 299–323, first published March 17, 2016, 10.1177/1094428116633872.

Boellstorff, T., Nardi, B., Pearce, C. and Taylor, T. (2012) *Ethnography and Virtual Worlds: A Handbook of Method*. Princeton, NJ: Princeton University Press.

Dutta, M. (2013) 'Communication, Power and Resistance in the Arab Spring: Voices of Social Change from the South', *Journal of Creative Communications*, 8(2–3), 139–55, first published February 19, 2014, 10.1177/0973258613512570.

Georgalou, M. (2015) 'Small Stories of the Greek Crisis on Facebook', *Social Media + Society*, first published October 8, 2015, 10.1177/2056305115605859.

Hine, C. (2015) *Ethnography for the Internet*. London: Bloomsbury Academic.

Keeley-Browne, E. (2010) 'Cyber-Ethnography: The Emerging Research Approach for 21st Century Research Investigation'. In Kurubacak, G. and Yuzer, T. (eds.) *Handbook of Research on Transformative Online Education and Liberation: Models for Social*

Equality, Hershey, PA: Information Science Reference, 330–38, 10.4018/978-1-60960-046-4.ch017.

Kozinets, R. (2010) *Netnography: Doing Ethnographic Research Online.* London: Sage.

Kozinets, R. (2015) *Netnography: Redefined.* London: Sage.

Kurtz, L., Trainer, S., Beresford, M., Wutich, A. and Brewis, A. (2017) 'Blogs as Elusive Ethnographic Texts: Methodological and Ethical Challenges in Qualitative Online Research', *International Journal of Qualitative Methods*, first published online April 28, 2017, 10.1177/1609406917705796.

McLelland, M. (2002) 'Virtual Ethnography: Using the Internet to Study Gay Culture in Japan', *Sexualities*, 5(4), 387–406, first published November 01, 2002, 10.1177/1363460702005004001.

Prior, D. and Miller, L. (2012) 'Webethnography: Towards a Typology for Quality in Research Design', *International Journal of Market Research*, 54(4), 503–20, first published July 1, 2012, 10.2501/IJMR-54-4-503-520.

Snelson, C. (2015) 'Vlogging about School on YouTube: An Exploratory Study', *New Media & Society*, 17(3), 321–39, first published September 26, 2013, 10.1177/1461444813504271.

Steinmetz, K. (2012) 'Message Received: Virtual Ethnography in Online Message Boards', *International Journal of Qualitative Methods*, 11(1), 26–39, first published February 01, 2012, 10.1177/160940691201100103.

CHAPTER 38

Online experiments

Overview

An online experiment is a scientific procedure that is carried out over the internet to test hypotheses, make discoveries or test known facts. It is an online method of investigation based on prediction, experimentation and observation. It can be used to validate results from offline field research or laboratory experiments, or used as an online research method in its own right. Technologies used in online experiments include email, text, modified web interfaces, bots (programmes or scripts) and add-ons (browser extensions or mobile apps, for example). Online experiments can also be referred to as web-based experiments or internet-based experiments, although technically the two terms are slightly different (the World Wide Web is a way of accessing information over the medium of the internet: it is a service, whereas the internet is the whole computer network that includes non-web based communication and networking such as email, newsgroups and instant messaging).

There are different types, terms and techniques used for online experiments and these tend to depend on academic discipline and/or researcher preference and training. Examples are provided below. It is important to note, however, that the following list provides a simplified overview and that types, terms and techniques can be categorised and sub-categorised in various or alternative ways, depending on discipline and preference:

- online controlled experiments that have an experiment group and a control group (Kohavi and Longbotham, 2017), which can include:

  ○ web-based A/B tests, online A/B tests or split tests, which direct users randomly to two or more variations of an interface to measure some outcomes (clicks, purchases or recommendations, for example) (Xu et al., 2015);

     ◦ web-based C/D experimentation (a term coined by Benbunan-Fich, 2017 to describe company-sponsored online experiments with unsuspecting users);

- interactive online experiments in which subjects interact for more than one repetition (Arechar et al., 2018);

- online field experiments that occur in the environment of interest (Parigi et al., 2017);

- online natural experiments (or online natural field experiments) where subjects do not know they are part of a research project (Lange and Stocking, 2009).

Online experiments are used in a variety of disciplines and fields of study including psychology, sociology, economics, political science, health and medicine, law, marketing, business studies, internet studies, gaming and game development. Examples of research projects that have used, assessed or critiqued online experiments include an investigation into the similarities and differences between interactive experiments conducted in the laboratory and online (Arechar et al., 2018); an assessment of the use of online field experiments in social psychology (Parigi et al., 2017); research into the use of online laboratories as a way to perform social psychological experiments with large numbers of diverse subjects (Radford et al., 2016); and a study using laboratory and online experiments to find out whether foreign language use reduces optimistic estimates of personal future (Oganian et al., 2018).

If you are interested in conducting online experiments and you are new to this topic, it is useful to find out first about how to conduct experiments. Field and Hole (2003) provide a comprehensive guide, taking the reader through each stage of the process from design to writing up, including statistical data analysis. For online experiments a good starting point is Kohavi and Longbotham (2017), who discuss controlled online experiments and A/B testing, covering issues such as estimating adequate sample size, gathering the right metrics and tracking the right users. There are a wide variety of organisations and commercial companies that provide services for researchers who are interested in conducting online experiments. Some of these provide software that enables researchers to design and carry out online experiments, others provide online participant recruitment and invitation services. Some provide open source code and experiment exchange services that enable researchers to create their own experiments. It is useful to visit some of these sites so that you can get an idea of services, functions, purposes and costs (if any). You may also find it useful to obtain more information about online panel research (Chapter 42), online questionnaires (Chapter 43) and online research communities (Chapter 44).

Questions for reflection

Epistemology, theoretical perspective and methodology

- Are online experiments neutral and objective? Is it possible that your online experiment might be influenced by personal and structural bias, politics, culture, networks and/or society? What other types of bias can influence online experiments (selection bias, design bias, measurement bias and reporting bias, for example)? Is it possible to eliminate such bias? Paying attention to issues of validity and reliability will help to address these issues: see below.

- Do you have a good understanding of issues concerning validity in online experiments? Validity refers to the accuracy of the measurement, asking whether the tests that have been used by the researcher are measuring what they are supposed to measure. There are different types of validity that are relevant to online experiments:

  o Face validity: are the tests that have been performed, or the questions that have been asked, effective in terms of the stated aims?

  o Content validity: does an instrument measure what it purports to measure? Is a specific question or test useful, essential or irrelevant?

  o Construct validity: how well does an instrument measure the intended construct? How well does the online experiment measure up to its claims?

  o Internal validity: how well is an online experiment performed (including the elimination of bias) and to what extent is a causal conclusion warranted?

  o External validity: to what extent can the results of an online experiment be generalised to other situations, settings, populations or over time? Has the study been replicated, perhaps with different subject populations or offline?

    ○   Predictive validity: to what extent can the measure being used make predictions about something that it should be able to predict theoretically (behaviour, performance or outcomes of an online experiment, for example)? Are the research results a useful predictor of the outcome of future experiments or tests (online or offline)?

    ○   Concurrent validity: how well do the results of a particular test or measurement correspond to those of a previously established measurement for the same construct?

- Do you have a good understanding of issues concerning reliability in online experiments? Reliability refers to the way that the research instrument is able to yield the same results in repeated trials. It asks whether other researchers would get the same results under the same conditions. The following methods can be used to determine the reliability of online experiments:

    ○   inter-observer (or inter-rater) reliability: used to show the degree of agreement, or consensus, between different observers or raters;

    ○   test-retest reliability: assesses the consistency of a measure from one time to another;

    ○   inter-method reliability: assesses the degree to which test scores are consistent when there are variations in methods or instruments;

    ○   internal consistency reliability: used to assess the consistency of results across items within a test (on one occasion).

Ethics, morals and legal issues

- Will taking part in the online experiment cause harm to subjects? This could be physical, psychological or financial harm, for example.

- How do you intend to obtain informed consent? How can you ensure that this consent is given voluntarily and freely by a person who is able to make the decision? Some online experiments do not ask for consent unless the data are to be used for research purposes and some unscrupulous organisations do not ask for consent, even though the data are used for research and business development (Benbunan-Fich, 2017).

- How do you intend to ensure anonymity, confidentiality and privacy, and store and manage data safely and securely? If data are to be shared, how will you make subjects aware of this?

- If you use a commercial company to recruit subjects, how can you be sure that the company treats them fairly (especially in terms of workload and amount of pay)?

Practicalities

- Do you have a good understanding of experimental processes and statistical techniques? Field and Hole (2003) provide comprehensive advice.

- How can you verify the identity of those participating in your online experiment?

- How can you ensure that instructions are followed correctly? This can involve making sure that instructions are simple, clear and not too long, for example.

- What influence might distractions have on those taking part in the experiment? This is of particular importance when individuals are using mobile devices and/or are in public spaces. Techniques that can be used to assess the impact of distractions can include mouse tracking, keystroke logging and background timing.

- How can you overcome problems that may occur with technology used by subjects (and researchers), such as frozen screens, loose network cables, loss of Wi-Fi, IP address conflicts, firewall malfunctions, router failures or bandwidth caps?

- How might your presence, behaviour and expectations influence the reactions or behaviour of online subjects? This is referred to as 'experimenter-expectancy effect' or 'observer effect': is this effect different for offline and online experiments? Also, some online subjects may not believe that they are interacting with an actual researcher: how might this influence their behaviour in the experiment?

- How can you reduce dropout? Arechar et al. (2018) provide advice on this in relation to interactive experiments online.

- How do you intend to answer questions raised by subjects? These can be technological questions, or questions about the research project, for example.

Useful resources

Examples of organisations and commercial companies that provide platforms, software, services or access to subjects for researchers interested in online experiments, available at time of writing, include (in alphabetical order):

- Experimental Tribe (http://xtribe.eu);
- Gorilla (https://gorilla.sc);
- IbexFarm (http://spellout.net/ibexfarm);
- Inquisit (www.millisecond.com);
- JATOS (www.jatos.org);
- LabVanced (www.labvanced.com);
- MTurk (www.mturk.com);
- oTree (www.otree.org);
- PlanOut (https://facebook.github.io/planout);
- psiTurk (https://psiturk.org);
- PsyToolkit (www.psytoolkit.org);
- WEXTOR (http://wextor.org/wextor/en).

There are a number of websites that provide access to online experiments for those who are interested in becoming a research subject. It is useful to visit some of these sites (and perhaps take part in a few experiments) so that you can find out more about the types of experiment that are available, learn more about how they are conducted and understand what it feels like to be a research subject (and assess the quality of the experiment). Examples that are available at time of writing include (in alphabetical order):

- Face Research (http://faceresearch.org);
- Online psychology research (https://experiments.psy.gla.ac.uk);
- Volunteer Science (https://volunteerscience.com).

A comprehensive list of current and past psychological Web experiments can be found at www.wexlist.net and links to known experiments on the internet that are psychology related can be found at https://psych.hanover.edu/research/exponnet.html.

Key texts

Arechar, A., Gächter, S. and Molleman, L. (2018) 'Conducting Interactive Experiments Online', *Experimental Economics*, 21(1), 99–131, first published May 9, 2017, 10.1007/s10683-017-9527-2.

Benbunan-Fich, R. (2017) 'The Ethics of Online Research with Unsuspecting Users: From A/B Testing to C/D Experimentation', *Research Ethics*, 13(3–4), 200–18, first published November 23, 2016, 10.1177/1747016116680664.

Field, A. and Hole, G. (2003) *How to Design and Report Experiments*. London: Sage.

Kohavi, R. and Longbotham, R. (2017) 'Online Controlled Experiments and A/B Testing', In Sammut, C. and Webb, G. (eds.) *Encyclopedia of Machine Learning and Data Mining*. Boston, MA: Springer.

Lange, A. and Stocking, A. (2009) 'Charitable Memberships, Volunteering, and Discounts: Evidence from a Large-Scale Online Field Experiment', *NBER*, Working Paper No. 14941, issued May 2009, retrieved from www.nber.org/papers/w14941.

Oganian, Y., Heekeren, H. and Korn, C. (2018) 'Low Foreign Language Proficiency Reduces Optimism about the Personal Future', *Quarterly Journal of Experimental Psychology*, first published May 22, 2018, 10.1177/1747021818774789.

Parigi, P., Santana, J. and Cook, K. (2017) 'Online Field Experiments: Studying Social Interactions in Context', *Social Psychology Quarterly*, 80(1), 1–19, first published March 1, 2017, 10.1177/0190272516680842.

Radford, J., Pilny, A., Reichelmann, A., Keegan, B., Welles, B., Hoye, J., Ognyanova, K., Meleis, W. and Lazer, D. (2016) 'Volunteer Science: An Online Laboratory for Experiments in Social Psychology', *Social Psychology Quarterly*, 79(4), 376–96, first published December 18, 2016, 10.1177/0190272516675866.

Xu, Y., Chen, N., Fernandez, A., Sinno, O. and Bhasin, A. (2015) 'From Infrastructure to Culture: A/B Testing Challenges in Large Scale Social Networks', in *Proceedings of the 21st ACM SIGKDD International Conference on Knowledge Discovery and Data Mining* 2227–36, Sydney, Australia, 10.1145/2783258.2788602.

CHAPTER 39

Online focus groups

Overview

Online focus groups are a collection of interacting individuals, with common characteristics or interests, holding an online discussion (verbal or text-based) on a particular topic that is introduced and led by a moderator. This can be through live chat rooms, video interaction, web cams, text-based forums, instant messaging or bulletin boards, for example. It is also possible to conduct focus groups in virtual worlds (Stewart and Shamdasani, 2017). Online focus groups are used as a research method within a variety of qualitative methodologies, including action research, grounded theory and ethnography, and in mixed methods approaches to help explain statistics that have been gathered in survey research, for example. Online focus groups can be used in the preliminary or planning stages of a research project, perhaps to help inform and develop a questionnaire, or they can be used as the sole data collection method in a research project (de Lange et al., 2018 and Fox et al., 2007 provide examples of research projects that use online focus groups as their only data collection method). The aim of an online focus group is not to reach consensus: instead, it is to gain a greater understanding of attitudes, opinions, beliefs, behaviour, perceptions and/or reactions to stimuli.

There are two types of online focus group: synchronous online focus groups take place in real-time by video, web conferencing or in a chat room, for example. Participants are able to communicate with each other, and with the moderator, listening, reacting, commenting and adding to the discussion as it unfolds (Fox et al., 2007). Asynchronous online focus groups (or bulletin board focus groups), on the other hand, take place over several days. Questions are posted at intervals and participants can log in over a period of time to answer questions and contribute to the discussion. The moderator does not have to be online for the duration of the focus group. Instead the discussion is reviewed and encouraged through further questioning and discussion at timed or periodic intervals (Williams et al., 2012).

Online focus groups are used in a wide variety of disciplines including education, sociology, politics, health and medicine, business and marketing. Examples of projects that have used online focus groups as a research method for data collection include a project on dementia case management (de Lange et al., 2018); research into young people with chronic skin conditions (Fox et al., 2007); research into young people's engagement with health resources online (Boydell et al., 2014); a study into the experiences of gay and bisexual men diagnosed with prostate cancer (Thomas et al., 2013); and research into the experiences of travel nurses (Tuttas, 2015). These projects illustrate that one of the main advantages of online focus group research is that it enables people to join an online group when they may be unwilling or unable to join a face-to-face group (due to the sensitive nature of the project, for example). Another advantage is that online focus groups can reach hard to access participants and enable researchers to include participants from a variety of locations across geographical boundaries with very few costs involved.

If you are interested in using online focus groups as a research method, it is important that you understand more about the focus group method in general. Barbour (2007), Kruegar and Casey (2015) and Liamputtong (2011) all provide valuable information for those interested in focus group research (Chapter 11 of Kruegar and Casey, 2015 discusses 'telephone and internet focus group interviewing'). Chapter 3 in Coulson (2015) also provides information about running online focus groups, along with chapters on other online methods. You can also visit You Tube (www.youtube.com) to search for online focus group demonstrations and tutorials. Software and online tools are available to help you to run focus groups and maximise group engagement including whiteboards, video sharing, demos, web browser sharing, desktop sharing, app sharing, drawing tools and polling tools, for example. Tools enable you to send open and closed ended questions, upload and present video clips, track time and monitor participation levels, annotate and receive instant transcriptions. There are also apps available for mobile devices that enable participants to take part in online focus groups at a place of their choosing or while on the move. It is useful to read about some of the software and tools available, if you are intending to use online focus groups as a research method. Some are simple to use and free, whereas others are more complex with significant costs attached. Examples are given below. You may also find it useful to obtain more information about qualitative data analysis software (Chapter 9) and video and audio analysis (Chapters 2 and 55).

Questions for reflection

Epistemology, theoretical perspective and methodology

- How will you ensure compatibility of method with underlying epistemological, theoretical and methodological standpoint? As we have seen above, online focus groups can be used within a variety of methodological and theoretical frameworks, but you must ensure that they are compatible.

- What sampling techniques do you intend to use? What about people who do not have computers or mobile devices, those who cannot read or write (or who type very slowly), or do not speak English, for example? Boydell et al. (2014) provide a useful discussion on sampling and recruitment strategies (and difficulties) for online focus groups.

- What analysis methods do you intend to use? This could include, for example, content analysis, grounded theory analysis, discourse analysis or narrative analysis. The method(s) that you choose is guided by epistemology, theoretical perspective and methodology: these will also guide your decisions about appropriate data analysis software (Chapter 9).

Ethics, morals and legal issues

- How might conventional power dynamics exist within the virtual or online environment? How might this have an influence on your focus group participants, outcomes and analyses?

- Have you identified a suitable and secure online venue for your focus group? Will you use an existing venue or create your own? How can you ensure that it is secure (password-protected access, for example)?

- Have you thought about developing a code of ethics or a code of behaviour and, if so, what is to be included (respecting each other, avoiding negative or derogatory comments, taking note of what others say/write and respecting confidentiality, for example)?

- How will you maintain a duty of care to participants? See Williams et al. (2012) for an enlightening discussion on participant safety, netiquette, anonymity and participant authentication.

- Are you going to provide incentives to encourage participants to take part? If so, what are they, are they appropriate and what costs are involved? What influence might incentives have on participant behaviour? Boydell et al. (2014) provide interesting reflections on recruitment and incentives for online focus groups.

- Is it possible that written communication can provide emotive detail that may cause strong emotive reactions in others (Williams et al., 2012)? Reflection on your research topic and methodology is important, in particular, in cases where online focus groups are seen to be a way to reach people who are unwilling to meet face-to-face, where there is a possibility that sensitive issues will be written about or discussed.

Practicalities

- The successful running and completion of online focus groups depend, in part, on the skill of the moderator. What skills are important, and how will you develop these skills? For example, in online focus groups you need experience with the required style of real-time discussion and the ability to react and type quickly, if your focus group is text-based (Fox et al., 2007). You will also need to develop an understanding of online group dynamics.

- How do you intend to recruit participants (online or offline, via websites, apps, or social media, for example)? How might your recruitment strategy influence participation rates? Stewart and Shamdasani (2017: 55) note that online commitments may not be as compelling as verbal commitments, for example.

- How many participants will you recruit, and how many will you over-recruit, assuming that all will not take part? Fox et al. (2007) found that they had an average attendance of three participants in each group after dropout, which meant that discussions were manageable, whereas de Lange et al. (2018) had an attendance of four to eleven over a total of thirteen online focus groups. You should ensure that you have enough participants to make data meaningful, but not too many that the group is difficult to run and control: seven to eleven participants is a good number for synchronous online focus groups, whereas it is possible to recruit up to thirty for asynchronous online focus groups.

- How will you deal with digressions, break-away conversations, insulting or derogatory comments or arguments, for example? It is important that you develop good skills as a moderator and that you understand how to deal with these problems.

- How can you minimise disturbances and distractions? In a face-to-face focus group these can be controlled, but in an online focus group you have no control over what happens in the physical location of the online participant (especially when participants may be on the move or in public spaces). Problems can be reduced by offering guidelines about choosing a suitable location and asking participants to ask others not to disturb them, for example.

- If you intend to use focus group software and data analysis software, what costs are involved and what type and level of support is offered by the provider? What features are available and do they meet your needs?

Useful resources

Examples of tools and software that are available for online focus groups at time of writing include (in alphabetical order):

- AdobeConnect (www.adobe.com/products/adobeconnect.html);
- e-FocusGroups (https://e-focusgroups.com/online.html);
- FocusGroupIt (www.focusgroupit.com);
- Google Hangouts (https://hangouts.google.com);
- GroupQuality (http://groupquality.com);
- itracks (www.itracks.com);
- Skype (www.skype.com/en/);
- TeamViewer (www.teamviewer.com/en);
- VisionsLive (www.visionslive.com/platform/online-focus-groups);
- Zoom (https://zoom.us).

Key texts

Abrams, K., Wang, Z., Song, Y. and Galindo-Gonzalez, S. (2015) 'Data Richness Trade-Offs between Face-to-Face, Online Audiovisual, and Online Text-Only Focus Groups', *Social Science Computer Review*, 33(1), 80–96, first published June 3, 2014, 10.1177/0894439313519733.

Barbour, R. (2007) *Doing Focus Groups*. London: Sage.

Boydell, N., Fergie, G., McDaid, L. and Hilton, S. (2014) 'Avoiding Pitfalls and Realising Opportunities: Reflecting on Issues of Sampling and Recruitment for Online Focus Groups', *International Journal of Qualitative Methods*, 206–23, first published February 1, 2014, 10.1177/160940691401300109.

Coulson, N. (2015) *Online Research Methods for Psychologists*. London: Palgrave.

de Lange, J., Deusing, E., van Asch, I., Peeters, J., Zwaanswijk, M., Pot, A. and Francke, A. (2018) 'Factors Facilitating Dementia Case Management: Results of Online Focus Groups', *Dementia*, 17(1), 110–25, first published February 24, 2016, 10.1177/1471301216634959.

Fox, F., Morris, M. and Rumsey, N. (2007) 'Doing Synchronous Online Focus Groups with Young People: Methodological Reflections', *Qualitative Health Research*, 7(4), 539–47, first published April 1, 2007, 10.1177/1049732306298754.

Kruegar, R. and Casey, M. (2015) *Focus Groups: A Practical Guide for Applied Research*, 5th edition. Thousand Oaks, CA: Sage.

Liamputtong, P. (2011) *Focus Group Methodology: Principle and Practice*. London: Sage.

Stewart, D. and Shamdasani, P. (2017) 'Online Focus Groups', *Journal of Advertising*, 46(1), 48–60, published online Nov 16, 2016, 10.1080/00913367.2016.1252288.

Stewart, K. and Williams, M. (2005) 'Researching Online Populations: The Use of Online Focus Groups for Social Research', *Qualitative Research*, 5(4), 395–416, first published November 1, 2005, 10.1177/1468794105056916.

Thomas, C., Wootten, A. and Robinson, P. (2013) 'The Experiences of Gay and Bisexual Men Diagnosed with Prostate Cancer: Results from an Online Focus Group', *European Journal of Cancer Care*, 22(4), 522–29, first published June 03, 2013, 10.1111/ecc.12058.

Tuttas, C. (2015) 'Lessons Learned Using Web Conference Technology for Online Focus Group Interviews', *Qualitative Health Research*, 25(1), 122–33, first published September 5, 2014, 10.1177/1049732314549602.

Williams, S., Clausen, M., Robertson, A., Peacock, S. and McPherson, K. (2012) 'Methodological Reflections on the Use of Asynchronous Online Focus Groups in Health Research', *International Journal of Qualitative Methods*, first published September 1, 2012, 10.1177/160940691201100405.

CHAPTER 40

Online interviews

Overview

Online interviews (or e-interviews) are structured, semi-structured or unstructured interviews that are carried out over the internet. These can be synchronous interviews that take place in real-time (e.g. by video, webcam, web conferencing, live chats, text, or instant messaging) or asynchronous interviews that are not in real-time and can take place over an extended period of time (e.g. email, pre-recorded video, microblogs, blogs, wikis or discussion boards). Within these two categories online interviews can be:

- video-based, enabling a full verbal and non-verbal exchange, including eye contact, facial expressions, posture, gestures and body movement;

- audio-based, enabling a full verbal and audio exchange, including pacing and timing of speech, length of silence, pitch, tone and emotion in speech;

- text-based, enabling a full written exchange that can also include emoticons, fonts, text size and colour;

- image-based, enabling a full visual exchange, including the generation and viewing of images, charts, graphs, diagrams, slides, maps, visual artefacts and virtual worlds and environments;

- multichannel-based, using a combination of some or all of the above.

Online interviews can be used as a method of data collection by researchers from a variety of epistemological and methodological standpoints. For example, structured and semi-structured online interviews can be used by those approaching their work from an objectivist standpoint: the questions are grouped into pre-determined categories that will help to answer the research question, or confirm/disconfirm the hypothesis. The assumption is that the respondent has experiential knowledge that can be transmitted to the interviewer. Data can be quantified and compared and contrasted with data from other interviews and, if correct procedures have been followed, generalizations can be made to the target population. Discursive, conversational and unstructured online interviews, on the other hand, can be employed by those approaching their work from a subjectivist standpoint: participants are free to tell their life stories in a way that they wish, with the researcher helping to keep the narrative moving forward. The emphasis is on finding meanings and acquiring a deep understanding of people's life experiences. This type of interview can take place in real-time or over a period of time with both the researcher and interviewee returning to the interview when they wish to probe for, or provide, further information.

Examples of research projects that have used, assessed or critiqued online interviews include a discussion about the experiences of two PhD researchers using Skype to interview participants (Deakin and Wakefield, 2014); a discussion on how to make meaning from thin data, based on email interviews with teachers (Kitto and Barnett, 2007); an evaluation of the use of conferencing software for synchronous online interviewing in relation to a project on 'cyberparents' (O'Connor and Madge, 2004); an assessment of the potential sexualisation of the online interview encounter when researching the users of a Nazi fetish Internet site (Beusch, 2007); and an evaluation into using the internet to conduct online synchronous interviews on gay men's experiences of internet sex-seeking (Ayling and Mewse, 2009).

If you are interested in using online interviews for your research, and you are new to interviewing, you might find it useful to obtain more information about interviewing techniques in social research. Brinkmann and Kvale (2015) provide a comprehensive guide, covering epistemological, ethical and practical issues, whereas Seidman (2013) provides a practical guide for those interested in phenomenological interviewing. Information specifically about online interviewing can be obtained from Salmons (2012). This book provides some interesting case studies, along with a useful first chapter that includes information about the different ways in which online interviews can be conducted, sampling and recruitment methods, the position of the researcher, selecting the right technology, conducting the interview and ethical issues. You may also find it useful to increase your understanding of technology available for online interviewing, some examples of which are listed in useful resources, below. Information about mobile phone interviewing can be found in Chapter 33, including a discussion on structured, semi-structured and unstructured interviews. Details of mobile phone survey methodology can be found in Chapter 34. Information about other types of computer-assisted interviewing is provided in Chapter 8 and details of computer-assisted qualitative data analysis software that can be used to analyse interview data is provided in Chapter 9.

Questions for reflection

Epistemology, theoretical perspective and methodology

- What type of online interview do you intend to conduct? This will depend on epistemology, theoretical perspective, methodology and topic, all of which will have an influence on the type of interview, the focus of the interview and the method. For example:

  - Positivists seek to understand and describe a particular phenomenon: structured online interviews can be used as part of a large scale survey to obtain attitudes and beliefs about a particular topic. Emphasis is placed on neutrality, technical detail and ensuring that experiential knowledge is transmitted from the interviewee to the interviewer.

  - Phenomenologists are interested in finding out how participants experience life-world phenomena: life-world online interviews (unstructured, in-depth interviews) can be used to obtain a detailed description of the lived world in a person's own words. The interview is reciprocal and informal.

  - Postmodernists focus on the construction of social meaning through language and discursive practices. Postmodern online interviewing is a conversation with diverse purposes, where interview roles are flexible and ever-changing. Different techniques are adopted, depending on postmodernist standpoint.

  - Feminists aim to conduct online interviews in a co-constructive, non-hierarchical, reciprocal and reflexive manner, although methods vary, depending on feminist standpoint. Women are encouraged to provide their own accounts in their own words with the interviewer only serving as a guide (or as a partner). Issues of rapport, trust, empathy and respect are highlighted. An awareness of gender relations is important during data collection and analysis.

- Do you intend to use online interviews as your only data collection method, or will you use other methods such as offline interviews or online questionnaires?

- Are you intending to interview participants about their online behaviour in an online setting, or are you intending to interview participants about real-life events, using online communication technologies? Salmons (2012) will help you to address these questions.

- What are the strengths and weaknesses of conducting synchronous online interviews for qualitative research? This question is discussed by Davis et al. (2004) in relation to their research on gay/bisexual men living in London.

- What sampling technique do you intend to use? Salmons (2012: 14–15) discusses this issue in relation to nomination (the identity of a participant is verified by another person), existing sampling frame (existing records such as membership lists) and constructed sample frames (created by the researcher).

Ethics, morals and legal issues

- How do you intend to obtain informed consent? Do all participants understand what is meant by informed consent? How can you ensure that participants understand the purpose, benefits and potential risks of taking part in your research and that they are given time to make their decision? What information do you need to provide (the goals of the research, the length and degree of commitment required, opt-in and opt-out clauses and issues of anonymity and confidentiality, for example)? This should encompass the whole research process including data collection, analysis and dissemination of results.

- How do you intend to protect data and keep data secure? Do you have a good understanding of relevant data protection legislation (such as the Data Protection Act 2018 in the UK)?

- Is the interview to take place in a public or private online space? If private, can you guarantee privacy? What happens about privacy when the interviewee chooses to undertake the interview in a physical public space on a mobile device, for example (Chapter 33)?

- How can you establish trustworthiness and credibility in online interviews? James and Busher (2006) provide a detailed discussion on this topic.

Practicalities

- Do you need to send equipment to participants, or will they use their own equipment (webcams, mobiles or laptops, for example).

- Are interviewers and interviewees competent in the use of equipment and software, or is training required? Have you (or other interviewers) practised with the equipment? This can include ensuring that you get the correct camera angle and are able to make 'eye contact' with your interviewee, or ensuring that speakers and microphones are placed in the right location and working correctly, for example.

- How can you ensure that both interviewer and interviewee have reliable technology that will not disrupt the interview process (good broadband connection, reliable recording and good sound and video quality, for example)? This is of particular importance when conducting synchronous interviews.

- What misunderstandings might occur in your online interviews (social, cultural, organisational or language-based misunderstandings, for example)? How can these be addressed?

- What issues are important when compiling an interview schedule? Chapter 34 lists pertinent issues on structure, length, relevance and wording of questions.

- How do you intend to establish rapport when you introduce yourself and begin the interviews? This can include issues of courtesy, respect, honesty, politeness, empathy and appropriate choice of communication technologies, for example.

Useful resources

There are a wide variety of digital tools and software packages available that can be used for different types of online interviewing. Some are free and are aimed at the general market, whereas others are aimed at businesses, market researchers and those interested in qualitative research. Some of these have significant costs attached. Examples available at time of writing include (in alphabetical order):

- BlogNog (www.blognog.com);

- Dub InterViewer (www.nebu.com/dub-interviewer-data-collection-software);

- FreeConference (www.freeconference.com);

- Google Hangouts (https://hangouts.google.com);

- GoToMeeting (www.gotomeeting.com/en-gb);

- Group Quality (http://groupquality.com/products/online-interviews);

- Isurus In-depth Interviews (http://isurusmrc.com/research-tools/qualitative-research-tools);

- ReadyTalk (www.readytalk.com);

- Skype (www.skype.com/en);

- VisionsLive V+ Online In-Depth Interviews (www.visionslive.com/platform/online-in-depth-interviews).

Key texts

Ayling, R. and Mewse, A. (2009) 'Evaluating Internet Interviews with Gay Men', *Qualitative Health Research*, 19(4), 566–76, first published April 1, 2009, 10.1177/1049732309332121.

Beusch, D. (2007) 'Textual Interaction as Sexual Interaction: Sexuality and/in the Online Interview', *Sociological Research Online*, 12(5), 1–13, first published September 1, 2007, 10.5153/sro.1569.

Brinkmann, S. and Kvale, S. (2015) *InterViews: Learning the Craft of Qualitative Research Interviewing*, 3rd edition. Thousand Oaks, CA: Sage Publications, Inc.

Davis, M., Bolding, G., Hart, G., Sherr, L. and Elford, J. (2004) 'Reflecting on the Experience of Interviewing Online: Perspectives from the Internet and HIV Study in London', *AIDS Care*, 16(8), 944–52, first published September 27, 2010, 10.1080/09540120412331292499.

Deakin, H. and Wakefield, K. (2014) 'Skype Interviewing: Reflections of Two PhD Researchers', *Qualitative Research*, 14(5), 603–16, first published May 24, 2013, 10.1177/1468794113488126.

Fritz, R. and Vandermause, R. (2018) 'Data Collection via In-Depth Email Interviewing: Lessons from the Field', *Qualitative Health Research*, 28(10), 1640–49, first published January 17, 2017, 10.1177/1049732316689067.

James, N. and Busher, H. (2006) 'Credibility, Authenticity and Voice: Dilemmas in Online Interviewing', *Qualitative Research*, 6(3), 403–20, first published August 1, 2006, 10.1177/1468794106065010.

Kitto, R. and Barnett, J. (2007) 'Analysis of Thin Online Interview Data: Toward a Sequential Hierarchical Language-Based Approach', *American Journal of Evaluation*, 28(3), 356–68, first published September 1, 2007, 10.1177/1098214007304536.

O'Connor, H. and Madge, C. (2004) 'Cyber-Mothers: Online Synchronous Interviewing Using Conferencing Software', *Sociological Research Online*, 9(2), 1–17, first published May 1, 2004, 10.1177/136078040400900205.

Salmons, J. (ed.) (2012) *Cases in Online Interview Research*. Thousand Oaks, CA: Sage Publications, Inc.

Seidman, I. (2013) *Interviewing as Qualitative Research: A Guide for Researchers in Education and the Social Sciences*, 4th edition. New York, NY: Teachers College Press.

CHAPTER 41

Online observation

Overview

Online observation is a research method that involves selective and detailed viewing, monitoring, acquisition and recording of online phenomena. This can include noticing facts, taking measurements and recording judgements and inferences. It involves access to the online phenomena that is being researched: textual encounters and exchanges in social networks, discussion groups, forums and micro-blogging sites; visual activities and behaviour in virtual worlds, videos or online games; and observations from controlled experiments that use questionnaires, visual images, websites and apps, for example. The way that online observation is used in research depends, in part, on theoretical perspective and methodology: positivists following the scientific method can use online observations to formulate and test hypotheses (Chapter 38), whereas ethnographers can use observations to describe, interpret and record online behaviour, relationships and interaction (Chapter 37), for example. In quantitative research where the goal is to provide quantifiable, reliable and valid data, online observation is structured, formal and systematic, following a set procedure or checklist. In qualitative research online observation does not follow a set, pre-defined procedure and can, instead, be open, unstructured, flexible and diverse. Careful and systematic recording of all online observation is required in both qualitative and quantitative research.

Online observation can be carried out overtly (direct or reactive) or covertly (unobtrusive or non-reactive). In overt observation participants know that they are part of a research project and have given informed consent. They might be observed during online experiments, during focus groups or within a virtual world or online community in which the researcher has declared their presence and purpose, for example. In covert observation people do not know that they are part of a research project and that their online behaviour, interaction, words and/or images are being observed. Researchers can observe search behaviour, key words, log files or interaction and behaviour in a virtual world or online community but do not declare their presence or purpose. There are ethical implications associated with both types of observation, and these are raised in the questions for reflection listed below.

There are different online observation methods, examples of which include:

- Online participant observation: this is a method used by ethnographers to study online environments. It enables them to provide interpretive and descriptive analyses of the symbolic and other meanings that inform the routine practices of online behaviour and activity. Participant observation can be carried out within any online community, network or group and can be either overt or covert (see Brotsky and Giles, 2007 for an example of covert online participant observation). Researchers immerse themselves into the community: the action is deliberate and intended to add to knowledge. Although most online participant observation is qualitative in nature, some researchers employ quantitative techniques (such as coding and counting responses) where appropriate, depending on methodological standpoint. Nørskov and Rask (2011) point out that there are four types of participant observer, (the complete participant, the participant-as-observer, the observer-as-participant and the complete observer), and go on to discuss each of these in relation to online participant observation.

- Eye-movement observation: this involves tracking eye movements of individuals during website use or when playing online games, for example. In this type of research participants know that they are taking part in a study: they have given informed consent, understand the purpose of the research and are aware of the eye-tracking tools that are being used by the researcher (integrated eye-trackers, peripheral eye-trackers or standalone eye trackers, for example). More information about eye-tracking research is provided in Chapter 19.

- Log file observation: this involves the observation and analysis of log file data (web server log files, mobile device logs, logs created by software installers and log files generated by operating systems or software utilities, for example). Log file observation and analysis is a covert, non-reactive method because participants are unaware that their search behaviour is being observed (unless they have agreed to install logging software for specific research purposes). More information about log file analysis can be found in Chapter 28.

- Observation of online focus groups: this involves the observation of interaction and behaviour in synchronous online focus groups (those taking place in real-time by video, web conferencing or in a chat room) and in asynchronous online focus groups (those taking place over several days such as bulletin board focus groups). Participants are aware that they are part of a research study: they have agreed to take part in the research and have provided informed consent. More information about online focus groups is provided in Chapter 39.

- Observation of online research communities and panels: this involves observation of the interaction, behaviour and activity of online research communities and panels. These are groups of people who have been targeted, invited and recruited into a public or private online community or panel to provide attitudes, opinions, ideas and feelings on specified issues, ideas or products over an extended period of time. Participants have given informed consent and are aware of the research and its purpose. More information about online research communities is provided in Chapter 44 and more information about online panel research is provided in Chapter 42.

- Observation in online experiments: this method of investigation is used to test hypotheses, make discoveries or test known facts. This type of observation can be overt (online controlled experiments that have an experiment group and a control group for which informed consent has been given, for example) or covert (online natural experiments or company-sponsored online experiments with unsuspecting users, for example). More information about online experiments is provided in Chapter 38.

- Observation of documents, text and images: this involves the systematic observation, acquisition and analysis of textual and visual online material. This can include social media communication, entries on micro-blogging sites, blogs, chat rooms, discussion groups, forums, email, photographs and video, for example. It can also include the use of digital archives relating to observation (see the Mass Observation Archive, below). This type of research tends to be covert or non-reactive, but can also include overt methods where specific text or images are produced, provided or requested for the research. Associated information can be found in

Chapter 12 (textual data mining), Chapter 23 (information retrieval), Chapter 16 (digital visual methods), Chapter 48 (researching digital objects) and Chapter 52 (social media analytics).

Examples of research projects that have used, assessed or critiqued online observation methods include an examination of the general and specific advantages and disadvantages of the observer roles in online settings, drawing on a study of online observation of a virtual community (Nørskov and Rask, 2011); a discussion about the re-engineering of participant observation for operation in an online pseudo-physical field (Williams, 2007); a study to examine the difficulties of obtaining informed consent online through a Facebook case study (Willis, 2017); a covert participant observation into the meanings of interaction in the 'pro-ana' online community to consider the beliefs of community members towards eating disorders and the processes of treatment and recovery (Brotsky and Giles, 2007); and a longitudinal observation of 20 individual Facebook users with type 1 and type 2 diabetes (Koteyko and Hunt, 2016).

If you are interested in online observation methods, it is useful to find out more about how observation techniques are used in research. Smart et al. (2013) provide a comprehensive guide, covering the historical development of observational methods and techniques, theoretical and philosophical understandings and assumptions and practical issues associated with conducting an observational study. You can also visit the relevant chapters listed above for more information related to specific online observation methods and techniques. It is important that you have a good understanding of ethical implications associated with online observation, and these are addressed in questions for reflection and useful resources, below.

Questions for reflection

Epistemology, theoretical perspective and methodology

- What observation techniques do you intend to adopt? These will be guided by theoretical perspective and methodology (a realist perspective that suggests there is a real world to observe, experimental research that manipulates and contrives in a systematic way or narrative approaches that tell a story, for example).

Online observation

- Do you intend to use online observation as your only data collection method, or do you intend to combine online methods with offline methods? Nørskov and Rask (2011) discuss this issue in relation to the level and type of researcher participation when observing online and offline.

- How can you understand the meaning of online observations and know how to interpret and analyse them? Are you interested in descriptive, evaluative or inferential observation, for example?

- In what way are online observations dependent on what online participants are willing to share? To what extent might online personas, identities, presentations and accounts be fabrications?

- How can you avoid observer bias in your research? Techniques can include setting aside preconceptions, personal judgement and personal bias when observing, for example.

Ethics, morals and legal issues

- Do you intend to undertake overt or covert online observation? What are the strengths and weaknesses, and ethical implications, of each? This can include, for example,

  o Overt:

    - helps to build trust and rapport;

    - displays honesty;

    - enables full and free informed consent to be obtained;

    - enables participants to opt in or out;

    - helps to protect participants from harm;

    - enables researchers to address and adhere to all rules, regulations and legal issues;

    - participants could change their behaviour if they know they are being observed.

  o Covert:

    - informed consent is not obtained;

    - there is the possibility of deception and/or harm;

    - participants cannot choose to opt in/out;

- groups that might not provide consent can be accessed;

- the observer effect can be avoided;

- there is a risk of detection with negative consequences (giving researchers a bad name and discouraging people from participating in future projects, for example).

- If informed consent is required, how do you intend to obtain it, in particular, if you intend to undertake covert online observation? Willis (2017) provides an interesting discussion on this topic in relation to obtaining observed data from Facebook.

- Do you intend to be honest about who you are and what you are doing? Are you prepared to lie if it means you can gain access to an online community? If so, how would you deal with any problems that may arise due to your deception?

- How much personal information are you willing to share, perhaps to encourage participants to trust you and share their own personal information?

- How can you ensure that there will not be repercussions for any member of the online community due to their involvement in your study? Could involvement lead to abuse, harassment or online trolling for you or your participants?

Practicalities

- Are there websites, text, images or log files that are not available for research purposes (due to their terms and conditions about data protection, for example)? What are the implications for your research?

- Do you have the required attributes and skills for successful online observation? Clear, structured and systematic observation is important for online experiments and immersion, concentration and focus are important for online participant observation, for example.

- How can you ensure that distractions do not disturb or influence your observations (technology, people and unexpected events, for example)?

- How will you deal with the unexpected when undertaking online observation? How can you ensure that you are open to new possibilities and unexpected results (setting aside expectations and prejudgments, for example)? Your reaction to, and recording of, unexpected events are important aspects of the observation process.

Useful resources

The British Psychological Society (BPS) (www.bps.org.uk) has produced useful ethical guidelines for internet-mediated research, which are useful for researchers who intend to carry out online observation (and other internet-based research): BPS (2017) *Ethics Guidelines for Internet-mediated Research*. British Psychological Society, Leicester. The document can be accessed on the BPS website.

The Mass Observation Archive (www.massobs.org.uk) preserves and presents to the public papers of the original Mass Observation research in Britain from 1937 to the 1950s and from the Mass Observation Project from 1981 onwards. It is kept and maintained by the University of Sussex and contains personal writing and observations of people's behaviour and conversations.

Key texts

Brotsky, S. and Giles, D. (2007) 'Inside the "Pro-Ana" Community: A Covert Online Participant Observation', *Eating Disorders*, 15(2), 93–109, 10.1080/10640260701190600.

Koteyko, N. and Hunt, D. (2016) 'Performing Health Identities on Social Media: An Online Observation of Facebook Profiles', *Discourse, Context & Media*, 12, 59–67, 10.1016/j.dcm.2015.11.003.

Nørskov, S. and Rask, M. (2011) 'Observation of Online Communities: A Discussion of Online and Offline Observer Roles in Studying Development, Cooperation and Coordination in an Open Source Software Environment', *Forum Qualitative Sozialforschung/ Forum: Qualitative Social Research*, 12(3), 10.17169/fqs-12.3.1567.

Smart, B., Peggs, K. and Burridge, J. (eds.) (2013) *Observation Methods*, (four volume set). London: Sage.

Williams, M. (2007) 'Avatar Watching: Participant Observation in Graphical Online Environments', *Qualitative Research*, 7(1), 5–24, first published February 1, 2007, 10.1177/1468794107071408.

Willis, R. (2017) 'Observations Online: Finding the Ethical Boundaries of Facebook Research', *Research Ethics*, first published November 11, 2017, 10.1177/1747016117740176.

CHAPTER 42

Online panel research

Overview

Online panel research is a research method that brings together a group of online research participants (or respondents) to answer questions on various topics over a period of time. The participants are pre-recruited and profiled: they volunteer to take part in surveys and provide demographic (and possibly lifestyle) information when they enrol for the panel. Some type of remuneration might be offered, which can include cash payments, entry into competitions, charitable donations, discount vouchers or gifts. There are various terms that can be used to describe this type of research including online panels, custom online panels, internet access panels, customer advisory panels, web panels and proprietary panels. Online research communities (Chapter 44) also recruit members for research purposes over an extended period of time, but differ in that they concentrate on qualitative methods and view individuals as community members, rather than as research participants.

Online panel research is used in market research, social/public attitude research, health research and psychological research. It is used to gain information about attitudes, opinions and behaviour of customers, potential customers, patients or members of the general public. In market research there are two main types of survey that are used in online panel research: B2C (business-to-consumer) and B2B (business-to-business). B2C surveys are targeted at the general consumer whereas B2B surveys are targeted at business professionals or those from specific occupations. In social and public attitude research a panel can consist of a large number of pre-recruited individuals who can be targeted altogether or in smaller groups, depending on the research question, aims and objectives. Health and psychological panels can consist of existing patients or clients, or of members of the public who express an interest in becoming a member of the panel.

Individuals can be selected for online panels in two ways: through probability sampling (probability-based panels) or through nonprobability sampling (volunteer nonprobability panels or opt-in panels). In probability samples, all individuals within the study population have a specifiable chance of being selected. These samples are used if the researcher wishes to provide explanations, predictions or generalisations. The sample serves as a model for the whole research population: therefore, it must be an accurate representation of this population. A popular method for online panels is to use address-based sampling, with candidates invited to join the panel. If candidates do not have internet access they are provided with a laptop or tablet, along with internet connectivity. Nonprobability samples (also referred to as purposive samples) are used if description rather than generalisation is the goal. In this type of sample it is not possible to specify the possibility of one individual being included in the sample. Individuals are recruited using methods such as online advertisements, recommendation from other websites or personal recommendation from friends and family who are already panel members. Once a pool of candidates exists, individuals are chosen for specific surveys based on their profile.

Once the sample has been worked out and individuals invited to participate, a number of data collection methods can be used. These include:

- online questionnaires (Chapter 43);

- mobile phone interviews and surveys (Chapters 33 and 34);

- online focus groups (Chapter 39);

- smartphone apps (Chapter 50);

- smartphone questionnaires (Chapter 51).

Examples of research projects that have used, assessed or critiqued online panel research include an assessment and analysis of the German Internet Panel, which is a longitudinal panel survey that collects political and economic attitudes (Blom et al., 2017); a study into online panel respondents' overall average relative response time (Vocino and Polonsky, 2014); a comparison of data quality between online panels and intercept samples (Liu, 2016); research into respondent conditioning in online panel surveys (Struminskaya, 2016); an investigation into the sampling discrepancies of the different stages of a probability-based online panel (Bosnjak et al., 2013); a study into the effects of lotteries and of offering study results on response behaviour (Göritz and Luthe, 2013); and research using web panels to understand whether online advertisement exposure influences information-seeking behaviour (Kim et al., 2012).

If you are interested in finding out more about online panel research, a good starting point is Callegaro et al. (2014). This edited volume is a comprehensive guide to the subject, aimed at opinion and market researchers, academic researchers interested in web-based data collection, governmental researchers, statisticians, psychologists and sociologists. Topics covered include coverage bias, nonresponse, measurement error, adjustment techniques to improve sample representativeness and motivations for joining an online panel. A useful book for market researchers is Postoaca (2006). Although this book could do with a little updating given rapid technological development, it still contains useful information on issues such as the language of online communication and the relationship between the researcher and online panellists.

There are a wide variety of commercial organisations providing online panel services. If you intend to use this type of service for your research it is important to visit some of their websites to find out more about services and costs. Also, you may find it beneficial to join a few panels, participate in some surveys and assess the quality of research service offered (poor services soon become apparent: panellists are invited to complete too many questionnaires; leading, ambiguous or double-barrelled questions are asked; questionnaires are given to panellists who cannot answer questions because they know nothing about the topic; questionnaires are far too long, for example). You may also find it useful to find out more about computer-assisted interviewing (Chapter 8), mobile phone interviews (Chapter 33), mobile phone surveys (Chapter 34), online focus groups (Chapter 39), online questionnaires (Chapter 43), online interviews (Chapter 40), smartphone app-based research (Chapter 50) and smartphone questionnaires (Chapter 51).

Questions for reflection

Epistemology, theoretical perspective and methodology

• Do you intend to use only online panels for your research, or could it benefit from the inclusion of offline panels? Blom et al. (2015) illustrate how the inclusion of offline households improves the representation of older people and women and, in another paper, illustrate that there are significant problems with coverage bias if offline households are excluded (Blom et al., 2017).

- What influence might self-selection bias have on your research? Individuals select themselves onto the panel and, once invited to take part in a survey, can decide whether or not to do so. Will self-selection bias affect your ability to represent your intended target population accurately (and make generalisations, if that is your goal)?

- What influence might sample composition bias have on your research? Bosnjak et al. (2013) consider this issue in relation to the first stage of the development of a probability-based online panel (selecting eligible participants), in the second stage (asking about willingness to participant in the panel) and in the third stage (participation in online surveys).

Ethics, morals and legal issues

- How can you ensure that online panellists are treated with respect and that membership does not cause harm or distress? This includes issues of data security, personal privacy, consent, confidentiality, anonymity, misrepresentation, survey content and dissemination of results, for example.

- If using commercial companies, how can you ensure that they are not selling a product under the guise of market research?

- If using a survey company, is it possible to ensure that they do not collect IP addresses of respondents (several companies have the facility for switching off this feature, for example)?

- Do you intend to offer some type of remuneration or material incentive to encourage panel members to take part? How might such incentives influence behaviour? Göritz (2008) discusses the long term effect of material incentives and Göritz and Luthe (2013) look at the effects of lotteries and offering study results to online panellists.

Practicalities

- Do you intend to use probability-based online panels or volunteer non-probability panels? Do you have a good understanding of the advantages and disadvantages of each? Daniel (2012) provides a useful introduction to those new to sampling, with Chapter 3 of his book offering advice about choosing between probability and nonprobability sampling.

- What data collection methods do you intend to use once your online panel has been established? If you are using an existing panel that can only be accessed through a commercial company, is there flexibility in method choice (including qualitative and quantitative techniques) and what costs are involved?

- How can you calculate response rate? In opt-in panels where there is no sampling frame (a list, source or device from which the sample is drawn) it is not possible to know the probability of selection from the population. 'Response rate' refers to the number of people who have taken part in the survey divided by the number of people in the sample. It is usually expressed as a percentage: the lower the percentage, the more likely some form of bias has been introduced into the research process. DiSogra and Callegaro (2016) provide useful advice on this topic.

- What data analysis methods do you intend to use? If you are working with a commercial company do they undertake the analysis and what costs are involved? If you are undertaking your own analysis, do you have access to relevant software and the required knowledge and understanding (or is training available)?

- How might your research be influenced by panel conditioning, where one wave of a survey has an effect on respondents' answers in subsequent waves? Halpern-Manners et al. (2017) provide an enlightening discussion on this issue.

Useful resources

There are a wide variety of commercial companies offering access to online panels, including market research companies and those specialising in attitude and opinion research. Some of these that are available at time of writing are listed below, in alphabetical order. It is useful to visit some of these sites so that you can find out a little more about the services that are offered, sampling techniques, data collection methods, data analysis techniques and costs.

- Borderless Access Panels (www.borderlessaccess.com);
- Critical Mix (www.criticalmix.com);

- OnePoll (www.onepoll.com/panel);

- Opinions 4 Good (Op4G) (http://op4g.com);

- Precision Sample (www.precisionsample.com);

- Quantability (www.quantability.com/solutions/field-data-collection);

- Research For Good (www.researchforgood.com);

- Skopos Insight Group (London) (http://skopos.london);

- SurveyMonkey Audience (www.surveymonkey.com/mp/audience);

- YouGov (https://yougov.co.uk).

Key texts

Blom, A., Gathmann, C. and Krieger, U. (2015) 'Setting up an Online Panel Representative of the General Population: The German Internet Panel', *Field Methods*, 27(4), 391–408, first published March 26, 2015, 10.1177/1525822X15574494.

Blom, A., Herzing, J., Cornesse, C., Sakshaug, J., Krieger, U. and Bossert, D. (2017) 'Does the Recruitment of Offline Households Increase the Sample Representativeness of Probability-Based Online Panels? Evidence from the German Internet Panel', *Social Science Computer Review*, 35(4), 498–520, first published June 2, 2016, 10.1177/0894439316651584.

Bosnjak, M., Haas, I., Galesic, M., Kaczmirek, L., Bandilla, W. and Couper, M. (2013) 'Sample Composition Discrepancies in Different Stages of a Probability-Based Online Panel', *Field Methods*, 25(4), 339–60, first published January 29, 2013, 10.1177/1525822X12472951.

Callegaro, M., Baker, R., Bethlehem, J., Goritz, A., Krosnick, J. and Lavrakas, P. (eds.) (2014) *Online Panel Research – A Data Quality Perspective*. Chichester: John Wiley & Sons Ltd.

Daniel, J. (2012) *Sampling Essentials: Practical Guidelines for Making Sampling Choices*. Thousand Oaks, CA: Sage Publications, Inc.

DiSogra, C. and Callegaro, M. (2016) 'Metrics and Design Tool for Building and Evaluating Probability-Based Online Panels', *Social Science Computer Review*, 34(1), 26–40, first published March 5, 2015, 10.1177/0894439315573925.

Göritz, A. (2008) 'The Long-Term Effect of Material Incentives on Participation in Online Panels', *Field Methods*, 20(3), 211–25, first published March 3, 2008, 10.1177/1525822X08317069.

Göritz, A. and Luthe, S. (2013) 'Lotteries and Study Results in Market Research Online Panels', *International Journal of Market Research*, 55(5), 611–26, first published September 1, 2013, 10.2501/IJMR-2013-016.

Halpern-Manners, A., Warren, J. and Torche, F. (2017) 'Panel Conditioning in the General Social Survey', *Sociological Methods & Research*, 46(1), 103–24, first published May 20, 2014, 10.1177/0049124114532445.

Kim, A., Duke, J., Hansen, H. and Porter, L. (2012) 'Using Web Panels to Understand whether Online Ad Exposure Influences Information-Seeking Behavior', *Social*

Marketing Quarterly, 18(4), 281–92, first published November 8, 2012, 10.1177/1524500412466072.

Liu, M. (2016) 'Comparing Data Quality between Online Panel and Intercept Samples', *Methodological Innovations*, first published October 12, 2016, 10.1177/2059799116672877.

Postoaca, A. (2006) *The Anonymous Elect: Market Research Through Online Access Panels*. Berlin: Springer.

Struminskaya, B. (2016) 'Respondent Conditioning in Online Panel Surveys: Results of Two Field Experiments', *Social Science Computer Review*, 34(1), 95–115, first published March 2, 2015, 10.1177/0894439315574022.

Vocino, A. and Polonsky, M. (2014) 'Career Overall Average Relative Response Time of Online Panellists', *International Journal of Market Research*, 56(4), 443–65, first published July 1, 2014, 10.2501/IJMR-2014-001.

CHAPTER 43

Online questionnaires

Overview

Online questionnaires are questionnaires that are administered over the internet to gather data about behaviour, experiences, attitudes, beliefs and values. They are used as a research method within a variety of methodologies including survey research (Mei and Brown, 2018), grounded theory (Jenkins et al., 2015) and ethnography (Murthy, 2008), and can be used as a standalone method (Cumming et al., 2007) or combined with other online or offline methods (Kamalodeen and Jameson-Charles, 2016). There are three basic types of online questionnaire: closed-ended (used to generate statistics in quantitative, survey research), open-ended (used in qualitative research to gather in-depth attitudes and opinions) or a combination of both (used to generate statistics and explore ideas, beliefs and/or experiences). Questionnaires can be embedded into websites or blogs, sent out as a bulk email to targeted groups (as attachments or unique links), shared on social networks or as QR Codes (Quick Response Codes) for respondents to scan with their mobile devices, for example. Online questionnaires can also be referred to as web-based questionnaires, web questionnaires, electronic questionnaires or e-questionnaires.

There are a wide variety of functions and features that can be incorporated into online questionnaires. These include:

- customisation: enabling a specific structure, style, content and layout for individual projects;
- collaboration: teams can work together through multi-user accounts (see Chapter 36);

- embedding: video, sound, photos, images, illustrations, logos and URL links can be embedded into the questionnaire;

- adaptability: the layout and size changes automatically to suit the device on which the questionnaire is being viewed;

- optimisation: for mobile devices (see Chapter 34);

- dynamic content: respondents can be addressed by name and personalised incentives offered;

- individualisation (customised logic): questionnaires can be tailored to users so that their experience is personalised;

- language: the questionnaire can be produced or translated into a number of different languages;

- branching (skipping or skip logic): irrelevant questions can be skipped based on previous answers;

- piping: selected answers from previous questions can be incorporated into other questions;

- randomisation: questions and options can be displayed in a different order each time;

- mandatory: a question must be answered before the respondent can move on to the next question;

- protection: password protection for participating in a survey, protection against multiple filling in of questionnaires and safe and secure protection of data (compliant with the General Data Protection Regulation, for example);

- notifications: researchers can be notified when a questionnaire is completed and provide instant feedback, if required;

- evaluation: enabling automatic, real-time evaluation, including cross-tabulations;

- filtering: results can be sorted and narrowed;

- downloading and exporting: answers can be downloaded or exported to a variety of statistical packages, spreadsheets and databases;

- sharing: files, answers or results can be shared with others through direct links;

- reporting: researchers can produce visualisations, build custom reports and generate web reports.

Online questionnaires are used in wide variety of disciplines and fields of study including health and medicine (Cumming et al., 2007), the social sciences (Murthy, 2008), education (Bebermeier et al., 2015), second language research (Wilson and Dewaele, 2010) and business and marketing (Brace, 2018). Examples of research projects that have used, assessed or critiqued online questionnaires include an assessment of the use of online question-naires to improve learning and teaching statistics (Bebermeier et al., 2015); a study to compare the length of responses to open-ended questions in online and paper questionnaires (Denscombe, 2008); research into women's attitudes to hormone replacement therapy, alternative therapy and sexual health, using questionnaires offered through a UK, patient-tailored and clinician-led meno-pause website (Cumming et al., 2007); an assessment of the use of web questionnaires in second language acquisition and bilingualism research (Wilson and Dewaele, 2010); and a study to evaluate five online questionnaire hosting platforms (Cândido et al., 2017).

If you are interested in using online questionnaires for your research, it is useful to view some of the websites of online survey and questionnaire design companies, as these give a good overview of styles, functions, design, optimi-sation and adaptability. Examples of these are listed below. It is also impor-tant that you have a thorough understanding of methodological issues surrounding survey research and an understanding of methods associated with questionnaire design, sampling, administration and analysis. More infor-mation about these issues can be found in Rea and Parker (2014), Brace (2018) and Gillham (2007). Two useful books that cover online questionnaires are Callegaro et al. (2015) and Toepoel (2016). Further relevant information can be found in Chapter 33, which discusses mobile phone interviews, and in Chapter 51, which discusses smartphone questionnaires. You might also find it useful to obtain more information about computer-assisted qualitative data analysis software (Chapter 9) and data visualisation (Chapter 13).

Questions for reflection

Epistemology, theoretical perspective and methodology

- What type of online questionnaire do you intend to use for your research? Your choice will be guided by epistemology and theoretical perspective. For example, objectivists who hope to describe an objective truth will employ a stimulus-response model, with standardised questions and answers in large scale surveys. Validity and reliability are extremely important in this type of survey (more information about the different types of questionnaire used in survey research can be found in Chapter 51 and more information about validity and reliability can be found in Chapter 38). On the other hand, constructionists suggest that meaning is constructed not discovered. Since it is meaning and the construction of meaning that are important, questionnaire design and the way they are administered is very different from survey research. Both researcher and respondent play an important role in helping to construct meaning and, as such, questionnaires can be fluid, flexible and adaptive. Issues of integrity, dependability, authenticity, credibility, trustworthiness and transparency are important in this type of research.

- How will you choose your sample? How will you take account of people without internet access, who cannot read, have literacy difficulties, are blind or partially sighted, do not have English as their first language or do not know how to use a computer, for example?

- How will you avoid or address issues of self-selection bias? This is bias that is introduced when respondents choose whether or not to complete an online questionnaire.

Ethics, morals and legal issues

- Have your respondents been told who the research is for and what will happen to the results? Have they been assured that you understand and will comply with relevant data protection legislation?

- Is your online questionnaire likely to cover sensitive topics? Some respondents may be more willing to answer these questions online, rather than face-to-face. However, you might be better asking an indirect question rather than a direct question in these circumstances. This could include asking the respondent to think about how other people might react to, or behave in, a given situation. Asking respondents to understand another person's perspective, or asking them how they see (or position) other people, can be a useful way to ask about sensitive topics in online questionnaires.

- What verification mechanisms are in place? Is it possible to know who has completed an online questionnaire and know whether they have completed more than one questionnaire (this might be a problem if cash or other valuable or desirable incentives are offered)? Sending out a unique link to the questionnaire in an email may help to address some of these issues: the link is unique to the email recipient, respondents are only able to complete one questionnaire, respondents can return to their questionnaire at a later date and the researcher can check which respondents have completed the questionnaire and send reminders, if required.

Practicalities

- What questionnaire design tools, templates, software packages or hosting platform do you intend to use? A brief evaluation of five hosting platforms is provided by Cândido et al. (2017).

- Is your online questionnaire relevant to the lives, attitudes and beliefs of the respondents? Are your intended respondents likely to complete the questionnaire (do they have the required technology, do they know how to use the technology, is the questionnaire well-constructed and does it work as intended, for example)?

- Are your instructions straightforward and have you been realistic about how long the online questionnaire will take to complete? Is there a logical flow to the questionnaire?

- Is your questionnaire adaptive? Will the layout and size change automatically to suit the size of screen and the device that is being used to view the questionnaire?

- Have you piloted (tested) your questionnaire with a sample of your research population? This will enable you to get an idea about expected response rates, data quality, data validity, data reliability, adaptability and the comprehensibility of questions. Your questionnaire may need to be modified as a result of your pilot study.

- What data analysis software, visualisation tools and methods of dissemination do you intend to use? Will your chosen software enable you to produce charts and graphs, export data to spreadsheets or your choice of statistics software and print results to PDFs, for example?

Useful resources

There are a wide variety of organisations that provide tools, software, templates and hosting platforms for designing and administering online questionnaires, some of which are open source and free to access, use and adapt, whereas others have significant costs attached. A snapshot of what is available at time of writing includes (in alphabetical order):

- Google Forms (www.google.com/forms/about);
- JD Esurvey (www.jdsoft.com/jd-esurvey.html);
- Kwick Surveys (https://kwiksurveys.com);
- LimeSurvey (www.limesurvey.org);
- Mobi-Survey (www.soft-concept.com/uk/mobi-survey-mobile-survey-system.htm);
- Qualtrics Research Core (www.qualtrics.com/research-core);
- Responster (www.responster.com);
- SnapSurveys (www.snapsurveys.com);
- Survs (https://survs.com);
- Typeform (www.typeform.com);
- Web Survey Creator (www.websurveycreator.com).

The WebSM website 'is dedicated to the methodological issues of Web surveys' (www.websm.org). You can find useful resources and information about web survey methodology and methods, including information about latest technology.

The European Survey Research Association (ESRA) is a membership organisation for people who are interested in survey research (www.europeansurveyresearch.org). Here you can access Survey Research Methods, which is the official journal of the ESRA. It is peer-reviewed and published electronically, with free and open access.

Key texts

Bebermeier, S., Nussbeck, F. and Ontrup, G. (2015) '"Dear Fresher…" – How Online Questionnaires Can Improve Learning and Teaching Statistics', *Psychology Learning & Teaching*, 14(2), 147–57, first published April 7, 2015, 10.1177/1475725715578563.

Brace, I. (2018) *Questionnaire Design: How to Plan, Structure and Write Survey Material for Effective Market Research*, 4th edition. London: Kogan Page.

Callegaro, M., Manfreda, K. and Vehovar, V. (2015) *Web Survey Methodology*. London: Sage.

Cândido, R., Perini, E., Menezes de Pádua, C. and Junqueira, D. (2017) 'Web-Based Questionnaires: Lessons Learned from Practical Implementation of a Pharmacoepidemiological Study', *F1000 Research*, 6(135), first published February 13, 2017, 10.12688/f1000research.10869.1.

Online questionnaires

Cumming, G., Herald, J., Moncur, R., Currie, H. and Lee, A. (2007) 'Women's Attitudes to Hormone Replacement Therapy, Alternative Therapy and Sexual Health: A Web-Based Survey', *Post Reproductive Health*, 13(2), 79–83, first published June 1, 2007.

Denscombe, M. (2008) 'The Length of Responses to Open-Ended Questions: A Comparison of Online and Paper Questionnaires in Terms of a Mode Effect', *Social Science Computer Review*, 26(3), 359–68, first published December 3, 2007, 10.1177/0894439307309671.

Gillham, B. (2007) *Developing a Questionnaire*, 2nd edition. London: Continuum.

Jenkins, K., Fakhoury, N., Marzec, M. and Harlow-Rosentraub, K. (2015) 'Perceptions of a Culture of Health: Implications for Communications and Programming', *Health Promotion Practice*, 16(6), 796–804, first published November 24, 2014, 10.1177/1524839914559942.

Kamalodeen, V. and Jameson-Charles, M. (2016) 'A Mixed Methods Research Approach to Exploring Teacher Participation in an Online Social Networking Website', *International Journal of Qualitative Methods*, first published March 1, 2016, 10.1177/1609406915624578.

Mei, B. and Brown, G. (2018) 'Conducting Online Surveys in China', *Social Science Computer Review*, 36(6), 721–34, first published September 8, 2017, 10.1177/0894439317729340.

Murthy, D. (2008) 'Digital Ethnography: An Examination of the Use of New Technologies for Social Research', *Sociology*, 42(5), 837–55, first published October 1, 2008, 10.1177/0038038508094565.

Rea, L. and Parker, A. (2014) *Designing and Conducting Survey Research: A Comprehensive Guide*, 4th edition. San Francisco, CA: Jossey-Bass.

Toepoel, V. (2016) *Doing Surveys Online*. London: Sage.

Wilson, R. and Dewaele, J. (2010) 'The Use of Web Questionnaires in Second Language Acquisition and Bilingualism Research', *Second Language Research*, 26(1), 103–23, first published March 31, 2010, 10.1177/0267658309337640.

CHAPTER 44

Online research communities

Overview

Online research communities are groups of people who have been targeted, invited and recruited into a public or private online community to provide attitudes, opinions, ideas and feelings on specified issues, ideas or products over an extended period of time. This research method is popular in market research, the service industries and product development. Online research communities differ from online research panels (Chapter 42) in that individuals are seen to be community members, rather than respondents, and the research adopts a greater variety of qualitative approaches, asking members to take part in community discussions, test hypotheses, develop ideas, pose questions, evaluate products and comment on service delivery, for example. Members interact with each other and the moderator via their computer, tablet and/or mobile phone. Researchers are able to observe this interaction, consider the ideas presented by members and record feelings about the topic(s) under discussion. They can also obtain real-time reactions to events and situations. The term 'online research community' can also refer to networking sites that enable research-ers to interact, collaborate and share findings: these and other online colla-boration tools are discussed in Chapter 36.

There are different types of online research community, which include:

- online customer communities (existing customers communicate and interact with each other, or commercial companies build a community of customers that can be targeted with specific advertisements, for example);

- online brand communities (enthusiasts of a brand communicate, interact, pose questions, and promote, share and submit ideas for brand development);

- market research online communities (a network of members take part in conversations and exercises over a period of time on specific projects or topics);

- online insight communities (targeted individuals give rapid and on-going feedback and actionable information to inform business decisions).

These online research communities can be open communities (public) or closed communities (private). Open communities are available for anyone who wishes to join the group: discussions are public and can be accessed by non-members. Those who wish to join can register their details and take part in discussions when they feel they wish to contribute. In closed communities members have been invited to join, usually based on previous answers to a questionnaire or through previous brand association. They may have completed a personal profile questionnaire to assess suitability and to provide information about motivational factors. Discussions are only available to members. Closed communities tend to be used for commercially-sensitive, private or confidential topics. Once an open or closed online research community has been established, a variety of methods can be used to engage members and collect data. This can include online focus groups (Chapter 39), online experiments (Chapter 38), online observation (Chapter 41) and online questionnaires (Chapter 43). It can also include discussion forums, ideas sessions, journals and diaries, blogs, live chat, polls, activity streaming, group-work, online collage and games (Chapter 47).

Examples of research projects that have used, assessed or critiqued online research communities include a study into consumers' motivations for participating in market research online communities (Bang et al., 2018); research into what constitutes the 'ideal participant' in closed online research communities (Heinze et al., 2013); a study into the relationship between customer and company in online communities (Paterson, 2009); research into why online brand communities succeed, the benefits that can be extracted from them and how negative information that emerges from these communities can be managed effectively (Noble et al., 2012); and a study into how viewing posts can affect visitors' intentions to join an online brand community (Zhou et al., 2013).

If you are interested in finding out more about online research communities Chapter 11 of Poynter (2010) provides a good starting point, covering the theory and practice of online research communities, quality and ethical issues, future directions and some interesting examples. If you are interested in online brand communities a comprehensive guide is provided by Martínez-López et al. (2016).

The book covers topics such as brand and the social web; conceptual approaches to community, virtual community and online brand community; types of virtual communities and online brand communities; and motivation for joining online brand communities. There are a wide variety of commercial organisations that provide online research community services. Some of these build market research online communities, whereas others provide platforms, software and apps that enable researchers or businesses to build their own online research communities. If you intend to use this type of service for your research it is important to visit some of their websites to find out more about services and costs. You may also find it useful to find out more about computer-assisted interviewing (Chapter 8), mobile phone interviews (Chapter 33), online focus groups (Chapter 39), online interviews (Chapter 40), smartphone app-based research (Chapter 50) and research gamification (Chapter 47).

Questions for reflection

Epistemology, theoretical perspective and methodology

- Do you intend to adopt qualitative, quantitative or a combination of approaches for your online community research? How will your theoretical perspective and methodological framework guide and inform these approaches? Is there scope for mixed methods and, if so, how will you combine and integrate qualitative and quantitative data? How can you ensure credibility of integration? How can you ensure that your mixed methods approach is transparent?

- Do you intend to concentrate only on online research communities or could your research benefit from the inclusion of offline research communities? Could you gain further insight by encouraging community members to meet researchers, attend events and get to know others in the community by meeting face-to-face? Would this help with motivation and engagement?

- What factors have motivated participants to become members of the online research community? How might these factors influence engagement, decisions and outcomes? Heinze et al. (2013) discuss typologies of open and closed online research community members, before going on to consider the social dynamics that drive these networks of individuals and the concept of social capital in relation to participant typology.

- What influence might self-selection bias have on your research? Individuals decide whether or not to become a member of an online research community (through invitation or through visiting a relevant site, for example) and, once invited to take part in a specific community activity, can decide whether or not to do so. Will self-selection bias affect your ability to represent your intended target population accurately?

Ethics, morals and legal issues

- If you intend to conduct research with an online research community that is operated by a market research company (and possibly a sponsoring company), how can you reconcile commercial with academic interests? Will commercially-sensitive data be withheld and how might this affect your research, for example?

- How can you improve the health of your online research community (encouraging and building loyalty, engagement and enjoyment and developing an emotional connection among members, for example)?

- How can you build trust among community members? Bilgrei (2018) discusses the issue of trust in relation to online drug communities (these are not online research communities, but the points raised in the paper have direct relevance to online research communities: collective identity, trustworthiness, competence and authenticity, for example).

- How should community members behave? Do you intend to produce a code of ethics or a code of behaviour? If you are using an existing online research community, what code of ethics/behaviour has been adopted by the operating company? Is it adequate (would it receive ethics committee approval, for example)?

- How secure is your online research community? How can you ensure privacy, in particular, in closed online research communities?

Practicalities

- What type of community do you need? How large a community is required (and what over-recruitment is required to account for dropout)? Over what duration is the community required (short-term communities can explore one particular issue, whereas a number of issues can be explored over the long-term)? Do you intend to establish your own community or use an existing community?

- If you are intending to build an online research community, how can you encourage members to be engaged and committed? Points to consider include:

  ○ simplicity: ensure that it is easy for members to take part (the technology is simple to use, communication strategies easy to adopt and there are clear instructions, for example);

  ○ communication: keep regular contact and encourage members to communicate with each other (within acceptable boundaries: see above);

  ○ interest: ensure that the right people are recruited and invited to attend (they are interested, have something to say, feel part of the community and can share their thoughts, ideas and feelings with like-minded community members, or with members who are interested in the topic);

  ○ encouragement: make sure that all contributions are acknowledged, thank members for contributions, provide gentle encouragement for non-contributors and understand how encouragement can be used as a motivator;

  ○ respect: listen and take opinions seriously and respect all views, opinions, wishes and feelings, encouraging others within the community to do the same (encourage respectful challenging of ideas and opinions, for example).

- How will you deal with inactive members? This can include emailing to say that membership will be removed after a certain period of time, and once members are removed, recruiting more members to take their place, for example.

- What training is required for new members (code of ethics/behaviour, introduction to technology and communication strategies, for example)? How will this training be undertaken? How can you ensure that new members feel part of the community and become active contributors?

Useful resources

There are a wide variety of commercial marketing and/or research companies that have existing online research communities that can be accessed by researchers or businesses. Other companies enable researchers to build online research communities using their platform, software or apps. Examples of both types that are available at time of writing are listed below, in alphabetical order.

- Angelfish Fieldwork (https://angelfishfieldwork.com/);

- Fresh Squeezed Ideas (www.freshsqueezedideas.com);

- Fuel Cycle (https://fuelcycle.com);

- Insight By Design (www.insight-by-design.com);

- Join the Dots (www.jointhedotsmr.com);

- KL communications (www.klcommunications.com);

- Mediacom (www.mediacom.com);

- My-Take (https://my-take.com);

- Recollective (www.recollective.com);

- Vision Critical (www.visioncritical.com).

Key texts

Bang, J., Youn, S., Rowean, J., Jennings, M. and Austin, M. (2018) 'Motivations for and Outcomes of Participating in Research Online Communities', *International Journal of Market Research*, 60(3), 238–56, first published May 22, 2018, 10.1177/1470785317744110.

Bilgrei, O. (2018) 'Broscience: Creating Trust in Online Drug Communities', *New Media & Society*, 20(8), 2712–27, first published September 7, 2017, 10.1177/1461444817730331.

Heinze, A., Ferneley, E. and Child, P. (2013) 'Ideal Participants in Online Market Research: Lessons from Closed Communities', *International Journal of Market Research*, 55(6), 769–89, first published November 1, 2013, 10.2501/IJMR-2013-066.

Martínez-López, F., Anaya-Sánchez, R., Aguilar-Illescas, R. and Molinillo, S. (2016) *Online Brand Communities: Using the Social Web for Branding and Marketing*. Cham: Springer.

Noble, C., Noble, S. and Adjei, M. (2012) 'Let Them Talk! Managing Primary and Extended Online Brand Communities for Success', *Business Horizons*, 55(5), 475–83, September–October 2012, 10.1016/j.bushor.2012.05.001.

Paterson, L. (2009) 'Online Customer Communities: Perspectives from Customers and Companies', *Business Information Review*, 26(1), 44–50, first published March 1, 2009, 10.1177/0266382108101307.

Poynter, R. (2010) *The Handbook of Online and Social Media Research: Tools and Techniques for Market Researchers*. Chichester: John Wiley & Sons.

Zhou, Z., Wu, J., Zhang, Q. and Xu, S. (2013) 'Transforming Visitors into Members in Online Brand Communities: Evidence from China', *Journal of Business Research*, 66 (12), 2438–43, December 2013, 10.1016/j.jbusres.2013.05.032.

CHAPTER 45

Predictive modelling

Overview

Predictive modelling refers to the process of creating a model that will forecast or predict outcomes, using techniques and methods such as data mining (Chapter 12), probability, mathematics and statistics. Models tend to predict future outcomes, although they can be used to make predictions about events that take place at any time. Applications of predictive modelling include weather forecasting; financial forecasting; predicting or detecting crimes; identifying suspects; identifying, targeting and interacting with customers or potential customers; identifying and hiring the 'best' employees; assessing risk in the insurance industry; and improving health care, for example. Predictive modelling is closely aligned to the fields of machine learning (Chapter 29), predictive analytics (Chapter 10) and computer modelling and simulation (Chapter 7).

Methods used to generate predictive models require two types of data: predictors (or inputs) and outcomes or outputs (the behaviour that is to be predicted). An appropriate statistical or mathematical technique is applied to the data to determine relationships, which are captured in the resulting model. This model can be applied to situations where the predictors are known, but where outputs are unknown. There are two broad classes of predictive models: parametric models that have a finite number of parameters, or known inputs, and nonparametric models that have a (potentially) infinite number of parameters, or many different inputs (parameters are flexible and not fixed in advance). It is also possible to have semi-parametric models that contain components from both of these models.

There are a wide variety of techniques and methods that can be used for predictive modelling. Examples include:

- simple linear regression that models the relationship between the independent (predictor) variable and the dependent (outcome or response) variable;

- multiple linear regression that models the relationship between one continuous dependent variable and two or more independent variables;

- decision trees that model the possible outcomes of a series of related choices;

- random forests that build an ensemble of decision trees for wider diversity;

- neural networks that review large volumes of labelled data to search for relationships and patterns (Chapter 29);

- Bayesian analysis that considers the probability of event occurrence and is suitable for predicting multiple outcomes (with associated propensity scores).

If you are interested in finding out more about these methods and techniques, along with a wide variety of others that can be used, comprehensive coverage can be found in Kuhn and Johnson (2013) and in Ratner (2017).

Predictive modelling is used by researchers from a number of disciplines and fields of study, including health and medicine (Cestari, 2013), social policy (Keddell, 2015), engineering (Van Buren and Atamturktur, 2012), the social sciences (Hofman et al., 2017), the environmental sciences (Hiestermann and Rivers-Moore, 2015), law and justice (Karppi, 2018), military and defence (King et al., 2016) and education (Essa and Ayad, 2012). Examples of research projects that have used, assessed or critiqued predictive modelling include a study into how predictive modelling promotes rethinking of the regulation of chemicals (Thoreau, 2016); a study that modelled and predicted the outcome of the 2016 referendum on the UK's continued membership of the EU across 380 local authorities (Manley et al., 2017); a comparison of the predictive capabilities of various models of wind turbine blades (Van Buren and Atamturktur, 2012); a study to assess the use of predictive models in urology (Cestari, 2013); the use of predictive models to predict counterinsurgent deaths (King et al., 2016); a critique concerning the ethics of predictive risk modelling in the Aotearoa/New Zealand child welfare context (Keddell,

2015); and a critique of the ethics, effectiveness and cultural techniques of predictive policing (Karppi, 2018).

If you are interested in finding out more about predictive modelling for your research, a good starting point is Kuhn and Johnson (2013), which is aimed at practitioners and covers the entire process of predictive modelling, from data pre-processing to measuring predictor importance. It also presents a comprehensive guide to statistical and mathematical techniques used for predictive modelling. Another comprehensive guide to the topic is provided by Ratner (2017) who makes the distinction between statistical data mining and machine learning data mining, and aims the book at data scientists (it may be a little complicated for those who are new to quantitative methodologies). A good understanding of data mining techniques (Chapter 12), data analytics (Chapter 10), big data analytics (Chapter 4) and data visualisation (Chapter 13) will help to build your understanding of predictive modelling. Other types of computer modelling are discussed in Chapter 7 and agent-based modelling and simulation is discussed in Chapter 1. The questions for reflection listed below will help you to think more about the suitability and use of predictive modelling for your research.

Questions for reflection

Epistemology, theoretical perspective and methodology

- Do you intend to use predictive modelling as a standalone method, or use with other methods by adopting a mixed methods approach? For example, predictive modelling is used to forecast outcomes (what, where and when), rather than provide an explanation (how and why). If you are interested in both explanation and prediction, a combination of methods would provide a greater depth of information.

- Do predictive models really work? When might models be wrong? How do we know that they are wrong, falsifiable or false?

- How neutral is modelling technology? How might predictive modelling (and data collected for modelling) be influenced by society, culture, history, politics and technology? Getting to grips with the field of critical data studies will help you to address this question. Richterich (2018) provides detailed information on ethics and critical data studies in relation to biomedical research and the *Big Data & Society* journal has produced a critical data studies special edition (see useful resources, below).

- Is it right to assume that historical data can be used to predict the future? What factors might affect validity and reliability of historical data and subsequent models (sampling errors, choice of variables, bias in labelling, choice of model and choice of statistical procedures, for example)?

- How might design choices (e.g. selection of the task, dataset and model) affect stated results? Hofman et al. (2017: 486), for example, illustrate 'that different researchers – each making individually defensible choices – can arrive at qualitatively different answers to the same question'.

Ethics, morals and legal issues

- Karppi (2018: 3), when discussing predictive policing, asks 'how ethical can a machine or a computational technique be? What is the role assigned for humans?' What relevance do these insights have to your research?

- How might predictive modelling reinforce power differentials and reduce trust? These issues, along with additional ethical concerns, are discussed by Keddell (2015) in relation to child abuse prevention in New Zealand.

- How might predictive modelling cause harm to individuals, groups or society, or lead to social injustice and discrimination? Richterich (2018) provides an enlightening discussion for researchers interested in these issues.

- How do individuals react to being modelled? Might they try to fool, thwart or disrupt modelling processes? If so, what are the implications for modelling research?

- How might predictive modelling outcomes and interpretations influence behaviour? When decisions are made based on models, how might this incentivise manipulation? Is it possible that your research might lead to changes in behaviour or to some type of manipulation?

Practicalities

- How can you ensure that data are accurate and reliable (if data are inaccurate or irrelevant then the model's results will be unreliable or misleading)? What methods can be used to clean data?

- Are you aware of common errors that can occur in predictive modelling, and do you know how to overcome such errors? This can include:
  - lack of knowledge about algorithms and techniques (ensure that you acquire the necessary knowledge and understanding: take courses, read, seek advice, practise);
  - bias in algorithm and technique choice (understand what is meant by bias, take time to reflect on choices, discuss choices with mentors/ peers, consider all alternatives);
  - mistakes in application and interpretation (ensure you have the appropriate training/practise; undertake rigorous cross-validation through partitioning, testing and evaluating; reflect on, and discuss, interpretations);
  - invalid or misleading predictions (ensure that inferences apply only to model variables/predictors, ensure correlation is not mistaken for causation, ensure that the outcomes are not perceived as more predictable than the model justifies).

Useful resources

There are a wide variety of digital tools and software packages available for those interested in predictive modelling. Some of these are automated predictive modelling tools designed for business analysts, whereas others provide platforms for developing predictive, descriptive and analytical models that are aimed at researchers from a variety of fields of study. Examples that are available at time of writing include (in alphabetical order):

- Board (www.board.com/en/capabilities-predictive-and-advanced-analytics);
- GMDH Shell (https://gmdhsoftware.com/predictive-modeling-software);
- IBM SPSS Modeler (www.ibm.com/uk-en/marketplace/spss-modeler);
- Infer (www.ignitetech.com/solutions/marketing/infer);
- MATLAB (https://uk.mathworks.com/products/matlab.html);
- Orange (https://orange.biolab.si);
- RapidMiner Auto Model (https://rapidminer.com/products/auto-model);
- Salford Predictive Modeler software suite (www.salford-systems.com/products/spm);
- scikit-learn (http://scikit-learn.org/stable);

- Weka (www.cs.waikato.ac.nz/~ml/weka).

Big Data & Society (http://journals.sagepub.com/home/bdsis) is an 'open access peer-reviewed scholarly journal that publishes interdisciplinary work principally in the social sciences, humanities and computing and their intersections with the arts and natural sciences about the implications of Big Data for societies'. A critical data studies special edition can be found at http://journals.sagepub.com/page/bds/collections/critical-data-studies [accessed January 21, 2019].

Key texts

Cestari, A. (2013) 'Predictive Models in Urology', *Urologia Journal*, 80(1), 42–45, first published February 14, 2013, 10.5301/RU.2013.10744.
Essa, A. and Ayad, H. (2012) 'Improving Student Success Using Predictive Models and Data Visualisations', *Research in Learning Technology*, 20, first published August 30, 2012, 10.3402/rlt.v20i0.19191.
Hiestermann, J. and Rivers-Moore, N. (2015) 'Predictive Modelling of Wetland Occurrence in KwaZulu-Natal, South Africa', *South African Journal of Science*, 111(7/8), first published July 27, 2015, 10.17159/sajs.2015/20140179.
Hofman, J., Sharma, A. and Watts, D. (2017) 'Prediction and Explanation in Social Systems', *Science*, 355(6324), 486–88, first published February 3, 2017, 10.1126/science.aal3856.
Karppi, T. (2018) '"The Computer Said So": On the Ethics, Effectiveness, and Cultural Techniques of Predictive Policing', *Social Media + Society*, first published May 2, 2018, 10.1177/2056305118768296.
Keddell, E. (2015) 'The Ethics of Predictive Risk Modelling in the Aotearoa/New Zealand Child Welfare Context: Child Abuse Prevention or Neo-Liberal Tool?', *Critical Social Policy*, 35(1), 69–88, first published July 28, 2014, 10.1177/0261018314543224.
King, M., Hering, A. and Aguilar, O. (2016) 'Building Predictive Models of Counter-insurgent Deaths Using Robust Clustering and Regression', *The Journal of Defense Modeling and Simulation*, 13(4), 449–65, first published May 6, 2016, 10.1177/1548512916644074.
Kuhn, M. and Johnson, K. (2013) *Applied Predictive Modeling*. New York, NY: Springer.
Manley, D., Jones, K. and Johnston, R. (2017) 'The Geography of Brexit – What Geography? Modelling and Predicting the Outcome across 380 Local Authorities', *Local Economy*, 32(3), 183–203, first published May 8, 2017, 10.1177/0269094217705248.
Ratner, B. (2017) *Statistical and Machine-Learning Data Mining: Techniques for Better Predictive Modeling and Analysis of Big Data*, 3rd edition. Boca Raton, FL: CRC Press.
Richterich, A. (2018) *The Big Data Agenda: Data Ethics and Critical Data Studies*. London: University of Westminster Press.
Thoreau, F. (2016) '"A Mechanistic Interpretation, if Possible": How Does Predictive Modelling Causality Affect the Regulation of Chemicals?', *Big Data & Society*, first published September 22, 2016, 10.1177/2053951716670189.
Van Buren, K. and Atamturktur, S. (2012) 'A Comparative Study: Predictive Modeling of Wind Turbine Blades', *Wind Engineering*, 36(3), 235–49, first published June 1, 2012, 10.1260/0309-524X.36.3.235.

CHAPTER 46

Qualitative comparative analysis

Overview

Qualitative comparative analysis is a systematic method used to analyse qualitative data that works through a process of logical comparison, simplification and minimisation. It is used to overcome problems that can occur when researchers try to make causal inferences based on a small number of cases. Although the technique was first described and named by Charles Ragin in 1987 (a newer edition of his book is available: Ragin, 2014) it has been included in this book due to the availability and use of software packages and digital tools for this type of analysis. QCA can be used to help researchers analyse data where numbers are too small for complex statistical analysis, but too large for in-depth qualitative analysis, or in situations where the number of cases is small, but the number of variables that might explain a given outcome is large. It enables researchers to discover patterns across cases, generate new themes and concepts, summarise and order qualitative data, while also allowing for generalisations (Pattyn et al., 2017; Rihoux and Ragin, 2009).

There are three variants of QCA:

- Crisp-Set Qualitative Comparative Analysis (csQCA): the original version of QCA in which conditions and outcomes are translated in binary terms (two values). A case can be in or out of a set.

- Fuzzy Set Qualitative Comparative Analysis (fsQCA): a development on the original version in which cases can have partial membership in a particular set. The two qualitative states of full membership and full non-membership are retained, but cases can be located in the interval between these two states (between 0 and 1).

- Multi-Value Qualitative Comparative Analysis (mvQCA): an alternative to the two versions described above that allows for multi-value conditions, where each category is represented by a natural number (see the TOS-MANA software package, listed below).

QCA was originally developed by Ragin for use in the political sciences. However, it is now used in a wide variety of disciplines including public health and medicine, leisure and tourism, environmental sciences, management studies, sports science, education, social sciences and psychology. Examples of research projects that have used, assessed or critiqued QCA include an analysis of the potential and challenges of applying QCA in an evaluation context (Pattyn et al., 2017); an exploration of the potential of QCA to assist with complex interventions (Thomas et al., 2014); a comprehensive evaluation of the use of QCA for the purpose of causal inference (Baumgartner and Thiem, 2017); research to measure and assess the strategic goals and potential determinants of performance among sport governing bodies from Belgium (Winand et al., 2013); a study into the economic recession, job vulnerability and tourism decision-making of Greek holiday makers (Papatheodorou and Pappas, 2017); and an assessment of how to address complex policy problems using an example of tackling high rates of teenage conceptions in England's most deprived local authority areas (Blackman et al., 2013).

If you are interested in finding out more about QCA for your research the best starting point is Ragin (2014) who provides comprehensive information on methodological issues concerning comparative social science, methods information about data reduction using Boolean algebra and examples of Boolean-based qualitative analysis. Although the book was originally written in 1987 and has not been updated fully to include modern tools and software, it provides essential background reading for those interested in QCA. You may find it useful to read Ragin (2000), Rihoux and Ragin (2009) and Ragin (2008) as these delve deeper into software packages and tools that are available for QCA. Schneider and Wagemann (2012) provide information about the basic principles of set theory and the applied practices of QCA, and provide a helpful chapter on potential pitfalls and suggestions for solutions. A useful practical guide to QCA with R is provided by Thiem and Duşa (2013a) and a comprehensive guide for those working within management studies is provided by Schulze-Bentrop (2013). There are various software packages and digital tools available for researchers who are interested in QCA: some of these are listed below. A more comprehensive guide can be obtained from COMPASSS (details below). More information about terms, techniques and methods involved in QCA is provided in questions for reflection, below. You may also find it useful to read more about other types of computer-assisted qualitative data analysis software and information about these can be found in Chapter 9.

Questions for reflection

Epistemology, theoretical perspective and methodology

- QCA is sometimes seen as a way to challenge the qualitative/quantitative divide in the social sciences (Cooper et al., 2012). Do you agree with this observation and, if so, why do you believe this to be the case?

- Lucas and Szatrowski (2014: 1) suggest that 'QCA fails even when its stated epistemological claims are ontologically accurate', concluding that 'analysts should reject both QCA and its epistemological justifications in favor of existing effective methods and epistemologies for qualitative research'. If you intend to use QCA for your research, how will you counteract this argument and justify your use of QCA? Rihoux and Ragin (2009) address critiques of QCA in Chapter 7 of their book.

- Is QCA correct for the purposes of causal inference? Baumgartner and Thiem (2017) undertake a comprehensive evaluation of the technique and conclude that 'QCA is correct when generating the parsimonious solution type' but that the method is 'incorrect when generating the conservative and intermediate solution type'. They suggest that researchers using QCA for causal inference in human sensitive areas such as public health and medicine 'should immediately discontinue employing the method's conservative and intermediate search strategies'.

- Ragin, in his introduction (2014: xxix), points out that there is a 'fundamental mathematical and conceptual difference, for example, between studying asymmetric set-relational connections and studying symmetric correlational connections'. What is this difference and how might this relate to your research?

- Do you intend to adopt an inductive/explorative mode of reasoning or integrate deductive elements? Do you intend to emphasize case knowledge or is your design condition oriented? How will you ensure internal and external validity? Thomann and Maggetti (2017) will help you to address these questions.

- How can you ensure that the strategies you use can be justified on empirical grounds (case-based knowledge) and/or on theoretical grounds and are not the result of opportunistic manipulation (Rihoux and Ragin, 2009: 49)?

- Which variant of QCA is most appropriate for your research? Do you have a good understanding of the strengths and weaknesses of each variant?

Ethics, morals and legal issues

- Do you have a good understanding of ethical (and procedural) practice within QCA? This can include:
  - avoiding manufacturing positive results and holding back negative results;
  - recognising and avoiding untenable assumptions;
  - recognising and avoiding measurement error;
  - addressing problems of limited diversity due to small samples;
  - recognising and avoiding pitfalls associated with necessity and sufficiency;
  - using QCA software with judgement and understanding;
  - paying attention to validity of analyses;
  - engaging in constructive QCA discussion and debate to improve methods.

- How reliable are QCA software packages or digital tools? Are they free of faults and bias? How can you check reliability and accuracy (using additional software, for example)? Can you trust the software?

- Do you have a good understanding of software licensing conditions and correct citing practices for any tools and software used for analyses?

Practicalities

- What tools and software do you intend to use for QCA? Thiem and Duşa (2013b) provide a useful review that highlights similarities and differences of QCA software.

- Do you have a good understanding of technical terms and jargon associated with QCA? Some examples of these include (Ragin, 2014):

- ○ truth table: the core element of the formal data analysis in QCA, which is a table of configurations (a given combination of conditions associated with a given outcome);

- ○ conjunctural causation: a condition will only produce a certain outcome when combined with other conditions;

- ○ equifinality: a certain outcome can be triggered by numerous, nonexclusive combinations of conditions;

- ○ contradictory configurations: identical configurations of conditions lead to different outcomes;

- ○ causal asymmetry: if the presence of a particular condition (or a combination of conditions) is relevant for the outcome, its absence is not necessarily relevant for the absence of the outcome;

- ○ Boolean minimisation: reducing a long, complex expression to a shorter, more restrained expression.

- Do you have a good understanding of the sequence of steps involved in the QCA method? This includes (Ragin, 2008, 2014):

  - ○ identifying an outcome of interest;

  - ○ producing a list of conditions that may be associated with that outcome;

  - ○ developing calibration metrics;

  - ○ calibrating the data;

  - ○ developing a truth table;

  - ○ identifying necessary conditions;

  - ○ using logic to minimise the truth table;

  - ○ arriving at pathways to the outcome;

  - ○ assessing pathways using two parameters of fit (consistency and coverage), for example.

Useful resources

Some of the digital tools and software packages available at time of writing for QCA are listed below, in alphabetical order.

- Fm-QCA (https://github.com/buka632/Fm-QCA);

- fs/QCA (www.socsci.uci.edu/~cragin/fsQCA/software.shtml);

- Kirq and acq (http://grundrisse.org/qca/download);

- QCApro (https://cran.r-project.org/web/packages/QCApro/index.html);

- QCA tools (https://CRAN.R-project.org/package=QCAtools);

- R QCA (https://cran.r-project.org/web/packages/QCA/index.html);

- Tosmana (www.tosmana.net).

COMPASSS (COMPArative Methods for Systematic cross-caSe analySis) is

a worldwide network bringing together scholars and practitioners who share
a common interest in theoretical, methodological and practical advancements in
a systematic comparative case approach to research which stresses the use of
a configurational logic, the existence of multiple causality and the importance of
a careful construction of research populations.

A useful bibliography of relevant books and journals can be found on this
site, along with a comprehensive list of QCA software and tools.

Key texts

Baumgartner, M. and Thiem, A. (2017) 'Often Trusted but Never (Properly) Tested:
    Evaluating Qualitative Comparative Analysis', *Sociological Methods & Research*, first
    published May 3, 2017, 10.1177/0049124117701487.
Blackman, T., Wistow, J. and Byrne, D. (2013) 'Using Qualitative Comparative Analysis to
    Understand Complex Policy Problems', *Evaluation*, 19(2), 126–40, first published
    April 12, 2013, 10.1177/1356389013484203.
Cooper, B., Glaesser, J., Gomm, R. and Hammersley, M. (2012) *Challenging the Qualita-
    tive-Quantitative Divide: Explorations in Case-Focused Causal Analysis.* London:
    Continuum.
Lucas, S. and Szatrowski, A. (2014) 'Qualitative Comparative Analysis in Critical
    Perspective', *Sociological Methodology*, 44(1), 1–79, first published July 25, 2014,
    10.1177/0081175014532763.
Papatheodorou, A. and Pappas, N. (2017) 'Economic Recession, Job Vulnerability,
    and Tourism Decision Making: A Qualitative Comparative Analysis', *Journal of
    Travel Research*, 56(5), 663–77, first published June 9, 2016, 10.1177/
    0047287516651334.
Pattyn, V., Molenveld, A. and Befani, B. (2017) 'Qualitative Comparative Analysis as an
    Evaluation Tool: Lessons from an Application in Development Cooperation', *American
    Journal of Evaluation*, first published August 28, 2017, 10.1177/1098214017710502.
Ragin, C. (2000) *Fuzzy-Set Social Science.* Chicago, IL: University of Chicago Press.
Ragin, C. (2008) *Redesigning Social Inquiry: Fuzzy Sets and Beyond.* Chicago, IL: Uni-
    versity of Chicago Press.

Ragin, C. (2014) *The Comparative Method: Moving beyond Qualitative and Quantitative Strategies.* Oakland, CA: University of California Press.

Rihoux, B. and Ragin, C. (2009) *Configurational Comparative Methods: Qualitative Comparative Analysis (QCA) and Related Techniques.* Thousand Oaks, CA: Sage.

Schneider, C. and Wagemann, C. (2012) *Set-Theoretic Methods for the Social Sciences: A Guide to Qualitative Comparative Analysis.* Cambridge: Cambridge University Press.

Schulze-Bentrop, C. (2013) *Qualitative Comparative Analysis (QCA) and Configurational Thinking in Management Studies.* Frankfurt am Main: Peter Lang.

Thiem, A. and Duşa, A. (2013b) 'Boolean Minimization in Social Science Research: A Review of Current Software for Qualitative Comparative Analysis (QCA)', *Social Science Computer Review*, 31(4), 505–21, first published March 6, 2013, 10.1177/0894439313478999.

Thiem, A. and Duşa, A. (2013a) *Qualitative Comparative Analysis with R: A User's Guide.* New York, NY: Spinger.

Thomann, E. and Maggetti, M. (2017) 'Designing Research with Qualitative Comparative Analysis (QCA): Approaches, Challenges, and Tools', *Sociological Methods & Research*, first published October 3, 2017, 10.1177/0049124117729700.

Thomas, J., O'Mara-Eves, A. and Brunton, G. (2014) 'Using Qualitative Comparative Analysis (QCA) in Systematic Reviews of Complex Interventions: A Worked Example', *Systematic Reviews*, 3(67), first published June 20, 2014, 10.1186/2046-4053-3-67.

Winand, M., Rihoux, B., Robinson, L. and Zintz, T. (2013) 'Pathways to High Performance: A Qualitative Comparative Analysis of Sport Governing Bodies', *Nonprofit and Voluntary Sector Quarterly*, 42(4), 739–62, first published May 16, 2012, 10.1177/0899764012443312.

CHAPTER 47

Research gamification

Overview

Research gamification is a research method that applies the principles of gaming, game mechanics, game features or elements of game design, to a non-game research situation or context. It involves modifying or redesigning existing digital methods (online questionnaires or mobile interviews and diaries, for example) or creating new methods (apps that utilise avatars, for example) to incorporate game mechanics or design elements. The intention is to increase motivation and engagement with research and improve quality and quantity of responses. In market research it is used increasingly by businesses in an attempt to improve marketing and increase profit (these intentions and uses can be seen as problematic: see questions for reflection and key texts, below). Gamification methods can also be used in a non-research context, such as by businesses and organisations to engage employees and increase performance, and in education to motivate and engage students and increase the enjoyment of learning (gamification of learning, gamified learning or game-based learning: Karagiorgas and Niemann, 2017; Landers, 2014; Markopoulos et al., 2015). A related area also covered in this chapter is gamification research, which seeks to assess, evaluate and critique gamification methods (see Kim and Werbach, 2016; Landers et al., 2018; Woodcock and Johnson, 2018 for examples).

Research gamification can involve the following techniques:

- changing the language, structure, layout and design of existing questionnaires;

- introducing competition (developing and sharing leader boards, for example);

- encouraging respondent interaction and sharing;

- introducing a reward system such as points or badges;

- creating levels or progress steps through which respondents move (with incentives or rewards for completing a level);

- using visual images or sound;

- using avatars and storylines that mimic videogames while collecting data;

- encouraging respondents to choose or create their own storylines or adventures;

- developing and using puzzles, word searches or card games;

- using projection techniques (respondents are asked to put themselves in the place of others or are given scenarios to work through);

- using personalisation techniques (respondents are asked to provide personal advice to the researcher or to others, for example);

- providing instant feedback such as informing respondents that an answer is 'correct' or providing a description of their 'personality type' after they have identified attributes from a given list.

Research gamification is used by researchers from a variety of disciplines and areas of study including business, marketing, customer satisfaction, civic engagement, psychology, sociology, politics and education. Examples of research projects that have used, assessed, critiqued or evaluated research gamification (or gamification) include an experiment to test four different styles of presentation of a single questionnaire: text only, decoratively visual, functionally visual and gamified (Downes-Le Guin et al., 2012); a review of peer-reviewed empirical studies on gamification to assess whether or not it works (Hamari et al., 2014); an assessment and critique of 'gamification-from-above', described as 'the optimization and rationalising of work practices by management', and 'gamification-from-below' described as 'a form of active resistance against control at work' (Woodcock and Johnson, 2018); research into how gamification can influence engagement on civic engagement platforms (Hassan, 2017); and a study into 'the motivational power of game design elements according to the theory of psychological need satisfaction' (Sailer et al., 2017).

If you are interested in research gamification for your research project it is useful first to find out more about gamification: a good starting point is Dymek and Zackariasson (2016). This book provides a collection of essays that cover both the theory and practice of gamification. Essays discuss the contribution that can be made to academic knowledge and consider the future of gamification in business settings. If you are interested in the practical application of gamification in learning, Kapp (2012) illustrates how gamification can be used to improve learning, retention and the application of knowledge. It can be read together with its companion guide (Kapp et al., 2014), which provides practical activities and worksheets aimed at students. A related area is that of 'serious games', which involves the creation of games that have a more 'serious' purpose and use learning theories and instructional design principles to maximise learning or training success: more information is provided in Chapter 20. If you intend to adapt, redesign and gamify existing questionnaires, a good understanding of questionnaire design and administration is important: Chapter 43 introduces online questionnaires and Chapter 51 introduces smartphone questionnaires. If you intend to develop, adapt and gamify mobile diaries or mobile phone interviews, more information about these methods can be found in Chapters 30 and 33.

Questions for reflection

Epistemology, theoretical perspective and methodology

- What are the philosophical underpinnings of modern gamification research? Landers et al. (2018) will help you to address this question.

- What is meant by gamification? Deterding (2018: 2–3) discusses the rhetoric of gamification. He first describes 'the rhetoric of choice architecture' that 'casts humans as strategic rational actors and games as information and incentive systems, informed by neoclassical and behavioral economics'. He goes on to describe 'the rhetoric of humanistic design' that 'views humans as inherently social, emotional, growth-oriented, meaning-making beings' and sees play as 'the paragon of human activity satisfying basic psychological needs such as competence, autonomy, relatedness, or meaning, which fuel motivation, enjoyment, and well-being'.

- How effective is gamification? Does it work? Is it successful in increasing motivation and engagement with research and improving quality and quantity of response? Hamari et al. (2014) will help you to address these questions.

- How can you adopt a more gamified approach without losing data integrity (the accuracy and consistency of data)? What are the implications for data validity? Bailey et al. (2015) address these issues in relation to market research.

Ethics, morals and legal issues

- How neutral is gamification? Woodcock and Johnson (2018: 543), for example, state that

  gamification is a tool … and it is therefore vitally important to identify who introduces it, uses it, and for what ends. Gamification is not a neutral tool, and as it is currently implemented has become complicit in supporting and even further developing the economic relations of neoliberal capitalism.

- What are the consequences of gamification on organisations, groups and individuals?

- Is there a possibility that gamification might manipulate or exploit? Can gamification cause harm or distress to participants, workers or learners? Kim and Werbach (2016) provide a detailed discussion on these issues.

- Is research gamification suitable for all members of the research population? Might attitudes towards, and effectiveness of, research gamification differ, depending on factors such as age, education, employment, literacy, technological and digital experience, ownership of technology, culture and country, for example?

Practicalities

- What gamification research methods do you intend to use for your research? Examples include:
  - mobile diaries that use gamification approaches (small sections, rewards and feedback) to encourage participants to record everyday behaviour, actions and thoughts (Chapter 30);
  - adaptive and interactive questionnaires (mobile or online) that use gamification features such as visual images, drag and drop, audio or video, slider scales and cards sorts, for example (Chapter 34, 43 and 51);
  - online interviews (Chapter 40) or mobile phone interviews (Chapter 33) featuring game elements that requires respondents to become more involved, use their imagination, seek and receive rewards or compete with others.

- Do you intend to introduce some type of reward system? If so, how effective might this be? Hamari (2013: 236), after conducting a field experiment on badges as rewards, concludes that the mere implementation of gamification mechanisms does not automatically lead to significant increases in use activity in the studied utilitarian service, however, those users who actively monitored their own badges and those of others in the study showed increased user activity.

- If you intend to offer instant feedback, what type will be offered and how will you do this? If respondents can answer your questions at any time, is someone available to respond, or do you intend to have automatic responses? As we have seen above, two feedback methods are positive feedback for correct answers and descriptions of personality type. How can answers be 'correct'? What is the model, or theoretical perspective used to develop a description of 'personality type'? How valid, reliable, accurate, independent and comprehensive are such theories?

Useful resources

There are various research gamification and gamification for learning platforms, tools and software packages available: some are aimed specifically at the business sector, some are aimed at the wider research community and some are aimed at teachers and trainers. Examples available at time of writing include (alphabetical order):

- Breakout Edu for critical thinking, collaboration, creativity and communication (www.breakoutedu.com);
- Datagame for turning surveys into games (http://datagame.io);
- Evoq Engage for online community engagement (www.dnnsoftware.com/products/evoq-engage/gamification);
- Gametize for engagement, advocacy and learning (https://gametize.com);
- Gamiware to increase motivation in software projects (http://gamiware.com);
- Influitive for customer engagement and advocacy (https://influitive.com);
- Mambo.IO for customer loyalty, employee engagement and learning (https://mambo.io);
- Qualifio for interactive marketing and data collection (https://qualifio.com);
- Socrative for engagement and assessment in the classroom (www.socrative.com);

- Tango Card for rewards and incentives (www.tangocard.com).

The Gamification Research Network (http://gamification-research.org) is a 'communication hub for researchers and students interested in studying the use of game design in non-game contexts'. You can access a repository of gamification publications, blogs and papers on this website.

Key texts

Bailey, P., Pritchard, G. and Kernohan, H. (2015) 'Gamification in Market Research: Increasing Enjoyment, Participant Engagement and Richness of Data, but What of Data Validity?', *International Journal of Market Research*, 57(1), 17–28, first published January 1, 2015, 10.2501/IJMR-2015-003.

Deterding, S. (2018) 'Gamification in Management: Between Choice Architecture and Humanistic Design', *Journal of Management Inquiry*, first published August 15, 2018, 10.1177/1056492618790912.

Downes-Le Guin, T., Baker, R., Mechling, J. and Ruyle, E. (2012) 'Myths and Realities of Respondent Engagement in Online Surveys', *International Journal of Market Research*, 54(5), 613–33, first published September 1, 2012, 10.2501/IJMR-54-5-613-633.

Dymek, M. and Zackariasson, P. (eds.) (2016) *The Business of Gamification: A Critical Analysis*. New York, NY: Routledge.

Hamari, J. (2013) 'Transforming Homo Economicus into Homo Ludens: A Field Experiment on Gamification in a Utilitarian Peer-to-Peer Trading Service', *Electronic Commerce Research and Applications*, 12(4), 236–45, 10.1016/j.elerap.2013.01.004.

Hamari, J., Koivisto, J. and Sarsa, H. (2014) 'Does Gamification Work? A Literature Review of Empirical Studies on Gamification', in *Proceedings of the 47th Hawaii International Conference on System Sciences*, Hawaii, USA, January 6–9, 2014.

Hassan, L. (2017) 'Governments Should Play Games: Towards a Framework for the Gamification of Civic Engagement Platforms', *Simulation & Gaming*, 48(2), 249–67, first published December 19, 2016, 10.1177/1046878116683581.

Kapp, K. (2012) *The Gamification of Learning and Instruction: Game-Based Methods and Strategies for Training and Education*. San Francisco, CA: John Wiley & Sons, Inc.

Kapp, K., Blair, L. and Mesch, R. (2014) *The Gamification of Learning and Instruction Fieldbook: Ideas into Practice*. San Francisco, CA: John Wiley & Sons, Inc.

Karagiorgas, D. and Niemann, S. (2017) 'Gamification and Game-Based Learning', *Journal of Educational Technology Systems*, 45(4), 499–519, first published May 17, 2017, 10.1177/0047239516665105.

Kim, T. and Werbach, K. (2016) 'More than Just a Game: Ethical Issues in Gamification', *Ethics and Information Technology*, 18(2), 157–73, first published May 12, 2016, 10.1007/s10676-016-9401-5.

Landers, R. (2014) 'Developing a Theory of Gamified Learning: Linking Serious Games and Gamification of Learning', *Simulation & Gaming*, 45(6), 752–68, 10.1177/1046878114563660.

Landers, R., Auer, E., Collmus, A. and Armstrong, M. (2018) 'Gamification Science, Its History and Future: Definitions and a Research Agenda', *Simulation & Gaming*, 49(3), 315–37, first published May 21, 2018, 10.1177/1046878118774385.

Research gamification

Markopoulos, A., Fragkou, A., Kasidiaris, P. and Davim, J. (2015) 'Gamification in Engineering Education and Professional Training', *International Journal of Mechanical Engineering Education*, 43(2), 118–31, first published June 18, 2015, 10.1177/0306419015591324.

Sailer, M., Hense, J., Mayr, S. and Mandl, H. (2017) 'How Gamification Motivates: An Experimental Study of the Effects of Specific Game Design Elements on Psychological Need Satisfaction', *Computers in Human Behavior*, 69, April 2017, 371–80, 10.1016/j.chb.2016.12.033.

Woodcock, J. and Johnson, M. (2018) 'Gamification: What It Is, and How to Fight It', *The Sociological Review*, 66(3), 542–58, first published August 21, 2017, 10.1177/0038026117728620.

CHAPTER 48

Researching digital objects

Overview

Digital objects are sequences of units of data (bits) that exist on a particular medium. They include properties (characteristics or attributes) and methods (procedures and techniques for performing operations on objects). Some of these are born digital, originating as digital objects, whereas others are digitised, presenting and recording attributes and characteristics of physical objects. Both are encoded information with which software interacts so that the bit sequences can be seen and understood as representations on screen. These representations of digital objects exist within digital space and can include accounts and narratives, songs, music, images, photographs, artworks, video games, virtual worlds and environments (Chapter 56), social networks (Chapter 53), interactive media, simulation and modelling (Chapter 7), commercial videos, journal articles, research reports and datasets. The term can also refer to materials used in the production of, or documentation related to, representations of digital objects (data visualisations or written descriptions of artwork, for example). 'Digital consumption objects' is a term that has been used to refer specifically to personal objects that exist in the digital space such as personal photographs, social network profiles, wish lists, playlists, avatars, text messages and blogs, for example (Watkins, 2015). 'Digital cultural objects' refers to objects that have been created by humans that are preserved digitally (Foster and Rafferty, 2016). As such, digital objects, digital consumption objects and digital cultural objects are made up of an abundance of voices, works and contributions that are stored, managed and preserved in the digital environment.

Researching digital objects is carried out in a number of disciplines and fields of study including library and information science, information retrieval (Chapter 23), digital curation and preservation, art history, the humanities, the social sciences, computer sciences, education and media studies. Examples of digital object research projects include an investigation into how digital contact networks can provide platforms for effective engagement and digital reciprocation of digital objects (Hogsden and Poulter, 2012); research into the affective qualities of digital objects and the impact of digitisation initiatives on museum practice and on people's engagements with the past (Newell, 2012); a comparison of descriptions of digital objects by multiple expert communities (Srinivasan et al., 2010); the development of a framework for 'the inventorying, cataloging, searching and browsing of multimedia digital objects related to ICH (Intangible Cultural Heritage)' (Artese and Gagliardi, 2015); an exploration of sentimental objects in a virtual world (Gibson and Carden, 2018); and an assessment of how different systems of communication generate different ethical codes concerning digital objects (Vuorinen, 2007).

If you are interested in finding out more about researching digital objects, Foster and Rafferty (2016) provide a collection of essays that are divided into three parts: 1) analysis and retrieval of digital cultural objects; 2) digitising projects in libraries, archives and museums (case studies); 3) social networking and digital cultural objects (if you are interested in social networking, more information about social network analysis is provided in Chapter 53). For deeper insight into the philosophy and theory of digital objects, and the existential structure of digital objects, see Hui (2016). Adams and Thompson (2016) provide an enlightening discussion on the relationship between humans and digital objects, and on the ethical and political implications of this relationship. A good understanding of information retrieval methods (Chapter 23) and digital visual methods (Chapter 16) will be of benefit to your research. If your chosen digital objects contain audio or video content that require analysis, further information can be found in Chapter 2 (audio analysis) and Chapter 55 (video analysis). Digital objects can be identified by a persistent identifier or handle called a Digital Object Identifier (DOI). These tend to be used for academic papers, books, ebooks, ejournals, datasets, spreadsheets, government publications, EU publications, tables and graphs produced by the Organisation for Economic Co-operation and Development (OECD) and some commercial video content. More information about the DOI system can be obtained from the International DOI Foundation website (details below). Other useful websites are listed below and in Chapter 23.

Questions for reflection

Epistemology, theoretical perspective and methodology

- What is a digital object? Hui (2012: 380) addresses this question and will help you to make sense of the 'media-intensive milieu comprising networks, images, sounds, and text'.

- What is your theoretical perspective and methodological framework when researching digital objects? Adams and Thompson (2016), for example, provide an interesting discussion on interviewing digital objects in which they draw from Actor Network Theory, phenomenology, postphenomenology, critical media studies and related sociomaterial approaches.

- What is the relationship of digital objects to physical (natural, technical or material) objects?

- What are the advantages and disadvantages of digitising objects for those who have different interests in such objects (historical researchers, museum visitors or clan members, for example)? Newell (2012) will help you to address this question.

- In what way can digital objects enable people to engage with the past? Does this differ, depending on the object and the viewer?

- How can curators present different and possibly conflicting perspectives in such a way that the tension between them is preserved? This question is raised and discussed by Srinivasan et al. (2010).

- What ravages might time wreak on digital data? What are the implications for trust in digital objects and the organisations that own or manage such data? These questions are raised and addressed by Macdonald (2006) in her paper relating to pharmaceuticals.

Ethics, morals and legal issues

- To what ethical codes are digital objects subject? Vuorinen (2007: 27) notes that 'a copy-protected digital object is simultaneously an utmost example of the hidden source code (the open/free system), a perfect artefact that can be owned and sold (the proprietary system) and a challenge to be cracked (the cracker system)'. These systems 'generate a diversity of ethical codes as they give different shapes to digital objects'.

- What are the ethical implications associated with the manipulation of digital objects? Mindel (2016: 179) points out that such manipulation is done to aid users in accessing information: it is well-intended and not meant to be misleading or deceptive. He goes on to discuss ethical considerations and start a 'larger conversation about post-processing standardization'.

- Does preservation of a digital object imply authenticity? Is preservation an endorsement of authenticity?

- In what ways can digital objects help to reveal, share and preserve non-Western perspectives?

Practicalities

- How will you find, access and understand digital objects? How can you assess authenticity and determine whether or not the digital object is the same as when it was created? This can include issues such as authorship, provenance and originality, for example.

- What impact might preservation processes have on digital objects that are accessed for your research? Do these processes affect the authenticity or accuracy of digital objects? See Chapter 11 for more information about data collection and conversion and problems that can occur within these processes (lost data or inexact data, for example).

- How can you respect the order, structure and integrity of digital objects (when copying representations, for example)?

- What impact might degradation, damage, vulnerability, obsolescence or loss of access to digital objects have on your research?

- Do you have a good understanding of digital object file types and formats? Examples include:
  - Still image files such as text documents and photographs from original copies:

    - DOC (document file);

    - EPS (Encapsulated PostScript);

    - GIF (Graphics Interchange Format);

    - JPEG (Joint Photographic Experts Group);

- PDF (Portable Document Format);
- PNG (Portable Network Graphics);
- TIFF (Tagged Image File Format).

○ Video files produced from either original analogue or digital video formats:

- ASF (Advanced System Format);
- AVI (Audio Video Interleaved);
- FLV (Flash Video);
- MOV (Apple QuickTime Movie);
- MP4 (MPEG-4 Part 14);
- MPEG-1 or MPEG-2 (Moving Picture Experts Group);
- WMV (Windows Media Video).

○ Audio files produced from original analogue or digital audio formats:

- AAC (Advanced Audio Coding);
- AIFF (Audio Interchange File Format);
- MP3 (MPEG-1 Audio Layer 3);
- WAV (Waveform Audio File Format);
- WMA (Windows Media Audio).

Useful resources

The Digital Curation Centre in the UK (www.dcc.ac.uk) is 'an internationally-recognised centre of expertise in digital curation with a focus on building capability and skills for research data management'. You can find the Curation Lifecycle Model on the website, which 'provides a graphical, high-level over-view of the stages required for successful curation and preservation of data from initial conceptualisation or receipt through the iterative curation cycle'.

The website of the International DOI Foundation (www.doi.org) provides Digital Object Identifier (DOI) services and registration. This is a 'technical and social infrastructure for the registration and use of persistent interoper-able identifiers, called DOIs, for use on digital networks'. DOIs can be assigned to any entity (physical, digital or abstract). A useful handbook and practical factsheets can be found on the website.

Cordra (www.cordra.org) 'provides techniques for creation of, and access to, digital objects as discrete data structures with unique, resolvable identifiers'. Useful documents and a 'getting started' guide can be found on the website, along with open source software. Cordra is a core part of the Corporation for National Research Initiatives Digital Object Architecture (www.cnri.reston.va.us), 'which offers a foundation for representing and interacting with information in the Internet'.

The Open Preservation Foundation (http://openpreservation.org) 'sustains technology and knowledge for the long-term management of digital cultural heritage'. You can find information about relevant products and software on this site, including information about fido (Format Identification for Digital Objects), which is a command-line tool to identify the file formats of digital objects.

Key texts

Adams, C. and Thompson, T. (2016) *Researching a Posthuman World: Interviews with Digital Objects.* London: Palgrave Pivot.

Artese, M. and Gagliardi, I. (2015) 'A Multimedia System for the Management of Intangible Cultural Heritage', *International Journal of Heritage in the Digital Era*, 4(2), 149–63, first published June 1, 2015, 10.1260/2047-4970.4.2.149.

Foster, A. and Rafferty, P. (eds.) (2016) *Managing Digital Cultural Objects: Analysis, Discovery and Retrieval.* London: Facet Publishing.

Gibson, M. and Carden, C. (2018) 'Sentimental Objects'. In *Living and Dying in a Virtual World: Digital Kinships, Nostalgia, and Mourning in Second Life*, 107–25. Cham: Palgrave Macmillan.

Hogsden, C. and Poulter, E. (2012) 'The Real Other? Museum Objects in Digital Contact Networks', *Journal of Material Culture*, 17(3), 265–86, first published September 11, 2012, 10.1177/1359183512453809.

Hui, Y. (2012) 'What Is a Digital Object?', *Metaphilosophy*, 43(4), 380–95, first published July 16, 2012, 10.1111/j.1467-9973.2012.01761.x.

Hui, Y. (2016) *On the Existence of Digital Objects.* Minneapolis, MN: University of Minnesota Press.

Macdonald, A. (2006) 'Digital Archiving, Curation and Corporate Objectives in Pharmaceuticals', *Journal of Medical Marketing*, 6(2), 115–18, first published March 1, 2006, 10.1057/palgrave.jmm.5050030.

Mindel, D. (2016) 'Approaches and Considerations Regarding Image Manipulation in Digital Collections', *IFLA Journal*, 42(3), 179–88, first published September 27, 2016, 10.1177/0340035216659300.

Newell, J. (2012) 'Old Objects, New Media: Historical Collections, Digitization and Affect', *Journal of Material Culture*, 17(3), 287–306, first published September 11, 2012, 10.1177/1359183512453534.

Srinivasan, R., Becvar, K., Boast, R. and Enote, J. (2010) 'Diverse Knowledges and Contact Zones within the Digital Museum', *Science, Technology, & Human Values*, 35(5), 735–68, first published May 21, 2010, 10.1177/0162243909357755.

Vuorinen, J. (2007) 'Ethical Codes in the Digital World: Comparisons of the Proprietary, the Open/Free and the Cracker System', *Ethics and Information Technology*, 9(1), 27–38, first published online November 23, 2006, 10.1007/s10676-006-9130-2.

Watkins, R. (2015) *Digital Possessions: Theorising Relations between Consumers and Digital Consumption Objects*. Doctoral Thesis, University of Southampton, Southampton Business School, retrieved from https://eprints.soton.ac.uk/id/eprint/381667.

CHAPTER 49

Sensor-based methods

Overview

Sensors are devices that are used to detect, measure and respond to some type of input (or change in input) from machinery, people or the environment (touch, pressure, heat, moisture or motion, for example). Although sensors have a long pre-digital history they are included in this book because of the rapid growth of interconnected devices, wireless sensor networks, intelligent sensor networks, smart sensors and developments in software-designed sensing that employs artificial intelligence, machine learning (Chapter 29) and predictive analytics (Chapter 45). Examples of different types of sensor include the following (sensors that are incorporated into wearable devices are listed in Chapter 57):

- accelerometers to measure acceleration, tilt and vibration;

- chemical sensors to measure chemical composition and/or activity;

- environment sensors to measure environmental conditions and phenomena (these are often multi-function sensors that include sensors to measure temperature, humidity and air flow, for example);

- fluid velocity or flow sensors to measure the flow rate, speed or quantity of liquid, gas or vapour;

- humidity sensors to measure moisture and air temperature, or relative humidity;

- infrared sensors (IR sensors) to measure, emit or detect infrared light;

- light sensors to measure light, light density or light intensity;

- optical sensors to measure and translate a physical quantity of light (reflective optical sensors detect objects and measure their distance);

- position sensors to measure position and movement (including linear, angular, tilt, rotary and multi-axis);

- pressure sensors to measure, control and monitor the pressure of gas or liquids;

- proximity sensors to measure the presence and distance of nearby objects;

- smoke, gas and alcohol sensors to detect and determine the presence of smoke, gas or alcohol;

- temperature sensors to detect, measure and maintain temperature;

- touch sensors to detect the location, force and pressure of touch;

- ultrasonic sensors to detect and measure distances and positions, and detect solids, powders or liquids.

There are a wide range of applications for sensors including habitat, infrastructure and climate monitoring (Polastre et al., 2004); temperature and pressure monitoring and metering for the oil, gas and water industry (Khan et al., 2016); and automotive engines and transmissions research, development and monitoring (Ranky, 2002). Sensors are also used to track and digitise human activity, providing information about people and the world in which we live (measuring environmental, behavioural, physiological and/ or psychological phenomena). For example, in social research sensors enable researchers to measure and interpret non-verbal behaviour that is difficult to observe (Pentland, 2008) and track and record movement and location (Chapters 21 and 27). In health and medical research sensors can be used to monitor health and provide detailed insight into health behaviour and risks to inform health promotion, self-care and medical intervention (Pagkalos and Petrou, 2016). They can also be used to measure contact networks for infectious disease transmission (Smieszek et al., 2014). In education sensors can be used to research language learning through spoken interaction (Seedhouse and Knight, 2016) and to encourage effective, efficient and flexible learning (Zhu et al., 2016).

If you are interested in finding out more about sensors for your research, Fraden (2016) provides a good starting point, covering theory, design and practicalities of sensor use and design in industry, science and consumer applications. If you are interested in chemical sensors and biosensors, Banica (2012) provides a comprehensive and detailed guide, covering theory and basic principles. There are a wide variety of sensors on the market, and care should be taken to choose the most appropriate for your research project, with close attention paid to reliability, accuracy, precision and suitability. Choices also depend on the type of sensor required for your research: wearables need to be small, unobtrusive, flexible and comfortable; medical imaging sensors require high resolution and the ability to capture fine detail; pressure sensors need an appropriate range (and the ability to cope with spikes). It is also important to ensure that interfaces between transmitted data, software for analysis and archiving systems are as seamless as possible, and that you have the necessary training and experience to use the equipment and software. Advice about choosing and using sensors is provided by the organisations and books listed below. More information about wearables-based research, including examples of wearable sensors and devices, is provided in Chapter 57. A discussion on location awareness and location tracking is provided in Chapter 27 and more information about mobile methods can be found in Chapter 32.

Questions for reflection

Epistemology, theoretical perspective and methodology

- Are sensor-based measurements objective and true?

- How accurate are sensor-based measurements? How can you ensure that sensors are measuring what they are supposed to be measuring?

- How reliable are sensor-based measurements? Would measurements be the same in repeated trials?

- How can you improve the validity of sensor-based measurements? How well do sensor-based methods measure what they purport to measure? See Geraedts et al. (2015) for examples of video validation and user evaluation of their sensor-based research.

- How can you ensure that inferences made on the basis of measurements are appropriate?

- Is it possible to relate sensor data to existing social theory?

- Do you intend to use sensor-based methods as a standalone research method, or adopt a mixed methods approach? Would a mixed methods approach provide a more complete understanding? This could include qualitative data gathered from participants and quantitative data gathered from sensors, for example. Is it possible to combine data from these different sources (checking sensor-detected movement against self-reported activity, for example)? If you do intend to adopt a mixed methods approach, how will you deal with what could, potentially, be large amounts of both qualitative and quantitative data? See Smieszek et al. (2014) for an example of how contact surveys and wireless sensor networks were used together to collect contact network data and Geraedts et al. (2015) for an example of how sensor-based methods were used together with video observation and a user evaluation questionnaire.

Ethics, morals and legal issues

- What data protection measures are in place? What encryption and security measures are in place for the sensors, software and wireless networks you intend to use? How can you protect against interception and insertion of rogue data? Zhong et al. (2019) provide comprehensive advice about security and privacy that will help you to address these issues.

- If you are intending to collect, transmit and process personal data, are you aware of, and able to comply with, current data protection legislation? More information about data protection can be obtained from the International Conference of Data Protection and Privacy Commissioners (https:// icdppc.org).

- If you are tracking and digitising human activity, how will you obtain informed consent? If you are dealing with vulnerable groups will they understand what is required and be able to make an informed decision about participation? More information about obtaining informed consent can be obtained from Chapter 40 and more information about the ethics of tracking location (location privacy, tracking vulnerable people, avoiding stigma and the protection of civil liberties, for example) can be found in Chapter 27.

Practicalities

- What factors are important when choosing sensors and devices for your research? This can include:
  - cost of equipment, including hardware and software;
  - standard of equipment (and avoiding sub-standard equipment, perhaps due to budget constraints);

- ○ availability of equipment;

- ○ skills and experience of the researcher;

- ○ understanding, perceptions, willingness and user acceptance (to use equipment correctly and appropriately).

- Where will sensors be placed? Will this affect accuracy?

- How can you mitigate against potential technological problems? This could include:

  - ○ atmospheric conditions (excessive heat or moisture, for example);

  - ○ possible wear and tear;

  - ○ battery/power source depletion (what power supply is used, how it is recharged and whether it is possible to prolong life without sacrificing data accuracy);

  - ○ network congestion;

  - ○ positioning (the sensor moves too far from the router as people or machinery move and travel around, for example);

  - ○ vibrations;

  - ○ calibration needs and requirements, and the disruption this may cause for participants and data collection;

  - ○ fluctuations and variation, and the period of time that will be required to take measurements or record data to account for fluctuations and variation.

- For how long will sensors be used in your research? When do sensor data become redundant?

Useful resources

Sensor village (www.bcu.ac.uk/business/digital/sensor-village) is a knowledge hub at Birmingham City University in the UK that provides real-time analytics on urban activity in the local area and develops high tech sensor systems.

CamBridgeSens (www.sensors.cam.ac.uk) is an interdisciplinary network at Cambridge University in the UK for academics, industry and students who are interested in sensor research and applications. A Sensor Day conference is organised each year, and collaboration projects are welcomed.

The Sensor Technology Research Centre, University of Sussex in the UK (www.sussex.ac.uk/strc) is a group of academics who are interested in developing and sharing interests in sensors and their applications. Details of ongoing research projects are available on their website.

An interesting example of a study that uses real-time sensor data in the home can be found at https://clinicaltrials.gov/ct2/show/study/NCT01782157. This study illustrates how sensors can be used to recognise changes in a resident's pattern of activity, which could signal changes in well-being, such as missed meals and long periods of inactivity, acute health problems such as injury or illness, or slower progressive decline due to conditions such as dementia.

Useful links to a variety of papers that discuss the use of sensors in scientific research can be found on the Nature website (www.nature.com/subjects/sensors). Here you can find the latest reviews and research along with news and comments about a vast array of sensor topics.

'Sensors and instrumentation' is one of the research areas funded by the Engineering and Physical Sciences Research Council (EPSRC) in the UK. More information about research area connections and funded projects can be obtained from www.epsrc.ac.uk/research/ourportfolio/researchareas/instrumentation.

The IEEE Instrumentation and Measurement Society (http://ieee-ims.org) is 'dedicated to the development and use of electrical and electronic instruments and equipment to measure, monitor and/or record physical phenomena'. Useful publications, conferences and symposiums can be found on this site.

Key texts

Banica, F.-G. (2012) *Chemical Sensors and Biosensors: Fundamentals and Applications*. Chichester: John Wiley & Sons.

Chaffin, D., Heidl, R., Hollenbeck, J., Howe, M., Yu, A., Voorhees, C. and Calantone, R. (2017) 'The Promise and Perils of Wearable Sensors in Organizational Research', *Organizational Research Methods*, 20(1), 3–31, first published November 30, 2015, 10.1177/1094428115617004.

Fraden, J. (2016) *Handbook of Modern Sensors: Physics, Designs, and Applications*, 5th edition. San Diego, CA: Springer.

Geraedts, H., Zijlstra, W., Van Keeken, H., Zhang, W. and Stevens, M. (2015) 'Validation and User Evaluation of a Sensor-Based Method for Detecting Mobility-Related Activities in Older Adults', *PLoS One*, 10(9), e0137668, 10.1371/journal.pone.0137668.

Khan, W., Aalsalem, M., Gharibi, W. and Arshad, Q. (2016) 'Oil and Gas Monitoring Using Wireless Sensor Networks: Requirements, Issues and Challenges', *2016 International Conference on Radar, Antenna, Microwave, Electronics, and Telecommunications (ICRAMET)*, 3–5 October 2016, 10.1109/ICRAMET.2016.7849577.

Sensor-based methods

Liu, M., Lai, C., Su, Y., Huang, S., Chien, Y., Huang, Y. and Hwang, P. (2015) 'Learning with Great Care: The Adoption of the Multi-Sensor Technology in Education'. In Mason, A., Mukhopadhyay, S. and Jayasundera, K. (eds.) *Sensing Technology: Current Status and Future Trends III. Smart Sensors, Measurement and Instrumentation*. Cham: Springer, 11.

Mombers, C., Legako, K. and Gilchrist, A. (2016) 'Identifying Medical Wearables and Sensor Technologies that Deliver Data on Clinical Endpoints', *British Journal of Clinical Pharmacology*, February 2016, 81(2), 196–98, first published December 26, 2015, 10.1111/bcp.12818.

Pagkalos, I. and Petrou, L. (2016) 'SENHANCE: A Semantic Web Framework for Integrating Social and Hardware Sensors in e-Health', *Health Informatics Journal*, 22(3), 505–22, first published March 10, 2015, 10.1177/1460458215571642.

Pentland, A. (2008) *Honest Signals: How They Shape Our World*. Cambridge, MA: Massachusetts Institution of Technology.

Polastre, J., Szewczyk, R., Mainwaring, A., Culler, D. and Anderson, J. (2004) 'Analysis of Wireless Sensor Networks for Habitat Monitoring'. In Raghavendra, C., Sivalingam, K. and Znati, T.(eds.) *Wireless Sensor Networks*, 399–423. Boston, MA: Springer.

Ranky, P. (2002) 'Advanced Digital Automobile Sensor Applications', *Sensor Review*, 22 (3), 213–17, 10.1108/02602280210433043.

Seedhouse, P. and Knight, D. (2016) 'Applying Digital Sensor Technology: A Problem-Solving Approach', *Applied Linguistics*, 37(1), 7–32, 10.1093/applin/amv065.

Smieszek, T., Barclay, V., Seeni, I., Rainey, J., Gao, H., Uzicanin, A. and Salathé, M. (2014) 'How Should Social Mixing be Measured: Comparing Web-Based Survey and Sensor-Based Methods', *BMC Infectious Diseases*, 14(136), first published March 10, 2014, 10.1186/1471-2334-14-136.

Topol, E., Steinhubl, S. and Torkamani, A. (2015) 'Digital Medical Tools and Sensors', *Journal of the American Medical Association*, January 27, 2015, 313(4), 353–54, 10.1001/jama.2014.17125.

Zhong, S., Zhong, H., Huang, X., Yang, P., Shi, J., Xie, L. and Wang, K. (2019) *Security and Privacy for Next-Generation Wireless Networks*. Cham: Springer.

Zhu, Z., Yu, M. and Riezebos, P. (2016) 'A Research Framework of Smart Education', *Smart Learning Environments*, 3(4), 10.1186/s40561-016-0026-2.

CHAPTER 50

Smartphone app-based research

Overview

Smartphone apps are pre-loaded, pre-installed or built-in software on smartphones and software or tools that are downloaded from websites or app stores on to smartphones. They are designed to enable, facilitate and simplify tasks and/or make them more suitable for mobile use. In research, smartphone apps can be used by the researcher on their own phones to collect, record, store, analyse and visualise data, or can be installed or downloaded on participants' phones to collect data in a wide variety of ways. Preloaded or downloaded apps, for example, enable the researcher to communicate via email, perhaps inviting someone to take part in a survey (see Chapter 34) or by undertaking an email interview (see Chapter 40). Or pre-loaded apps can be used to enable participants to take notes, photos or videos that can be collected and analysed in a qualitative research study (García et al., 2016). Downloadable apps enable researchers to undertake survey research and collect statistical data (see Chapter 51), or undertake ethnographic studies (see Chapter 31). Alternatively, researchers can design an app for a specific research project, which can be downloaded by participants, used and deleted on completion (Hughes and Moxham-Hall, 2017). A related area of study includes the assessment, evaluation or critique of apps that have been developed for specific functions, such as those in health and medicine (Dempster et al., 2014) or learning and teaching (Jonas-Dwyer et al., 2012).

There are various smartphone apps available for research purposes. These include:

- ethnographic apps for studies into human activity, behaviour and/or movement (Chapter 31);

- survey or mobile data collection apps to collect data about thoughts, opinions, attitudes, behaviour and feelings (Chapters 34 and 51);

- panel apps that enable researchers to engage with pre-recruited and profiled panellists over a period of time (Chapter 42);

- mapping, tracking and location awareness apps for recording, visualising and monitoring location (Chapter 27);

- sensor-based apps, including those that utilise motion, environmental and position sensors (Chapters 49 and 57 and Herodotou et al., 2014);

- diary and time use data apps to record everyday behaviour, actions and thoughts (Chapter 30 and Sonck and Fernee, 2013);

- recording, photo and video apps that enable people to communicate, share their lives and illustrate their lived realities, from their perspective (García et al., 2016; Montgomery, 2018);

- data analytics apps to examine datasets, or raw data, to make inferences and draw conclusions (Chapter 10);

- web and mobile analytics apps for the collection, analysis and reporting of web data (Chapter 58);

- data visualisation apps to help researchers better understand data, see patterns, recognise trends and communicate findings (Chapter 13);

- apps for searching, retrieving, managing and citing information (Chapter 23);

- collaboration apps that enable individuals, groups and teams to work together in research, teaching and learning (Chapter 36).

Researchers from a wide variety of disciplines use smartphone apps for research purposes, including adult education (Roessger et al., 2017), psychology (Harari et al., 2016; Miller, 2012), the social sciences (García et al., 2016; Sugie, 2018) and health and medicine (Dempster et al., 2014; Heintzman, 2016). Examples of research projects that have used, assessed or critiqued smartphone apps include a project to design and evaluate a sensor-based mobile app for citizen science projects (Herodotou et al., 2014); a study into the role of football in participants' lives (García et al., 2016); and a study into the use of data collection apps to better understand learners' lived experiences and the effects of transformative learning in adult education (Roessger et al., 2017).

If you are interested in using smartphone apps for your research, an enlightening discussion about the potential for their use in behavioural research, along with interesting methodological and ethical challenges, is provided by Harari et al. (2016). Examples of apps that can be used for research purposes are given below (a more comprehensive list can be found at https://tqr.nova.edu/apps: see useful resources). If you are interested in designing your own app for your research project there are various websites available to help with your design (see for example, http://appinventor.mit.edu/explore, www.appmakr.com and http://researchkit.org). It is also important that you take note of the questions for reflection, listed below and in the relevant chapters listed above.

Questions for reflection

Epistemology, theoretical perspective and methodology

- How do theoretical perspective and methodological framework influence the type of smartphone app(s) that is used for research purposes? For example, an objectivist standpoint presumes that things exist as meaningful entities, independent of consciousness and experience, that they have truth and meaning in them as 'objects', whereas a subjectivist standpoint might maintain that reality as related to a given consciousness is dependent on that consciousness. These different standpoints will have an influence on the app used and the type of data collected: statistics from survey apps that use questionnaires with closed-ended questions (see Chapter 34 and 51) or apps that enable the creation and use of fluid, flexible and adaptable techniques for the collection of qualitative data, for example.

- Can smartphone app-generated data be considered naturalistic, given that, in most cases, respondents are aware that they are part of a research project? How might this awareness influence what is collected? Can this be offset through the use of sensors and the collection of data of which respondents are unaware (see Chapters 49 and 28)? Is this ethically acceptable?

- Why use an app rather than a web-based resource (the benefits of offline access, for example)?

Ethics, morals and legal issues

- When contacting people for your research, have you considered potential health and safety issues? This could include, for example, problems with contacting people by smartphone when they are driving, operating dangerous machinery or undertaking hazardous sports or activities. It can also include contacting vulnerable people, children (where permission has not been sought) and those who are in a situation where it is difficult to talk privately.

- Have you taken into consideration any air-time, roaming or data costs that your participants may be charged? Is it possible to design your study so that there is no cost for participants? If this is not possible, how will you compensate participants (and reach agreement on compensation)?

- How will you address issues of data privacy and security in cases where data are stored on a participant's phone (SMS, videos, photos or personal notes, for example)? Data are potentially vulnerable if a phone is stolen, lost or used by another person. How will you address these issues when contacting and recruiting participants, and when a research project comes to an end (ensuring that participants delete sensitive research data, for example)?

- If you are intending to use recording apps do you understand the legal issues surrounding the accidental recording of third parties (people who are unrelated to your participants, who have not given informed consent)? You can find an interesting discussion on this issue from The Conversation (https://theconversation.com/whos-listening-the-ethical-and-legal-issues-of-developing-a-health-app-69289) [accessed April 17, 2018]. More information about legislation related to mobile phone recording can be found in Chapter 33.

- How can users be protected from smartphone apps? This can include malware (viruses, Trojans and worms, for example) and greyware (adware, trackware and spyware, for example). Enck (2011) will help you to address this question.

Practicalities

- Have you considered possible advantages and disadvantages to using smart-phone apps for your research? Advantages could include the ability to contact respondents at any time, on devices that are familiar, used frequently and kept on person. Apps can be user-friendly, interactive, non-invasive, customisable and efficient, which can improve response and completion rates. They enable respondents to be spontaneous and become engaged in the research project, and they enable researchers to monitor participation and progress. Disadvantages could include problems with technology and connections, battery life, misunderstanding or misuse of technology, limited and self-selecting samples, and concerns about privacy, confidentiality and data security, for example. García et al. (2016) provide an enlightening discussion on these issues when smartphone-based apps are used for qualitative research.

- What costs are involved, especially if you have to work with a software company to design or modify a specific app for your research? Can costs be weighed up against savings in others areas (researcher time, for example)?

- Will your app work on all devices? What procedures do you have in place for reporting and overcoming technical difficulties that could be encountered? How will you limit participant drop-out through technical problems? Piloting your app on all possible devices is essential to iron-out potential problems.

- Is your chosen app suitable for your research? Will participants be able to use the app and feel comfortable doing so? Are text, language and images suitable for your intended participants? Jonas-Dwyer et al. (2012) will help you to consider these questions.

More questions for reflection that are relevant to those who are interested in smartphone app-based research can be found in Chapter 31, 32, 33, 34 and 51.

Useful resources

There are a wide variety of apps available for research purposes. Examples available at time of writing include:

- Dedoose for analysing qualitative and mixed methods research (www.dedoose.com);

- FieldNotes for geo-referenced data sharing and collection (http://fieldnotesapp.info);

- Formhum for data collection and analysis in the field (http://unfao.net);

- Grid Diary for recording daily thoughts, feelings and activities (http://griddiaryapp.com/en);

- Momento to capture and record experiences, events and activities (https://momentoapp.com);

- MyInsights for mobile qualitative market research (www.mobilemarketresearch.com/en/myinsights);

- Paco to observe and experiment with behaviour (www.pacoapp.com);

- PushForms for form creation, data collection and data sharing (www .getpushforms.com);

- SoundNote for taking and tracking notes while recording audio (http:// soundnote.com).

The Qualitative Report Guide to Qualitative Research Mobile Applications, curated by Ronald J. Chenail (https://tqr.nova.edu/apps) is a comprehensive list of mobile research apps, including those that can be used for note-taking, recording, collecting, analysing and visualising data [accessed March 19, 2018].

ResearchKit (http://researchkit.org) 'is an open source framework introduced by Apple that allows researchers and developers to create powerful apps for medical research'. Visit the site for more information about functions and capabilities. Jardine et al. (2015) provide a useful short summary of the software and its application.

Key texts

Dempster, N., Risk, R., Clark, R. and Meddings, R. (2014) 'Urologists' Usage and Perceptions of Urological Apps', *Journal of Telemedicine and Telecare*, 20(8), 450–53, first published October 14, 2014, 10.1177/1357633X14555622.

Enck, W. (2011) 'Defending Users against Smartphone Apps: Techniques and Future Directions'. In Jajodia, S. and Mazumdar, C.(eds.) *Information Systems Security, ICISS 2011. Lecture Notes in Computer Science*. Berlin: Springer, 7093.

García, B., Welford, J. and Smith, B. (2016) 'Using a Smartphone App in Qualitative Research: The Good, the Bad and the Ugly', *Qualitative Research*, 16(5), 508–25, first published August 3, 2015, 10.1177/1468794115593335.

Harari, G., Lane, N., Wang, R., Crosier, B., Campbell, A. and Gosling, S. (2016) 'Using Smartphones to Collect Behavioral Data in Psychological Science: Opportunities, Practical Considerations, and Challenges', *Perspectives on Psychological Science*, 11(6), 838–54, first published November 28, 2016, 10.1177/1745691616650285.

Heintzman, N. (2016) 'A Digital Ecosystem of Diabetes Data and Technology: Services, Systems, and Tools Enabled by Wearables, Sensors, and Apps', *Journal of Diabetes Science and Technology*, 10(1), 35–41, first published December 20, 2015, 10.1177/ 1932296815622453.

Herodotou, C., Villasclaras-Fernández, E. and Sharples, M. (2014) 'The Design and Evaluation of a Sensor-Based Mobile Application for Citizen Inquiry Science Investigations'. In Rensing, C., de Freitas, S., Ley, T. and Muñoz-Merino, P. (eds.) *Open Learning and Teaching in Educational Communities, EC-TEL 2014, Lecture Notes in Computer Science*. Cham: Springer, 8719.

Hughes, C. and Moxham-Hall, V. (2017) 'The Going Out in Sydney App: Evaluating the Utility of a Smartphone App for Monitoring Real-World Illicit Drug Use and Police

Encounters among Festival and Club Goers', *Substance Abuse: Research and Treatment*, first published June 28, 2017, 10.1177/1178221817711419.

Jardine, J., Fisher, J. and Carrick, B. (2015) 'Apple's ResearchKit: Smart Data Collection for the Smartphone Era?', *Journal of the Royal Society of Medicine*, 108(8), 294–96, first published August 12, 2015, 10.1177/0141076815600673.

Jonas-Dwyer, D., Clark, C., Celenza, A. and Siddiqui, Z. (2012) 'Evaluating Apps for Learning and Teaching', *International Journal of Emerging Technologies in Learning (Ijet)*, 7(1), 54–57, Kassel, Germany: International Association of Online Engineering, retrieved December 3, 2018 from, www.learntechlib.org/p/44890/.

Miller, G. (2012) 'The Smartphone Psychology Manifesto', *Perspectives on Psychological Science*, 7(3), 221–37, first published May 16, 2012, 10.1177/1745691612441215.

Montgomery, R. (2018) *Smartphone Video Storytelling*. New York, NY: Routledge.

Roessger, K., Greenleaf, A. and Hoggan, C. (2017) 'Using Data Collection Apps and Single-Case Designs to Research Transformative Learning in Adults', *Journal of Adult and Continuing Education*, 23(2), 206–25, first published September 18, 2017, 10.1177/1477971417732070.

Sonck, N. and Fernee, H. (2013) *Using Smartphones in Survey Research: A Multifunctional Tool. Implementation of a Time Use App; a Feasibility Study*. The Hague: The Netherlands Institute for Social Research.

Sugie, N. (2018) 'Utilizing Smartphones to Study Disadvantaged and Hard-To-Reach Groups', *Sociological Methods & Research*, first published January 18, 2016, 10.1177/0049124115626176.

CHAPTER 51

Smartphone questionnaires

Overview

Smartphone questionnaires are questionnaires that are disseminated, administered and completed on smartphones, by the person who owns, uses or is given, the smartphone. Smartphones differ from ordinary mobile or cell phones in that they have a user-friendly, mobile operating system, internet access, location tracking, a touchscreen interface, a virtual store of apps for download and, on some models, a built-in personal assistant. Therefore, although smartphones are mobile phones, with many of the issues that have been raised in Chapters 33 and 34 pertinent to their use in research, a separate entry has been included in this book because there are specific features, functions and methods that are applicable to smartphone questionnaires. These include:

- different modes of communication such as voice, text or video (see Conrad et al., 2017 for a discussion on four modes: Human Voice, Human Text, Automated Voice, and Automated Text);

- alerts, notifications or pop-up questions over a period of time when the respondent is in a specific location or undertaking specific types of communication, search behaviour or interaction (this can be referred to as a daily diary method or experience sampling: more information about mobile diaries is provided in Chapter 30);

- screen rotation to rotate between portrait and landscape for more effective viewing of questions and grids, and for easier scrolling;

- automatic synchronisation that enables data stored in a respondent's smartphone to be sent to the main database every time there is internet connectivity;

- collection of auxiliary, supplementary or complementary data such as GPS trackers to record location (Chapter 27), sensors to record movement and interaction (Chapters 49 and 57) and log files for metadata about communication and search behaviour (Chapter 28).

There are two types of questionnaire that can be disseminated, administered and completed on smartphones. The first are online questionnaires that are accessed via a mobile browser. If questionnaires are adaptive, the layout and presentation of content change automatically to suit the screen size of the device, enabling researchers to design questionnaires for regular browsers that will adapt to work on smartphones (and tablets, if required). These can be individualised, customised and optimised. They can contain specific features that enable irrelevant questions to be skipped based on previous answers, or display questions and options in different orders, for example. Video and audio content can also be embedded into the questionnaire. A detailed list of features and functions of online questionnaires is provided in Chapter 43. The second type includes questionnaires that are embedded in a smartphone app for download by respondents, which can overcome the need for internet connectivity while completing a questionnaire: more information about smartphone apps for research can be found in Chapter 50. Questionnaires, within these two types, can be closed-ended (used to generate statistics in quantitative, survey research), open-ended (used in qualitative research to gather in-depth attitudes and opinions) or a combination of both (used to generate statistics and explore ideas, beliefs and/or experiences).

Smartphone questionnaires are used in a wide variety of disciplines including education, health and medicine, the social sciences and business and marketing. They can be used as standalone questionnaires, in mixed methods research or to supplement sensor-based data, for example. Smartphone questionnaires can accommodate various question types including rating scales, response matrices (with comments if required), multiple choice questions, Likert-scale items, timed exposure of images and questions with images, GIFs, photos and video. Examples of research projects that have used, assessed or critiqued smartphone questionnaires include a study that merged smartphone data collection apps with single-case research designs for a study into transformative learning in adults (Roessger et al., 2017); a study of respondent mode choice in a smartphone survey (Conrad et al., 2017); and a study that used an array of behavioural measures and respondent-reported information to study men who had recently been released from prison (Sugie, 2018).

If you are interested in using smartphone questionnaires for your research, it is useful to view some of the websites, worksheets or help documents provided by software survey or questionnaire design companies, as these give a good overview of styles, functions, design, optimisation and adaptability. Some of these are listed below. It is also important that you have a thorough understanding of methodological issues associated with survey research or the use of questionnaires for qualitative research; an understanding of methods associated with questionnaire design, sampling, administration and analysis; and an understanding of methodological implications of mobile phone research. More information about these issues can be found in Rea and Parker (2014); Brace (2018); Gillham (2007); Callegaro et al. (2015) and Häder et al. (2012). The questions for reflection listed below will help you to think more about using smartphone questionnaires for your research, and Chapter 43 provides more information about online questionnaires.

Questions for reflection

Epistemology, theoretical perspective and methodology

*   If you are intending to use smartphone questionnaires in survey research, what type of survey do you intend to undertake? For example, questionnaires can be used in cross-sectional surveys, the aim of which is to describe attitudes and behaviour at one point in time. These tend to be exploratory and descriptive, but can also be used to find correlations and make predictions. They cannot assess changes over time. Questionnaires can also be used in longitudinal surveys that set out to explore and describe what is happening over time (over months or years, through the use of two or more discreet surveys). There are two main types of longitudinal survey: cohort surveys (or panel surveys) follow the same group of individuals, with a common characteristic or experience, over time (see Chapter 42); trend surveys ask the same questions to different respondents within the study population over a period of time. If you require participants to take part in longitudinal surveys there are a number of practical implications that you need to consider such as device upgrades, compatibility and phone numbers becoming inactive or discontinued.

*   What type of questionnaire do you intend to use for your research? As we have seen above, questionnaires can be online or embedded into apps, and they can be closed-ended, open-ended or a combination of both. Closed-ended questionnaires are used to generate statistics in quantitative

research, open-ended questionnaires are used in qualitative research to explore ideas, beliefs and/or experiences and combination questionnaires combine both approaches. All three types can be used in smartphone-based research. However, there are specific issues associated with design, layout, structure and format, such as scrolling, grids, the use of images and navigation. For example, grids with answers from closed-ended questions may require vertical and horizontal scrolling, and could encounter problems with respondents working quickly through the grid, merely tapping on the same response for each item in the grid. The WebSM website (see below) provides some interesting papers and discussion on this topic and Roessger et al. (2017) provide a good example of the practical application of data collection apps and questionnaires.

- When undertaking a smartphone survey, how can you conduct your survey in a way that will yield unbiased, reliable and valid data (see Chapter 34)?

- How will you overcome potential selection bias (how to take account of people who do not own, cannot use or are unwilling to use a smartphone)?

Ethics, morals and legal issues

- How can you ensure that participants understand ethical issues such as confidentiality, anonymity, privacy and informed consent? If these issues need to be explained in writing, how can you ensure that they are read, and understood, by participants? Smartphone and online users can ignore, or merely tick, user terms and conditions without reading them. How will you make sure that this does not happen when important ethical issues are to be considered?

- How will you ensure that data collection is transparent? This is of particular importance if you intend to collect behavioural data from smartphones, such as location awareness or location tracking tools (see Chapter 27) or from sensors in the smartphone (see Chapters 49 and 57).

- How can you ensure anonymity? Is it possible that location traces might lead to the identification of research participants, for example?

- How can you keep data secure and protect research participants against unwanted surveillance (adtrackers and spyware, for example)?

- If you intend to use digital tools, software packages or panels provided by commercial companies, how can you ensure that the commercial company conforms to the required ethical standards?

Practicalities

- What features do you want to include in your smartphone questionnaire? Iqbal et al. (2017) provide a usability evaluation of the effectiveness, efficiency and satisfaction of screen rotation, voice commands, LED notifications and kid mode, which will help you to think about and evaluate relevant features for your questionnaire.

- How can you encourage smartphone users to remain focussed, concentrate on and complete your questionnaire, especially when smartphones are used in short bursts and on the move? This could include, for example, not using images that take a long time to download, avoiding features that require a high degree of dexterity and ensuring that questions (and the questionnaire) are kept as short as possible. It can also include the decision whether or not to use progress bars. See Antoun et al. (2018) for a literature review of research projects from 2007 to 2016 that consider the effects of smartphone questionnaire design features on quality of response.

- How do you intend to optimise your questionnaire to make it easier to complete on smartphones? How can you avoid tapping errors, the need for scrolling and differences in the way questionnaires are displayed, for example? Antoun et al. (2018) provide an interesting discussion on optimisation and go on to identify five design heuristics for creating effective smartphone questionnaires: readability, ease of selection, visibility across the page, simplicity of design elements and predictability across devices.

- Will your questionnaire adapt automatically to the capabilities of your participants' devices? For example, some smartphones may have limited screen space, a slow speed mobile internet connection and small keyboards. Mobile metatags can be incorporated to enable questionnaires to look and perform better on handheld devices, specific smartphone templates can be used, or a 'publish all' function supplied by the survey hosting company can be used to deliver the appropriate questionnaire according to the size of device. Once you have developed your questionnaire, view it on a number of devices to check adaptability before piloting.

- How will you take account of problems that could occur due to lack of internet access, or intermittent connection, while respondents are completing the questionnaire? Using downloaded apps that work offline may help to overcome this problem: see Chapter 50.

Useful resources

There are a variety of digital tools and software packages available at time of writing that can be used for designing questionnaires for smartphones:

- Client Heartbeat (www.clientheartbeat.com);

- LimeSurvey (www.limesurvey.org);

- Mobi-Survey (www.soft-concept.com/uk/mobi-survey-mobile-survey-system.htm);

- Qualtrics Research Core (www.qualtrics.com/research-core);

- SnapSurveys (www.snapsurveys.com);

- SoGoSurvey (www.sogosurvey.com);

- SurveyMonkey (www.surveymonkey.com);

- Survey Planet (https://surveyplanet.com);

- Typeform (www.typeform.com);

- Web Survey Creator (www.websurveycreator.com).

The WebSM website 'is dedicated to the methodological issues of Web surveys' (www.websm.org). This website contains useful papers and discussions on using smartphone designed questionnaires, such as Hanson, T. (2017) *Adapting Questionnaires for Smartphones: An Experiment on Grid Format Questions*, which provides interesting reading about grid questions on smartphones [accessed April 16, 2018].

The European Survey Research Association (ESRA) is a membership organisation for people who are interested in survey research (www.europeansurveyresearch.org). Here you can access Survey Research Methods, which is the official journal of the ESRA. It is peer-reviewed and published electronically, with free and open access.

Key texts

Antoun, C., Katz, J., Argueta, J. and Wang, L. (2018) 'Design Heuristics for Effective Smartphone Questionnaires', *Social Science Computer Review*, 36(5), 557–74, first published September 27, 2017, 10.1177/0894439317727072.
Brace, I. (2018) *Questionnaire Design: How to Plan, Structure and Write Survey Material for Effective Market Research*, 4th edition. London: Kogan Page.
Callegaro, M., Manfreda, K. and Vehovar, V. (2015) *Web Survey Methodology*. London: Sage.

Smartphone questionnaires

Conrad, F., Schober, M., Antoun, C., Yan, H., Hupp, A., Johnston, M., Ehlen, P., Vickers, L. and Zhang, C. (2017) 'Respondent Mode Choice in a Smartphone Survey', *Public Opinion Quarterly*, 81(S1), 307–37, published online April 13, 2017, 10.1093/poq/nfw097.

Gillham, B. (2007) *Developing a Questionnaire*, 2nd edition. London: Continuum.

Häder, S., Häder, M. and Kühne, M. (eds.) (2012) *Telephone Surveys in Europe: Research and Practice*. Heidelberg: Springer.

Iqbal, M., Ahmad, N. and Shahzada, S. (2017) 'Usability Evaluation of Adaptive Features in Smartphones', *Procedia Computer Science*, 112, 2185–94, first published online September 1, 2017, 10.1016/j.procs.2017.08.258.

Rea, L. and Parker, A. (2014) *Designing and Conducting Survey Research: A Comprehensive Guide*, 4th edition. San Francisco, CA: Jossey-Bass.

Roessger, K., Greenleaf, A. and Hoggan, C. (2017) 'Using Data Collection Apps and Single-Case Designs to Research Transformative Learning in Adults', *Journal of Adult and Continuing Education*, 23(2), 206–25, first published September 18, 2017, 10.1177/1477971417732070.

Sugie, N. (2018) 'Utilizing Smartphones to Study Disadvantaged and Hard-to-Reach Groups', *Sociological Methods & Research*, 47(3), 458–91, first published January 18, 2016, 10.1177/0049124115626176.

CHAPTER 52

Social media analytics

Overview

Social media analytics (or social media data analytics) refers to the process of scraping, capturing, extracting and examining data from social media platforms to make inferences, draw conclusions and provide actionable insight. It includes social media such as social networking, microblogging, blogs, photo sharing, video sharing, music sharing, crowdsourcing and messaging, for example. It also encompasses the tools, techniques and software that are used to capture and analyse both qualitative and quantitative social media data. Social media analytics can be guided by a number of theoretical and methodological frameworks, including critical theory and critical data studies (Felt, 2016); digital ethnography and narrative analysis techniques (Brooker et al., 2018); Distributional Lexical Semantics (Basili et al., 2017); visual analytics grounded in an 'abductive ontological perspective' (Brooker et al., 2016); and 'interface methods' (Marres and Gerlitz, 2016).

There are a number of qualitative and quantitative data analysis methods that can be used by those undertaking social media analytics and those that are chosen depend on epistemology, theoretical perspective, methodological framework, purpose of research and research question. Choices can also depend, in part, on available tools and software and on researcher understanding and experience of tools and software (see below). Examples of data analysis methods used in social media analytics include (in alphabetical order):

- comparative analysis to identify, analyse and explain similarities across groups;

- content analysis to provide an objective, systematic and quantified description of data;
- conversation analysis to study social interaction within conversation;
- co-occurrence analysis to extract and display rules and patterns;
- corpus analysis to investigate linguistic phenomena;
- discourse analysis to study and analyse the use of language;
- rhetorical analysis to identify how an author writes;
- sentiment analysis to identify and categorise opinions (this can also be referred to as opinion mining: see Chapter 12);
- social network analysis to investigate social structure through networks (see Chapter 53);
- temporal analysis to study events, behaviour or sentiments over time;
- textual analysis to consider content, structure and functions of messages within text, which can include:
  - collocation (words commonly appearing near to each other);
  - common phrases;
  - commons sets of words;
  - concordance (the context of a word or sets of words);
  - place and name recognition;
  - time and date recognition;
  - word frequency.

Social media analytics is used as a research method in various disciplines and fields of study including computer sciences, communication studies, sociology, psychology, politics, education, business and health and medicine. Examples of research projects that have used, assessed or critiqued social media analytics include a project analysing Twitter data from a 24-hour period following The Sisters in Spirit Candlelight Vigil, sponsored by the Native Women's Association of Canada (Felt, 2016); studies into flu outbreaks, Ebola epidemic and marijuana legalisation (Yang et al., 2016); the use of timeline data to interrogate users' socio-political attitudes over time (Brooker et al., 2018); and research into using social media data to enhance public health analytics (Ji et al., 2017).

If you are interested in finding out more about social media analytics for your research, Ganis and Kohirkar (2016) provide a good overview. Although this book is aimed at executives and marketing analysts, it nevertheless provides an interesting introduction and overview of practical tools for researchers from other disciplines and fields of study who are interested in social media analytics. Marres and Gerlitz (2016) and Brooker et al. (2016) provide relevant and interesting discussions for those interested in methods and methodology of social digital media. A method that is closely related to, and can be part of, social media analytics, is social media mining: a useful introduction to concepts, principles and methods for social media mining is provided by Zafarani et al. (2014). There are various digital tools and software packages available if you are interested in social media analytics and relevant websites are listed below. It is useful to visit some of these sites so that you can get a better idea of functions, capabilities, purposes, user-friendliness (some require programming skills, for example) and costs. You may also find it useful to read Yang et al. (2016) who describe the development of a web-based social media analytics and research testbed (SMART), which is designed 'with the goal of providing researchers with a platform to quickly test hypotheses and to refine research questions'. This will help you to think more about how social media analytics will help to answer your own research question. It is important that you also understand more about data analytics (Chapter 10), data mining (Chapter 12) and data visualisation (Chapter 13) and the ethics of social media research (see questions for reflection, below).

Questions for reflection

Epistemology, theoretical perspective and methodology

- How can you ensure that sensitivities to epistemology, theory and methodology are not neglected or subsumed by the intricacies of technology when undertaking social media analytics? How can you bring together potential complexities of algorithms, metrics and machine learning with the requirements and methodologies of the social sciences? These issues, and others, led to the development of the Chorus project (http://chorusanalytics.co.uk), which is discussed in Brooker et al. (2016).

- Do you intend to use social media analytics as your sole data collection and analysis method, or do you intend to use other methods, perhaps to tackle reliability, verifiability and validity challenges, for example?

- What types of data do you intend to capture: goal-driven data, data-driven data, user-driven data or semantic-driven data, for example? This will depend on your research question and methodology.

- How will you handle potential complexities? Brooker et al. (2016: 1), for example, point out that 'social media provides a form of user-generated data which may be unsolicited and unscripted, and which is often expressed multi-modally (i.e. through combinations of text, hyperlinks, images, videos, music, etc.)'. Does your methodology allow for such challenges?

- What are the possible benefits of social media analytics? Felt (2016: 3), for example, notes that Twitter creates 'an alternative voice accessible to researchers who are interested in contrasting dominant mass media frames with those produced through microblogging of the general public' and that Twitter allows access to 'subordinated voices'. She goes on to point out that 'it answers the social network questions, the global discourse questions, and the public sentiment questions'.

- How will you recognise and acknowledge the diversity of contexts within which social media occurs? Brooker et al. (2018) provide an interesting discussion on this topic that will help you to answer this question.

Ethics, morals and legal issues

- Are online postings public or private data? Are all equally available for research?

- Is it possible to obtain informed consent, in particular, in cases where participants do not know that they are part of a research project?

- How can you deal with issues of anonymity (when platforms insist that data are only reported in their original form and attributed to the original poster, or when reproducing social media comments that are traceable online, for example)? A useful publication titled *Social Media Research: A Guide to Ethics* that addresses all aspects of social media research ethics has been produced by Dr Leanne Townsend and Prof Claire Wallace at the University of Aberdeen. It can be retrieved from www.dotrural.ac.uk/socialmediaresearchethics.pdf [accessed May 2, 2018].

- What influence might large technology, internet or social media companies have on social media analytics (access, restrictions, collection and storage of data, for example)? What impact might such influence have on academic research? Richterich (2018) provides an interesting discussion on these issues relating to biomedical studies.

Practicalities

- Do you intend to focus on one social media platform, or do you intend to compare data across platforms? If so, is access to data equally available?

- Do you have a good understanding of the application programming interface (API) policies of relevant social media platforms? These are becoming increasingly restrictive, in particular after problems with recent breaches and abuses of personal data. Access to data may be limited or only available for purchase. What impact might restrictions and costs have on your research project?

- Are you aware and able to acknowledge limitations of research that relies on API-generated datasets? Felt (2016: 3) points out that these methods provide 'the most accessible free entry points to Twitter data', but that researchers are unable to reproduce data exactly due to the 'unknown logic of the algorithm used to produce them'. McKelvey (2015) provides an interesting discussion on this and the need for democratic methods of internet inquiry.

- We have seen above that there are various ways to analyse social media data. What technique(s) do you intend to use? Are you familiar with software and tools? Are they freely available or are significant costs involved? Do you know how to use the tools, or can you gain the required understanding?

Useful resources

There are various digital tools and software packages available for social media analytics. Examples available at time of writing include (in alphabetical order):

- Brandwatch (www.brandwatch.com);
- Chorus (http://chorusanalytics.co.uk);
- DiscoverText (https://discovertext.com);
- Echosec (www.echosec.net);
- Facebook Analytics (https://analytics.facebook.com);

- Followthehashtag (www.followthehashtag.com);

- Leximancer (https://info.leximancer.com);

- Netlytic (https://netlytic.org);

- SocioViz (http://socioviz.net/SNA/eu/sna/login.jsp);

- The Digital Methods Initiative Twitter Capture and Analysis Toolset (DMI-TCAT) (https://github.com/digitalmethodsinitiative/dmi-tcat/wiki);

- Trendsmap (www.trendsmap.com);

- Twitonomy (www.twitonomy.com);

- Webometric Analyst 2.0 (http://lexiurl.wlv.ac.uk).

The Algorithmic Media Observatory (www.amo-oma.ca/en) studies the changing media (social media platforms, mobile phones and home routers) that influence participation in politics and culture, and looks into 'the patterns and habits of these algorithmic media to detect their subtle influence'. Useful publications, posts and archives can be found on this website.

The Social Media Research Association (https://smra-global.org) is a 'global trade association dedicated to the advocacy, promotion and development of best practices for using social media as a source for insights and marketing efficiency'. Useful podcasts, news and events can be found on the website.

Key texts

Basili, R., Croce, D. and Castellucci, G. (2017) 'Dynamic Polarity Lexicon Acquisition for Advanced Social Media Analytics', *International Journal of Engineering Business Management*, first published December 20, 2017, 10.1177/1847979017744916.

Brooker, P., Barnett, J. and Cribbin, T. (2016) 'Doing Social Media Analytics', *Big Data & Society*, first published July 8, 2016, 10.1177/2053951716658060.

Brooker, P., Barnett, J., Vines, J., Lawson, S., Feltwell, T., Long, K. and Wood, G. (2018) 'Researching with Twitter Timeline Data: A Demonstration via "Everyday" Socio-Political Talk around Welfare Provision', *Big Data & Society*, first published April 3, 2018, 10.1177/2053951718766624.

Felt, M. (2016) 'Social Media and the Social Sciences: How Researchers Employ Big Data Analytics', *Big Data & Society*, first published April 29, 2016, 10.1177/2053951716645828.

Ganis, M. and Kohirkar, A. (2016) *Social Media Analytics: Techniques and Insights for Extracting Business Value Out of Social Media*. Indianapolis, IN: IMB Press.

Ji, X., Chun, S., Cappellari, P. and Geller, J. (2017) 'Linking and Using Social Media Data for Enhancing Public Health Analytics', *Journal of Information Science*, 43(2), 221–45, first published February 12, 2016, 10.1177/0165551515625029.

Marres, N. and Gerlitz, C. (2016) 'Interface Methods: Renegotiating Relations between Digital Social Research, STS and Sociology', *The Sociological Review*, 64(1), 21–46, first published February 1, 2016, 10.1111/1467-954X.12314.

McKelvey, F. (2015) 'Openness Compromised? Questioning the Role of Openness in Digital Methods and Contemporary Critical Praxis'. In Langlois, G., Redden, J. and Elmer, G. (eds.) *Compromised Data: From Social Media to Big Data.* New York, NY: Bloomsbury, 126–46.

Richterich, A. (2018) *The Big Data Agenda: Data Ethics and Critical Data Studies.* London: University of Westminster Press.

Schroeder, R. (2014) 'Big Data and the Brave New World of Social Media Research', *Big Data & Society*, first published December 17, 2014, 10.1177/2053951714563194.

Yang, J., Tsou, M., Jung, C., Allen, C., Spitzberg, B., Gawron, J. and Han, S. (2016) 'Social Media Analytics and Research Testbed (SMART): Exploring Spatiotemporal Patterns of Human Dynamics with Geo-Targeted Social Media Messages', *Big Data & Society*, first published June 24, 2016, 10.1177/2053951716652914.

Zafarani, R., Abbasi, M. and Liu, H. (2014) *Social Media Mining: An Introduction.* New York, NY: Cambridge University Press.

Social network analysis

Overview

Social network analysis enables researchers to map, analyse and represent patterns of social connections and flows between individuals, groups, organisations, computers and/or websites. It seeks to describe, explain, predict and/or model the structure of relationships among social entities and helps to build an understanding of how interdependence and social networks influence behaviour (constrain or enable certain behaviours and actions, for example). It can also be applied to non-human entities such as concepts, text and artefacts. Full or complete network analysis studies the complete network, which includes all relationships or interactions (ties) within a defined population. Egocentric social network analysis studies the structure, function and composition of ties around the individual (the ego). Examples of key terms and concepts associated with these approaches are given below. Social network analysis has a long, pre-digital history (see Prell, 2012 for detailed information on the history, theory and methodology of social network analysis). However, it has been included in this book because there are now a wide variety of software packages and digital tools available to aid analysis and there are numerous digital platforms, media and networks available to provide data for analysis.

Social network analysis is primarily a quantitative method that borrows from graph theory (the theoretical study of graphs, their structure and application) and sociometry (the study and measurement of relationships within a group of people). It also borrows from qualitative methods used by ethnographers interested in kinship, friendship, communities and interpersonal relations. If you are interested in finding out more about qualitative social network analysis,

Edwards (2010) provides an interesting and enlightening methodological discussion, describing qualitative methods such as walking interviews, participatory mapping, observation, network maps and in-depth interviewing.

Social network analysis is used in a wide range of disciplines and subject areas, examples of which include environmental governance (Scott, 2015), health and medicine (Hether et al., 2016), health and international development (Drew et al., 2011), social psychology (Clifton and Webster, 2017), applied criminology (Bichler et al., 2017), evaluation (Popelier, 2018) and the sociology of sport (Dobbels et al., 2018). Examples of research projects that have used, analysed or critiqued social network analysis include research into how small states acquire status (Baxter et al., 2018); research to identify stakeholders' influence on energy efficiency of housing in Australia (Zedan and Miller, 2017); a study that considers the potential for using affiliation networks to aid serial homicide investigation (Bichler et al., 2017); and a review of the current and potential use of social network analysis for evaluation purposes (Popelier, 2018).

If you are interested in finding out more about social network analysis, Scott (2017) provides a comprehensive account, including information about the history of social network analysis, methods of data collection, terminology, visualising and modelling. Borgatti et al. (2018) and Prell (2012) also provide a comprehensive account of social network analysis, including history, methods and techniques. If you are interested in a conceptual rather than mathematical approach to social network analysis, Yang et al. (2017) provide interesting reading. Edwards (2010) will help you to consider whether a mixed methods approach to social network analysis is appropriate for your research, giving examples of previous research projects and providing an enlightening theoretical discussion on a mixed methods approach. There are a wide variety of digital tools and software packages available that enable researchers to visualise and measure networks and some examples of these are given below. You may also find it useful to find out more about link analysis (Chapter 25), data mining (Chapter 12), log file analysis (Chapter 28), digital ethnography (Chapter 14), social media analytics (Chapter 52) and data visualisation (Chapter 13).

Questions for reflection

Epistemology, theoretical perspective and methodology

- On what theoretical assumptions do social network analyses rely? Prell (2012) will help you to consider this question.

- Have you considered adopting a mixed methods approach to social network analysis? Edwards (2010: 2) notes that 'mixing quantitative and qualitative approaches can enable researchers to explore the structure (or form) of networks from an "outsider's" view, and the content and processes of networks from an "insider's" view'. She also observes that a qualitative approach enables 'analysts to consider issues relating to the construction, reproduction, variability and dynamics of complex social ties'. How might these observations relate to your research?

- Social media platforms and networks are subject to continuous change, perhaps to improve customer experience or raise additional revenue. This can change user behaviour (algorithms can be used to assess, classify, enhance and shape behaviour, for example). What impact might this have on your research design?

- Have you assumed that complete network information is available? How might missing information about people/objects and their connections bias your analysis?

- What bias might be present in visual perception and what effect might this have on social network analysis? How can this type of bias be eliminated, reduced or acknowledged?

- Might there, on occasions, be a need to consider non-network data to help answer a particular question or explain patterns or relationships? Does your methodological framework allow for this?

- How might you move beyond descriptive network analysis? Popelier (2018) discusses how this can be done using alternative sources of network data, qualitative methods and inferential statistics, for example.

Ethics, morals and legal issues

- How will you deal with issues of informed consent and, if you decide to include an opt-out clause, what impact might this have on your analysis? In cases where it is not possible to obtain informed consent (where thousands of posts are subject to analysis, for example) what legal issues and fair use policies are relevant (the UK Data Protection Act 2018 and the EU General Data Protection Regulation, for example)? Hutton and Henderson (2015) will help you to consider this issue in more detail.

- How can you protect individual privacy? This is of particular importance when asking people to identify part of their network (it also has implications for informed consent: the person identifying their social network may have given consent, but others within the network may not have given consent).

- How can you ensure anonymity? Is it possible for third parties to guess the name of individuals due to their location in the network?

- Is it possible that your social network analysis could cause harm to individuals? Could conclusions be drawn that might lead to disciplinary action or dismissal of employees, for example? Could the analysis be used to make changes that could cause psychological harm for individuals?

Practicalities

- How do you intend to select your networks? How will you take account of overlapping, multiple structures in social life? How will you justify your network boundary?

- Do you have the required understanding of statistical techniques to undertake social network analysis? Scott (2017) and Borgatti et al. (2018) provide comprehensive information that will help you to develop these techniques.

- Do you understand what is meant by key terms and/or concepts associated with social network analysis? This can include:
  o nodes (individual people, items, objects or web pages within the network);
  o ties (relationships or interactions: they can be strong, weak, absent, positive or negative, for example);
  o arc (a tie with direction);
  o edge (a tie without direction);
  o centrality (a range of measures to quantify the importance or influence of a person, object or web page within a network);
  o density (the proportion of direct ties in a network relative to the total number possible);
  o distance (the minimum number of ties required to connect two people, objects or web pages);
  o structural holes (the absence of ties between two parts of a network);
  o bridge (a weak tie that fills a structural hole and connects two individuals, objects or web pages).

- Is visualisation the best way to interpret node and graph properties, or are these best interpreted through quantitative analysis (can visual representations misrepresent structural properties, for example)?

Useful resources

There are a variety of digital tools and software packages available that enable researchers to undertake social network analysis or undertake statistical modelling of networks. Some are aimed at researchers in disciplines such as the social sciences, health and medicine and physics, whereas others are aimed at business and the search engine optimisation (SEO) sector. Examples available at time of writing are given below, in alphabetical order.

- Harmari Web Identity Search Tool (www.harmari.com/web-identity-search-tool-wist);
- InFlow (www.orgnet.com/software.html);
- Jung (http://jung.sourceforge.net);
- ModeXL (http://nodexlgraphgallery.org/Pages/AboutNodeXL.aspx);
- Network Workbench (http://nwb.cns.iu.edu);
- Pajek (http://mrvar.fdv.uni-lj.si/pajek/);
- Sentinel Visualizer (www.fmsasg.com);
- Social Network Visualizer (http://socnetv.org);
- statnet (http://statnetproject.org);
- UCINET 6, NetDraw and Krackplot (www.analytictech.com/products.htm);
- Voson (http://vosonlab.net/VOSON).

The International Network for Social Network Analysis (www.insna.org) is 'the professional association for researchers interested in social network analysis'. You can obtain details of conferences and events and find useful publications on their website, including the *Journal of Social Structures*.

Key texts

Baxter, P., Jordan, J. and Rubin, L. (2018) 'How Small States Acquire Status: A Social Network Analysis', *International Area Studies Review*, first published May 22, 2018, 10.1177/2233865918776844.

Bichler, G., Lim, S. and Larin, E. (2017) 'Tactical Social Network Analysis: Using Affiliation Networks to Aid Serial Homicide Investigation', *Homicide Studies*, 21(2), 133–58, first published September 27, 2016, 10.1177/1088767916671351.

Borgatti, S., Everett, M. and Johnson, J. (2018) *Analyzing Social Networks*, 2nd edition. London: Sage.

Clifton, A. and Webster, G. (2017) 'An Introduction to Social Network Analysis for Personality and Social Psychologists', *Social Psychological and Personality Science*, 8 (4), 442–53, first published June 15, 2017, 10.1177/1948550617709114.

Dobbels, L., Voets, J., Marlier, M., De Waegeneer, E. and Willem, A. (2018) 'Why Network Structure and Coordination Matter: A Social Network Analysis of Sport for Disadvantaged People', *International Review for the Sociology of Sport*, 53(5), 572–93, first published September 6, 2016, 10.1177/1012690216666273.

Drew, R., Aggleton, P., Chalmers, H. and Wood, K. (2011) 'Using Social Network Analysis to Evaluate a Complex Policy Network', *Evaluation*, 17(4), 383–94, first published October 11, 2011, 10.1177/1356389011421699.

Edwards, G. (2010) *Mixed-Methods Approaches to Social Network Analysis*. Southampton: ESRC National Centre for Research Methods, NCRM/015.

Hether, H., Murphy, S. and Valente, T. (2016) 'A Social Network Analysis of Supportive Interactions on Prenatal Sites', *Digital Health*, first published February 4, 2016, 10.1177/2055207616628700.

Hutton, L. and Henderson, T. (2015) '"I Didn't Sign up for This!" Informed Consent in Social Network Research', in *Proceedings of the 9th International AAAI Conference on Web and Social Media (ICWSM)*, 178–87, Oxford, United Kingdom, May 26, 2015.

Peseckas, R. (2016) 'Collecting Social Network Data from Mobile Phone SIM Cards', *Field Methods*, 28(2), 194–207, first published February 26, 2016, 10.1177/1525822X15627809.

Popelier, L. (2018) 'A Scoping Review on the Current and Potential Use of Social Network Analysis for Evaluation Purposes', *Evaluation*, 24(3), 325–52, first published July 19, 2018, 10.1177/1356389018782219.

Prell, C. (2012) *Social Network Analysis: History, Theory and Methodology*. London: Sage.

Scott, J. (2017) *Social Network Analysis*, 4th edition. London: Sage.

Scott, M. (2015) 'Re-Theorizing Social Network Analysis and Environmental Governance: Insights from Human Geography', *Progress in Human Geography*, 39(4), 449–63, first published October 20, 2014, 10.1177/0309132514554322.

Yang, S., Keller, F. and Zheng, L. (2017) *Social Network Analysis: Methods and Examples*. Thousand Oaks, CA: Sage.

Zedan, S. and Miller, W. (2017) 'Using Social Network Analysis to Identify Stakeholders' Influence on Energy Efficiency of Housing', *International Journal of Engineering Business Management*, first published June 12, 2017, 10.1177/1847979017712629.

Zhang, M. (2010) 'Social Network Analysis: History, Concepts, and Research'. In Furht, B. (ed.) *Handbook of Social Network Technologies and Applications*, 3–21. New York, NY: Springer.

CHAPTER 54

Spatial analysis and modelling

Overview

Spatial analysis refers to the process of (and set of techniques used for) acquiring, analysing and presenting spatial data. Spatial modelling refers to the analytical process of representing spatial data (cartographic models, network models, spatio-temporal models or optimisation models, for example). Spatial analysis and modelling can also be referred to as spatial data analysis (Lloyd, 2010; Haining, 2003) or spatial analysis modelling (Patterson, 2016). Some organisations also use the term 'geospatial analysis' when referring to spatial analysis (this can be seen on some of the websites that provide spatial analysis tools and software, listed below). However, 'spatial' relates to the position, area, shape and size of places, spaces, features, objects or phenomena, whereas 'geospatial' relates to data that contain a specific geographic component (locational information). This chapter discusses spatial analysis and modelling and Chapter 21 discusses geospatial analysis, along with methods and techniques such as locational analysis, distance analysis, surface analysis and geovisualisation.

Spatial analysis and modelling involve the exploration of, and interaction with, spatial data, maps, tables, charts, graphs and multimedia. The methods can be used to discover patterns and trends, test hypotheses, describe and summarise, explain, measure, transform, optimise, problem-solve, predict and plan. Although spatial analysis has a long pre-digital history (see Simpson, 1989 for a discussion of the conceptual and historical development of spatial analysis in relation to landscape planning), it has been included in this book because of rapid technological development and increased use of software and digital tools for spatial analysis (and for spatial modelling). Also, the development, expansion and use of geographic information systems (GIS), remote-sensing technologies (Chapter 49), location awareness and tracking technologies (Chapter 27) and mobile technologies (Chapter 32) is providing access to many and varied sources of spatial data available for research purposes.

Spatial analysis and modelling is used by researchers approaching their work from various disciplines and fields of study including geography, ecology, sociology, history, planning and development, architecture, urban studies, transportation, waste management and criminology. Examples of research projects that have used, assessed or critiqued spatial analysis and modelling include a study into the urban occupation of Portuguese municipalities (Abrantes et al., 2017); a spatial analysis of motor vehicle theft and motor vehicle recovery in Colorado Springs, USA (Piza et al., 2017); an investigation into the use of spatial analysis modelling procedures based on space syntax to investigate the potential to model aggregate traffic flows at an urban scale (Patterson, 2016); the presentation of a framework to help practitioners and researchers choose the most appropriate spatial method of measuring alcohol outlet density (Fry et al., 2018); and an evaluation of the effect of varying spatial parameters on the economic performance of hierarchical structures within one waste transfer station (Tanguy et al., 2016).

If you are interested in finding out more about spatial analysis and modelling for your research a good starting point is Behnisch and Meinel (2018). This is a collection of papers divided into four parts covering the research of settlements and infrastructure; geographic data mining; spatial modelling, system dynamics and geosimulation; and multi-scale representation and analysis. It is aimed at those interested in the spatial sciences, urban planning and the planning of human settlement and infrastructure. If you are interested in ecology, Dale and Fortin (2014) provide a comprehensive and practical guide to spatial analysis, covering issues such as spatial and spatio-temporal graph theory, scan statistics, fibre process analysis and Hierarchical Bayesian analysis. Lloyd (2010) provides a practical guide for those interested in finding out more about specific techniques for analysing spatial data, covering issues such as key concepts, combining data layers, network analysis, exploring spatial point patterns, exploring data patterning in data values and spatial interpolation. Haining (2003) aims his book at those interested in scientific and policy-related research, with chapters divided into five sections: 1) the context for spatial data analysis, 2) spatial data: obtaining data and quality issues, 3) the exploratory analysis of spatial data, 4) hypothesis testing and autocorrelation and 5) modelling spatial data. More information about computer modelling and simulation is provided in Chapter 7 and more information about data visualisation is provided in Chapter 13. You may also find it useful to obtain more information about geospatial analysis (Chapter 21), location awareness and location tracking (Chapter 27), zoning and zone mapping (Chapter 60) and predictive modelling (Chapter 45).

Questions for reflection

Epistemology, theoretical perspective and methodology

- How do you intend to approach spatial analysis and modelling in your research? This could be an exploratory/descriptive approach where your aim is to explore and describe spatial patterns and relationships, or it could be an explanatory/confirmatory approach where your aim is to test your hypothesis about spatial patterns and relationships, for example.

- Is there scope for combining spatial analysis with qualitative approaches? Rucks-Ahidiana and Bierbaum (2017), for example, discuss the triangulation of spatial and qualitative data and illustrate how spatial analysis enhances depth and rigor in qualitative work across the social sciences.

- What is the utility or applicability of a spatial model to a given situation? Are models reliable and will they produce identical answers? Are algorithms mathematically robust and reliable? Are visual representations and analytical results accurate and reliable?

- How can you avoid errors in the design, analysis and interpretation of your spatial analysis? Ocaña-Riola (2010) provides insight into these issues, in relation to a study on small-area epidemiological studies.

Ethics, morals and legal issues

- How can you ensure location privacy of individuals? This can involve obscuring techniques such as deleting, randomising, anonymising, substituting, shuffling or encrypting, for example. Data masking or data obfuscation techniques of this type relate also to the protection of personal identifiable data, personal sensitive data or commercially-sensitive data.

- How can you avoid risking participant confidentiality in your analyses? Haley et al. (2016: 49) consider the issue of participant confidentiality through providing a 'scoping review of sexual and reproductive health literature published and indexed in PubMed between January 1, 2013 and September 1, 2015' and providing 'readers with a primer on key confidentiality considerations when utilizing linked social-spatial data for visualizing results'.

- What are the ethical issues that surround the use of privately-harvested data by publically-funded academics? This issue is addressed by Lovelace et al. (2014) in their paper that asks whether social media data can be useful in spatial modelling.

- Are strengths and weaknesses of models understood (or misunderstood) by stakeholders or decision-makers? How can proper use be made of models? How can you avoid misuse (or potential abuse) of models?

Practicalities

- What spatial data type do you intend to work with? For example, there are two primary types of spatial data in GIS: raster data, made up of pixels (or grid cells) and vector data, made up of vertices and paths (points, lines, networks and polygons). The natural environment such as rainfall, temperature, land cover and elevation is represented in raster grids: a discreet raster has a distinct theme or category, whereas a continuous raster has gradual change (these models are useful for data that vary and change on a continual basis). Features within the built environment such as roads, canals, pipelines and building footprints, and administrative features such as boundaries, districts and urban centres, are represented as vector data. If required, it is possible to covert one data type to the other (see Chapter 11). When making your decision you need to consider whether data are continuous or discreet, the required level of accuracy, the complexity of algorithms, flexibility, speed of processing and data storage, for example.

- Do you need to convert scanned images into vector-based feature layers? If so, visit http://desktop.arcgis.com/en/arcmap/latest/extensions/arcscan/what-is-arcscan-.htm, which give details of ArcScan software that enables you to covert raster data to vector features (vectorisation). More information about data conversion is provided in Chapter 11.

- Do you have a good understanding of spatial analysis methods and techniques? This can include, for example:
  - factor analysis (used to describe variability among observed, correlated variables);
  - spatial regression (used to model spatial relationships);

- ○ spatial interaction (used to estimate the flow of people, objects or information between locations);

- ○ spatial autocorrelation (used to measure the degree of dependency among observations in a particular location or space).

Useful resources

There are a variety of digital tools and software packages available that enable researchers to undertake spatial analysis and modelling. Some are aimed at researchers in disciplines such as the social sciences, ecology and geography, whereas others are aimed at business and the search engine optimisation (SEO) sector. Examples available at time of writing are given below, in alphabetical order. Additional platforms, tools and software that may be of interest can be found in Chapter 21.

- ArcGIS (www.esri.com/arcgis);
- Biodiverse (http://shawnlaffan.github.io/biodiverse);
- CARTO (https://carto.com);
- Croizat (https://croizat.sourceforge.io);
- eSpatial (www.espatial.com);
- FRAGSTATS (www.umass.edu/landeco/research/fragstats/fragstats.html);
- Google Earth (www.google.com/earth);
- Map Comparison Kit (http://mck.riks.nl);
- PySAL (http://pysal.org);
- Saga (www.saga-gis.org/en/index.html);
- SANET (http://sanet.csis.u-tokyo.ac.jp);
- SaTScan (www.satscan.org);
- Spatial Ecology Program (www.cnfer.on.ca/SEP);
- TerrSet (https://clarklabs.org/terrset).

R Spatial (www.rspatial.org) is a website that provides detailed information about spatial data analysis and modelling with R. It includes comprehensive guidance on spatial data manipulation with R, spatial data analysis and remote sensing image analysis, along with some interesting case studies.

Key texts

Abrantes, P., Rocha, J., Marques da Costa, E., Gomes, E., Morgado, P. and Costa, N. (2017) 'Modelling Urban Form: A Multidimensional Typology of Urban Occupation for Spatial Analysis', *Environment and Planning B: Urban Analytics and City Science*, first published March 29, 2017, 10.1177/2399808317700140.

Behnisch, M. and Meinel, G. (eds.) (2018) *Trends in Spatial Analysis and Modelling: Decision-Support and Planning Strategies*. Cham: Springer.

Dale, M. and Fortin, K. (2014) *Spatial Analysis: A Guide for Ecologists*, 2nd edition. Cambridge: Cambridge University Press.

Fry, R., Orford, S., Rodgers, S., Morgan, J. and Fone, D. (2018) 'A Best Practice Framework to Measure Spatial Variation in Alcohol Availability', *Environment and Planning B: Urban Analytics and City Science*, first published May 11, 2018, 10.1177/2399808318773761.

Haining, R. (2003) *Spatial Data Analysis: Theory and Practice*. Cambridge: Cambridge University Press.

Haley, D., Matthews, S., Cooper, H., Haardörfer, H., Adimora, A., Wingood, G. and Kramer, M. (2016) 'Confidentiality Considerations for Use of Social-Spatial Data on the Social Determinants of Health: Sexual and Reproductive Health Case Study', *Social Science & Medicine*, 166, 49–56, first published August 08, 2016, 10.1016/j.socscimed.2016.08.009.

Lloyd, C. (2010) *Spatial Data Analysis: An Introduction for GIS Users*. Oxford: Oxford University Press.

Lovelace, R., Birkin, M. and Malleson, N. (2014) 'Can Social Media Data Be Useful in Spatial Modelling? A Case Study of "Museum Tweets" and Visitor Flows'. In *GISRUK 2014*, 16–18, April 2014, University of Glasgow, Scotland.

Ocaña-Riola, R. (2010) 'Common Errors in Disease Mapping', *Geospatial Health*, 4(2), 139–54, first published May 1, 2010, 10.4081/gh.2010.196.

Patterson, J. (2016) 'Traffic Modelling in Cities – Validation of Space Syntax at an Urban Scale', *Indoor and Built Environment*, 25(7), 1163–78, first published July 4, 2016, 10.1177/1420326X16657675.

Piza, E., Feng, S., Kennedy, L. and Caplan, J. (2017) 'Place-Based Correlates of Motor Vehicle Theft and Recovery: Measuring Spatial Influence across Neighbourhood Context', *Urban Studies*, 54(13), 2998–21, first published August 24, 2016, 10.1177/0042098016664299.

Rucks-Ahidiana, Z. and Bierbaum, A. (2017) 'Qualitative Spaces: Integrating Spatial Analysis for a Mixed Methods Approach', *International Journal of Qualitative Methods*, first published May 26, 2017, 10.1177/160940691501400208.

Simpson, J. (1989) 'A Conceptual and Historical Basis for Spatial Analysis', *Landscape and Urban Planning*, 17(4), 313–21, 10.1016/0169-2046(89)90085-6.

Tanguy, A., Glaus, M., Laforest, V., Villot, J. and Hausler, R. (2016) 'A Spatial Analysis of Hierarchical Waste Transport Structures under Growing Demand', *Waste Management & Research*, 34(10), 1064–73, first published July 20, 2016, 10.1177/0734242X16658544.

CHAPTER 55

Video analysis

Overview

Video analysis is a term that is used to describe the study of visual recordings that have been produced from video cameras, camcorders, mobile phones, tablets, drones and video surveillance systems, for example. Video analysis can be used to investigate human behaviour, action and interaction; to analyse movement and motion; and to determine environmental, temporal and spatial events, for example. There are different types, terms and/or methods of video analysis that are used within different disciplines and fields of study and for different purposes. The more common of these are listed below.

- Video analysis is a generic term that tends to be used in the social sciences, education, science and technology, arts and the humanities to describe various methods that are used to study visual recordings. These methods are framed and guided by epistemology, theoretical perspective and methodology, and can include the following.
  - Content analysis of video that quantifies visual images (and text) using reliable and previously defined categories that are exhaustive and mutually exclusive. Using this method the researcher works through visual data identifying, coding and counting the presence of images, themes, characters, words, phrases and sentences, for example. See Basch et al. (2017) for an example of this type of video analysis. Inductive content analysis of video, instead of having predefined categories, uses an inductive approach to develop categories, identify themes and develop theory from video.

○ Video interaction analysis involves the study of social action and interaction on video, with a systematic consideration of the context in which the recorded interaction is embedded. Using this method the researcher analyses data from the perspective of those in the video, using techniques that analyse the turns involved in talk and action (conversation analysis or sequential analysis, for example). It is an interpretative method that has its roots in hermeneutics. Knoblauch and Schnettler (2012) provide an interesting methodological discussion that covers this type of video analysis.

○ Discourse analysis of video involves an interpretative and deconstructive reading of video but does not provide definitive answers or reveal the truth. Using this method researchers systematically deconstruct (and reconstruct) the various layers of meaning derived from different contexts of use. Researchers ask how videos are given particular meanings, how meanings are produced and how a particular video works to persuade (with a focus on claims to truth), for example. See Schieble et al. (2015) for an example of this type of video analysis.

○ Visual rhetoric of video considers the message that is being portrayed in a video and the meaning that is being communicated. It is related to visual semiotics, which looks at the way that videos communicate a message through signs, the relationship between signs (including systems of meaning that these create) and patterns of symbolism. However, it is a broader method of video analysis in that, in addition to considering signs and symbols, it also considers other forms of human communication through video such as colours, form, medium, scale, location, audience, context and purpose. Every part of a video has significance in the message that is being communicated. Hill and Helmers (2009) provide detailed information about visual rhetoric.

○ Psychoanalysis of video can guide the interpretation of videos and their effect on individual spectators. Researchers work with psychoanalytic concepts to explore how they are articulated through a particular video. They are interested in issues such as space, gaze, spectator, difference, resistance and identification, for example. Holliday (2004) provides an example of this type of research in her paper on the role of video dairies in researching sexualities.

• Video motion analysis involves the investigation of movement and motion in video. It can be used by sports scientists to study and improve performance (gait analysis or task performance analysis, for example); by physicists to study speed, velocity and acceleration; by biologists and veterinary scientists to study animal movement and behaviour; and by medical researchers to investigate orthopaedic, neurological and neuro-muscular dysfunctions, for example. A related field is video traffic flow analysis that extracts traffic flow data from traffic monitoring cameras.

- Video content analysis (or video content analytics: VCA) is becoming widespread in the security and retail sectors, where it is used for purposes such as facial recognition, people counting, licence plate reading and tracking a moving target. It enables researchers to capture, index and search faces, colours and objects, for example. It is used for marketing and development purposes and to increase and improve security.

- Intelligent video analysis (or intelligent video analytics) is a term that is sometimes used interchangeably with VCA. Specifically it refers to sys-tems that perform automatic analyses of surveillance streams to extract useful information. Applications include motion detection, video pattern matching and auto tracking (following a moving object), for example.

Examples of research projects that have used, assessed or critiqued video analysis as a method include a study into how the research practice of video analysis can lead to methodological insight (Knoblauch and Schnettler, 2012); research into YouTube videos related to prostate cancer (Basch et al., 2017); a study that uses discourse analysis and positioning theory to enable teachers to analyse videos of their practice during student teaching (Schieble et al., 2015); and a study to 'analyse the use of weapons as part of the body posturing of robbers as they attempt to attain dominance' (Mosselman et al., 2018).

If you are interested in using video analysis for your research project, two good books to start with are Harris (2016) and Heath et al. (2010). These books will help you to understand, design, conduct, analyse and disseminate video-based research. There are a wide variety of digital tools and software packages available for video analysis: it is useful to visit some of the sites listed below so that you can get an idea of functions, capabilities and purpose. You may also find it useful to read Chapter 9, which provides information about computer-assisted qualitative data analysis software and Chapter 2, which discusses audio analysis. If you are interested in video game analysis, or virtual world analysis, more information is provided in Chapter 56.

Questions for reflection

Epistemology, theoretical perspective and methodology

- Are videos a representation of reality? Can they be accurate and true? Do they record phenomena or social behaviour in an objective way (indeed, is it possible to be objective)? Hassard et al. (2018: 1403) will help you to address this question in their paper where they 'supplant the objectivist and realist philosophy underpinning traditional documentary filmmaking with sociologically interpretive and reflexive arguments for undertaking ethnography in organizations'.

- How might videos be influenced by culture, history and politics (of the producer, viewer or viewed)?

- How does framing of a video (the literal presentation and the interpretative markers) influence our interpretation of a video?

- What role does the gaze of the viewer have in constructing meaning about a video? Who is the viewer? How and why are they viewing the video?

- Mengis et al. (2018: 288) point out that 'while video has a number of general affordances, the research practices with which we use it matter and have an impact both on the analytical process and the researcher's findings'. They go on to consider how video-based research methods 'have a performative effect on the object of inquiry and do not simply record it'. How might such observations relate to your research?

- Knoblauch and Schnettler (2012: 342) point out that 'even though video provides very rich and overly abundant data, audio-visual recordings have obvious technical limitations'. What might these limitations be and how can they be addressed in your research?

Ethics, morals and legal issues

- How will you address ethical issues such as gaining the necessary permissions, data protection and issues of anonymity, confidentiality and informed consent? Do these vary, depending on the type of video and production methods?

- How will you address issues of privacy when it is difficult or almost impossible to disguise individuals appearing on video? Are masking techniques necessary and what impact might these have on analysis, sharing and reporting?

Video analysis

- How will you ensure that participants understand and know about the purpose and nature of the video recording? This should include issues such as:
  - where and how it will be stored;
  - whether it will be copied and, if so, how and by whom;
  - how it will be labelled;
  - how it will be analysed;
  - who will be able to see the video;
  - how long it will be kept for;
  - how and when it will be deleted.

- Is there any bias present when videos are made and viewed (observer bias or selection bias, for example)? If so, how does this bias influence the video, the making of the video or how it is viewed?

- Do the people being recorded know that they are being recorded? Does this influence their behaviour and, if so, in what way?

- How can you ensure that editing does not obscure, misrepresent or introduce bias?

Practicalities

- If you are producing your own videos for analysis, what hardware is required? When choosing hardware you will need to consider some or all of the following:
  - handheld device, shoulder mounted camera or tripod;
  - weight and size;
  - internal or external microphone;
  - lighting and light reflectors;
  - battery life and recording time;
  - lens type (wide angle, clear, protective and filter, for example);
  - image stabilisation;
  - frame rate (a higher frame rate provides a more complete video and improves the viewing experience);

- ○ high definition (and the ability to send a clean high definition signal through video output ports);

- ○ optical zoom for higher image quality (or digital zoom if image quality is not so important);

- ○ recording medium (video tape, memory card, hard drive or DVD, for example);

- ○ storage (external, portable hard drive, for example);

- ○ cost and warranty.

- When choosing software, do you have a good understanding of what is available and how it will be of use to your research? This can include measurement, timing, editing and analysis software, for example (see below).

- If you are using videos produced by others, how and why was the video produced? Who produced it? What was their motivation for producing it? If you are interested in participatory video research, Whiting et al. (2018: 317) provide a 'reflexive analysis of the paradoxical nature of the relationships and roles produced by participatory video research', which will help you to think more deeply about these issues. A related area of research uses video diaries and these are discussed in Chapter 30.

- At what point in your research project does the video analysis begin? Is production part of the analysis, for example?

Useful resources

There are a variety of digital tools and software packages available for video analysis, with differing functions, capabilities, purposes and costs. Some are aimed at the professional sports coaching market, some at teachers and educators and some at social researchers, for example. The following list provides a snapshot of what is available at time of writing (in alphabetical order):

- Dartfish Prosuite (www.dartfish.com/pro);

- iOmniscient (http://iomniscient.com);

- Kinovea (www.kinovea.org);

- Mintrics (https://mintrics.com);

- MotionView™ video analysis software (https://allsportsystems.eu);

- OpenCV Python (https://pythonprogramming.net);

- PhysMo (http://physmo.sourceforge.net);

- Simple Video Coder (https://sourceforge.net/projects/simplevideocoder);

- Tracker (https://physlets.org/tracker);

- Transana (www.transana.com);

- Vidooly (https://vidooly.com).

Key texts

Basch, C., Menafro, A., Mongiovi, J., Hillyer, G. and Basch, C. (2017) 'A Content Analysis of YouTube Videos Related to Prostate Cancer', *American Journal of Men's Health*, 11 (1), 154–57, first published September 29, 2016, 10.1177/1557988316671459.

Harris, A. (2016) *Video as Method*. Oxford: Oxford University Press.

Hassard, J., Burns, D., Hyde, P. and Burns, J.-P. (2018) 'A Visual Turn for Organizational Ethnography: Embodying the Subject in Video-Based Research', *Organization Studies*, 39(10), 1403–24, first published September 22, 2017, 10.1177/0170840617727782.

Heath, C., Hindmarsh, J. and Luff, P. (2010) *Video in Qualitative Research*. London: Sage.

Hill, C. and Helmers, M. (eds.) (2009) *Defining Visual Rhetorics*. New York, NY: Routledge.

Holliday, R. (2004) 'Filming "The Closet": The Role of Video Diaries in Researching Sexualities', *American Behavioral Scientist*, 47(12), 1597–616, first published August 1, 2004, 10.1177/0002764204266239.

Knoblauch, H. and Schnettler, B. (2012) 'Videography: Analysing Video Data as a 'Focused' Ethnographic and Hermeneutical Exercise', *Qualitative Research*, 12(3), 334–56, first published June 6, 2012, 10.1177/1468794111436147.

Margolis, E. and Pauwels, L. (eds.) (2011) *The SAGE Handbook of Visual Research Methods*. London: Sage.

Mengis, J., Nicolini, D. and Gorli, M. (2018) 'The Video Production of Space: How Different Recording Practices Matter', *Organizational Research Methods*, 21(2), 288–315, first published September 29, 2016, 10.1177/1094428116669819.

Mosselman, F., Weenink, D. and Rosenkrantz Lindegaard, M. (2018) 'Weapons, Body Postures, and the Quest for Dominance in Robberies: A Qualitative Analysis of Video Footage', *Journal of Research in Crime and Delinquency*, 55(1), 3–26, first published January 16, 2018, 10.1177/0022427817706525.

Nassauer, A. and Legewie, N. (2018) 'Video Data Analysis: A Methodological Frame for A Novel Research Trend', *Sociological Methods & Research*, first published May 17, 2018, 10.1177/0049124118769093.

Schieble, M., Vetter, A. and Meacham, M. (2015) 'A Discourse Analytic Approach to Video Analysis of Teaching: Aligning Desired Identities with Practice', *Journal of*

Teacher Education, 66(3), 245–60, first published February 23, 2015, 10.1177/0022487115573264.

Whiting, R., Symon, G., Roby, H. and Chamakiotis, P. (2018) 'Who's behind the Lens? A Reflexive Analysis of Roles in Participatory Video Research', *Organizational Research Methods*, 21(2), 316–40, first published September 29, 2016, 10.1177/1094428116669818.

CHAPTER 56

Virtual world analysis

Overview

Virtual world analysis (or virtual environment analysis) is an umbrella term that is used to describe the various methods and techniques that are used to investigate human action, interaction, relationships and behaviour (carried out usually by means of a virtual self or avatar) in computer-based simulated environments. Examples of methods and techniques that can be used for virtual world analysis include discourse analysis (Gee, 2014), comparative case study analysis (Grimes, 2015), multi-level modelling (LaPorte et al., 2012), content analysis (Waddell et al., 2014), social network analysis (Stafford et al., 2012), statistical analysis of survey data (Mavoa et al., 2018) and hypothesis testing and experimentation (Burrows and Blanton, 2016).

There are various virtual worlds or environments that can be analysed by researchers. Some of these are listed below. It is important to note, however, that these categories are fluid and flexible, with some overlap and with terms used interchangeably (multi-player games can be both collaborative and role-playing, for example).

- Collaborative virtual environments (CVEs) used for collaboration between many members, often over large distances, to share experiences and generate ideas. These environments can be digital or a mixture of the virtual and real world (augmented reality). Examples include multi-player games, interactive art installations, surgery simulation, distance learning, architectural walkthroughs and consumer interaction for marketing purposes, for example. CVEs provide the opportunity for analyses from a wide variety of disciplines, including sports science, behavioural science, market research, health and medicine and management and organisation studies, for example. See Cantrell et al. (2010) for an example of research into CVEs from a medical and nursing perspective and Koles and Nagy (2014) for an example of research into CVEs from an organisational psychology perspective.

- Massive multi-player online (MMO) that enable a large number of players to participate simultaneously and communicate with each other in a shared world. These provide the opportunity for analyses from various disciplines and field of study, including demography, the social and behavioural sciences, gaming and communication studies. See Waddell et al. (2014) for an example of MMO demographic research and Burrows and Blanton (2016) for an example of MMO communications research. Within this category there are various genres, including:
  ○ role-playing games (RPG) or massively multiplayer online role-playing games (MMORPGs) such as World of Warcraft (https://worldofwar craft.com);

  ○ first-person shooter (FPS) or massively multiplayer online first-person shooter (MMOFPS) such as Call of Duty (www.callofduty.com);

  ○ real-time strategy (RTS) or massively multiplayer online real-time strategy (MMORTS) such as League of Legends (https://play. na.leagueoflegends.com).

- Simulation games (or real-life games) that imitate the real world or simulate a fictional reality, such as Second Life (https://secondlife.com). These provide the opportunity for analyses from various fields of study including early childhood research (Marsh, 2010), education (Nussli and Oh, 2015), gaming history and culture (Veerapen, 2013) and social psychology (Yee et al., 2011).

- Sports games that simulate sports (driving, football or sports-based fighting games, for example). These virtual worlds can be of interest to researchers approaching their work from the field of geriatrics, sports science, psychology and the behavioural sciences, for example. LaPorte et al. (2012) provide an example of sports game research within the field of geriatrics.

Examples of research projects that have used, assessed or critiqued virtual world analysis include a content analysis of representations of gender and race in MMO (Waddell et al., 2014); an examination of motor skill acquisition in a virtual world for older adults (LaPorte et al., 2012); an investigation into young children's play in online virtual worlds (Marsh, 2010); a study of patterns of identity development in virtual worlds (Nagy and Koles, 2014); an examination of the expression of personality in virtual worlds (Yee et al., 2011); an evaluation of virtual worlds teacher training (Nussli and Oh, 2015); an investigation into how boys play in a tween-centric virtual world (Searle and Kafai, 2012); and a study into the development of virtual worlds (Veerapen, 2013).

If you are interested in finding out more about virtual world analysis for your research project, a useful starting point is Hunsinger and Krotoski (2014) who provide a collection of essays covering a wide variety of topics on learning and research within virtual worlds. For those interested in an ethnographic approach Boellstorff et al. (2012) provide a comprehensive and practical guide to the study of game and nongame virtual environments. If you are interested in virtual worlds for children, Burke and Marsh (2013) provide an interesting collection of papers covering topics such as play, friendships and online identities; the power of brain-computer games; and digital play structures. For those who are interested in the law of virtual and augmented reality, Barfield and Blitz (2019) provide a comprehensive guide that covers the USA, Europe and 'other jurisdictions'. Depending on your chosen theoretical perspective and methodological standpoint, you may find it useful to obtain more information about digital ethnography (Chapter 14), online ethnography (Chapter 37), online observation (Chapter 41), social network analysis (Chapter 53), video analysis (Chapter 55) and audio analysis (Chapter 2).

Questions for reflection

Epistemology, theoretical perspective and methodology

- What theoretical perspective and methodological standpoint will guide your virtual world analysis? El Kamel and Rigaux-Bricmont (2011), for example, consider the 'contributions of postmodernism' in their examination of the virtual life of Second Life avatars, whereas Boellstorff et al. (2012) use an ethnographical approach to guide their research (see Chapter 14 for more information on digital ethnography and Chapter 37 for more information about online ethnography).

- How might selection of site and presentation of self in a virtual world influence study outcomes? Martey and Shiflett (2012) will help you to address this question and think further about methodological frameworks for virtual world research.

- How might biases in the real world translate to (or follow individuals into) the virtual world? How can you address and/or avoid bias in your virtual world analysis?

- Virtual worlds are subject to continuous change, perhaps to improve customer experience or raise additional revenue. This can change user behaviour. What impact might this have on your research design?

Ethics, morals and legal issues

- How can you ensure that you do not infringe intellectual property laws or copyright when analysing and reporting virtual world research (the use of images or patentable inventions within a virtual world, for example)?

- How might a researcher's appearance and behaviour within a virtual world influence the behaviour of others within the virtual world?

- Is it possible that your research into, or within, virtual worlds may uncover harmful virtual acts? If so, what are the legal consequences of such acts? What are your responsibilities for reporting such acts (and to whom) and what are the implications for issues of anonymity and confidentiality?

- What are the ethical implications when the virtual world and the real world merge (players collecting real badges or paying money for online currency, for example)? Can such merging cause harm or distress? If such problems arise during your research, how will you deal with them?

Practicalities

- What do you intend to analyse within the virtual world? This can include:
  - Images recorded as screenshots, screen captures or video captures, which can include images of how avatars interact with each other and how they use or navigate the objects around them; constructions of buildings, bases or cities and their development over time; and maps of terrain, environments or movement, for example. Information about different video analysis techniques is provided in Chapter 55 (video content analysis, video interaction analysis and video motion analysis, for example) and information about digital visual methods is

provided in Chapter 16 (visual rhetoric, visual semiotics and visual hermeneutics, for example).

○ Textual data such as log files (Chapter 28), textual avatars, textual descriptions of avatars, avatar text-chat, virtual world chat rooms, discussion posts, storylines and game narratives, for example. Textual data can be analysed in a variety of ways, including content analysis, discourse analysis, conversation analysis and thematic analysis (Chapter 52).

○ Audio recording such as exchanges between players, pieces of music, songs, narratives and stories. Information about the different techniques that can be used for analysing audio data is provided in Chapter 2 (multimodal analysis, psychoacoustic analysis, semantic audio analysis and sound analysis, for example).

- How do you intend to deal with what could, potentially, be huge amounts of data? How do you intend to weed out data that are not relevant to your study? What filter methods will you use? Do you need to spend time within the virtual environment to get to know more about interaction, behaviour and characters so that you can understand more about what data should and should not be collected and recorded, for example?

Useful resources

The *Journal of Virtual Worlds Research* (www.jvwresearch.org) is a 'transdisciplinary journal that engages a wide spectrum of scholarship and welcomes contributions from the many disciplines and approaches that intersect virtual worlds research'. Past issues of the journal and details of events can be found on the website.

Key texts

Barfield, W. and Blitz, M. (eds.) (2019) *Research Handbook on the Law of Virtual and Augmented Reality*. Cheltenham: Edward Elgar Publishing Ltd.

Boellstorff, T., Nardi, B., Pearce, C. and Taylor, T. (2012) *Ethnography and Virtual Worlds: A Handbook of Method*. Princeton, NJ: Princeton University Press.

Burke, A. and Marsh, J. (eds.) (2013) *Children's Virtual Play Worlds: Culture, Learning, and Participation*. New York, NY: Peter Lang.

Burrows, C. and Blanton, H. (2016) 'Real-World Persuasion From Virtual-World Campaigns: How Transportation Into Virtual Worlds Moderates In-Game Influence', *Communication Research*, 43(4), 542–70, first published December 9, 2015, 10.1177/0093650215619215.

Cantrell, K., Fischer, A., Bouzaher, A. and Bers, M. (2010) 'The Role of E-Mentorship in a Virtual World for Youth Transplant Recipients', *Journal of Pediatric Oncology Nursing*, 27(6), 344–55, first published October 21, 2010, 10.1177/1043454210372617.

El Kamel, L. and Rigaux-Bricmont, B. (2011) 'The Contributions of Postmodernism to the Analysis of Virtual Worlds as a Consumption Experience. The Case of Second Life', *Recherche et Applications en Marketing (English Edition)*, 26(3), 71–92, first published June 23, 2016, 10.1177/205157071102600304.

Gee, J. (2014) *Unified Discourse Analysis: Language, Reality, Virtual Worlds and Video Games*. Abingdon: Routledge.

Grimes, S. (2015) 'Playing by the Market Rules: Promotional Priorities and Commercialization in Children's Virtual Worlds', *Journal of Consumer Culture*, 15(1), 110–134, first published July 12, 2013, 10.1177/1469540513493209.

Hunsinger, J. and Krotoski, A. (eds.) (2014) *Learning and Research in Virtual Worlds*. Abingdon: Routledge.

Koles, B. and Nagy, P. (2014) 'Virtual Worlds as Digital Workplaces: Conceptualizing the Affordances of Virtual Worlds to Expand the Social and Professional Spheres in Organizations', *Organizational Psychology Review*, 4(2), 175–95, first published October 22, 2013, 10.1177/2041386613507074.

LaPorte, L., Collins McLaughlin, A., Whitlock, L., Gandy, M. and Trujillo, A. (2012) 'Motor Skill Acquisition in a Virtual World by Older Adults: Relationships between Age, Physical Activity, and Performance', *Proceedings of the Human Factors and Ergonomics Society Annual Meeting*, 56(1), 2084–88, first published December 20, 2016, 10.1177/1071181312561442.

Marsh, H. (2010) 'Young Children's Play in Online Virtual Worlds', *Journal of Early Childhood Research*, 8(1), 23–39, first published February 22, 2010, 10.1177/1476718X09345406.

Martey, R. and Shiflett, K. (2012) 'Reconsidering Site and Self: Methodological Frameworks for Virtual-World Research', *International Journal of Communication*, 6(2012), 105–26, retrieved from https://ijoc.org/index.php/ijoc/article/view/971 [accessed January 24, 2019].

Mavoa, J., Carter, M. and Gibbs, M. (2018) 'Children and Minecraft: A Survey of Children's Digital Play', *New Media & Society*, 20(9), 3283–03, first published December 19, 2017, 10.1177/1461444817745320.

Nagy, P. and Koles, B. (2014) 'The Digital Transformation of Human Identity: Towards a Conceptual Model of Virtual Identity in Virtual Worlds', *Convergence*, 20(3), 276–92, first published May 7, 2014, 10.1177/1354856514531532.

Nussli, N. and Oh, K. (2015) 'A Systematic, Inquiry-Based 7-Step Virtual Worlds Teacher Training', *E-Learning and Digital Media*, 12(5–6), 502–29, first published October 12, 2016, 10.1177/2042753016672900.

Searle, K. and Kafai, Y. (2012) 'Beyond Freedom of Movement: Boys Play in a Tween Virtual World', *Games and Culture*, 7(4), 281–304, first published August 17, 2012, 10.1177/1555412012454219.

Stafford, G., Luong, H., Gauch, J., Gauch, S. and Eno, J. (2012) 'Social Network Analysis of Virtual Worlds', in Huang, R., Ghorbani, A., Pasi, G., Yamaguchi, T., Yen, N. and Jin, B. (eds.) *Active Media Technology*, AMT 2012, Lecture Notes in Computer Science, 7669, Springer, Berlin, Heidelberg, 10.1007/978-3-642-35236-2_41.

Veerapen, M. (2013) 'Where Do Virtual Worlds Come From? A Genealogy of Second Life', *Games and Culture*, 8(2), 98–116, first published February 26, 2013, 10.1177/1555412013478683.

Virtual world analysis

Waddell, T., Ivory, J., Conde, R., Long, C. and McDonnell, R. (2014) 'White Man's Virtual World: A Systematic Content Analysis of Gender and Race in Massively Multiplayer Online Games', *Journal of Virtual Worlds Research*, 7(2), 1–14, published online May, 2014, 10.4101/jvwr.v7i2.7096.

Yee, N., Harris, H., Jabon, M. and Bailenson, J. (2011) 'The Expression of Personality in Virtual Worlds', *Social Psychological and Personality Science*, 2(1), 5–12, first published August 6, 2010, 10.1177/1948550610379056.

CHAPTER 57

Wearables-based research

Overview

Wearables are computational devices and/or digital sensors that are worn (or carried) on the body, used to track and digitise human activity and/or behaviour. They can be worn as hats, badges, waistbands, armbands, around the neck, in eyeglasses, embedded into textiles, or in other devices such as smartphones, headphones or wristwatches, for example. Wearables are used to sense environmental, behavioural, physiological and/or psychological phenomena from the user or from the environment in which they are worn. Some are single function such as monitors for a specific health condition or audio or video recording devices, whereas others are multifunction, tracking motion, brain activity, heart activity and muscle activity, for example. Data are transmitted to the nearest gateway or processing unit (a smartphone, PDA or computer, for example) where they are processed, displayed and/or transmitted to a distant server (a health facility or research lab, for example). The user can receive processed information in the form of alerts, feedback, visual or textual content, and can interact with wearables via the gateway. 'Non-wearable' sensors can also be used to track and digitise human activity and these are discussed with other types of sensor in Chapter 49 (Muro-de-la-Herran et al., 2014 illustrate how both wearables and non-wearables can be used together in medical research).

Wearables are used in the social and behavioural sciences to measure and interpret social behaviour (Pentland, 2008). In health and medicine they are used to monitor health and provide detailed insight into health behaviour and risks, help to improve quality of life, improve drug treatment and monitor disease and treatment (Mombers et al., 2016). They can provide protection and peace of mind for people with epilepsy (see www.empatica.com for an example) and enable patients and doctors to monitor arrhythmias once they have been discharged from hospital (see www.preventicesolutions.com for an example). They can also be used for remote health monitoring (Majumder et al., 2017). In market research they are used to provide real-time data from real-world environments (Radhakrishnan et al., 2016) and in education they are used to enhance learning and student engagement (Borthwick et al., 2015).

Devices and sensors that can be integrated into wearable products include the following (non-wearable sensors are listed in Chapter 49):

- accelerometers that measure acceleration of the body (physical activity or whether a person is sitting, standing or walking, for example);
- built-in microphones or cameras;
- GPS for displaying location (and whether a person is inside or outside);
- gyroscopic sensors for measuring gait or posture;
- infrared sensors that can detect whether the wearer is facing other people wearing similar sensors;
- pedometers that measure steps taken when walking or running;
- piezoelectric pressure sensors for measuring heart rate and respiration rate;
- radio-frequency identification (RFID) tags for storing information, security, authentication and brand protection, for tracking people and goods, and for contactless payment, for example;
- temperature sensors for monitoring body temperature.

There are a wide variety of wearables on the market and care should be taken to choose the most appropriate for your research project, with close attention paid to reliability, accuracy and suitability. Godfrey et al. (2018) provide a useful A–Z guide for those new to wearables-based research that will help with your choices. It is also important to consider demographics, experience and perceptions: who is comfortable wearing what type of device; the type and look of device and how it is perceived (Adapa et al., 2018); and the potential benefits to participating in a research study and how this might influence participant behaviour and actions (a free fitness wearable for taking part in the study and how this might influence the amount of physical activity undertaken, for example).

If you are interested in using wearables for your research project, consult Chaffin et al. (2017), who provide an interesting discussion covering different studies of wearable channels (Bluetooth, infrared and microphone) and the potential for these devices in behavioural research. Other useful information can be obtained from Homesense (http://cress.soc.surrey.ac.uk/web/projects/homesense), which is an ESRC-funded research project in the UK that aims to develop and demonstrate the use of digital sensors in social research. You will also find it useful to read Chapter 49, which discusses non-wearable sensor-based methods and provides pertinent questions for reflection and key resources that are relevant to both types of device and sensor.

Questions for reflection

Epistemology, theoretical perspective and methodology

- Can wearables-based research provide a true picture of objective reality? If an objective reality exists to be measured, how can you ensure that a true picture is obtained (taking into account issues such as accuracy, range, precision and calibration, for example)?

- Is there any place for subjective reality in wearables-based research? If so, in what way (taking account of the different lived realities of those using wearables, for example)?

- How can data collected by wearables complement data collected by humans?

- What effect might levels of social acceptability have on wearables research? Kelly and Gilbert (2018) provide an interesting discussion that will help you to address this question.

Ethics, morals and legal issues

- How will you deal with issues of informed consent, especially among vulnerable groups in health research, for example? If wearables contain recording devices and the wearer records others who are not part of the study, what are the implications for informed consent? Also, participants may find it easy to consent to providing information of which they are aware, but sensors may generate information of which participants are unaware (how their body is functioning, for example). How will you deal with this issue in your research?

- How do you intend to address issues of participant privacy, anonymity and confidentiality, in particular, when using wearables to record video or sound (see Chapter 55 for more information about video and Chapter 2 for more information about audio)?

- If you intend to use wearables with vulnerable people, how will you approach issues such as protection of civil liberties, avoiding stigma and effective response? These issues, along with other ethical issues, are addressed in a paper produced by the Alzheimer's Society in the UK, which can be accessed at www.alzheimers.org.uk/about-us/policy-and-influencing/what-we-think/safer-walking-technology [accessed January 24, 2019].

- What data protection measures are in place where data are in transition (transmitted from gateway to server over the internet, for example)? What encryption, security and authentication measures are in place? How can you protect against interception, insertion of rogue data or theft of data?

- How are you going to check for, and protect against, reactions that could occur when wearables are placed against skin, such as allergies or dermatitis? How can you ensure that wearables are comfortable and unobtrusive?

- Do you have a good understanding of current regulations relating to wearable devices? Erdmier et al. (2016) provide a useful overview for those interested in the healthcare industry.

Practicalities

- How will wearables be chosen? How can you ensure that the right wearable is chosen and that sub-standard equipment is avoided?

- How much will equipment cost (hardware and software)? Is it possible that accuracy will be compromised due to budget restraints?

- Is equipment readily available and do you, and research participants, have the necessary skills and experience to use the equipment?

- If multiple devices (types and models, for example) are to be used, are they compatible?

- Do you have a good understanding of the methods that can be used to transmit data from wearables? This can include, for example, wireless communication technology such as WiFi (www.wi-fi.org), Bluetooth (www.bluetooth.com), Zigbee (www.zigbee.org) or Ant (www.thisisant.com).

- Are you aware of potential technological problems and do you know how to find solutions (the influence of atmospheric conditions, battery/power source depletion, vibrations and calibration needs and requirements, for example)?

- Will your participants want to wear wearables? Sethumadhavan (2018) provides a useful discussion on this topic.

- Will participants be able to wear, and use, devices correctly (will people with memory loss or limited experience of this type of technology struggle, for example)?

- How might your participants' behaviour change as a result of taking part in wearables-based research (increasing their activity levels, for example)? Is it possible that participants might manipulate devices (shaking pedometers to obtain more 'steps' for example)? What can you do to avoid these problems?

Useful resources

There are a wide variety of wearables on the market, available for different purposes. Some examples that are available at time of writing are given below, in alphabetical order. Additional wearables are listed and discussed by Godfrey et al. (2018).

- ActiGraph for medical-grade wearable activity and sleep monitoring solutions (www.actigraphcorp.com);

- Apple watches for activity, health and connection (www.apple.com/uk/watch);

- Fitbit range of devices for tracking activity, sleep, weight and exercise (www.fitbit.com);

- Grace bracelet for automated tracking and cooling of hot flushes (www.gracecooling.com);

- Microsoft Band for health monitoring, communication and organisation (www.microsoft.com/en-us/band);

- Moodmetric ring for monitoring and managing stress (www.moodmetric.com);

- Muvi™ wearable camcorders and cameras (www.veho-muvi.com);

- MYLE wearable for recording, organising and transcribing (http://getmyle.com);

- Senstone bracelets, clips and pendants for recording voice, organising and transcribing (http://senstone.io);

- The BioNomadix® system of wearable wireless devices for scientific research (www.biopac.com/product-category/research/bionomadix-wireless-physiology);

- Vue Smartglasses for audio, phone calls and fitness (www.enjoyvue.com).

Further information about the use of digital sensors and wearables in social research can be obtained from http://sensors-in-social-research.net/. Here you can find details of short courses, research projects and publications covering the use of digital sensors for observing social life.

ResearchKit (http://researchkit.org) enables researchers and developers to create apps for medical research. These apps, once developed, can access and use data from thousands of enrolled consumers (collected from sensors), along with heart rate data from HealthKit (https://developer.apple.com/healthkit).

Key texts

Adapa, A., Fui-Hoon Nah, F., Hall, R., Siau, K. and Smith, S. (2018) 'Factors Influencing the Adoption of Smart Wearable Devices', *International Journal of Human–Computer Interaction*, 10.1080/10447318.2017.1357902.

Borthwick, A., Anderson, C., Finsness, E. and Foulger, T. (2015) 'Personal Wearable Technologies in Education: Value or Villain?', *Journal of Digital Learning in Teacher Education*, 31(3), 85–92, published online July 02, 2015, 10.1080/21532974.2015.1021982.

Chaffin, D., Heidl, R., Hollenbeck, J., Howe, M., Yu, A., Voorhees, C. and Calantone, R. (2017) 'The Promise and Perils of Wearable Sensors in Organizational Research', *Organizational Research Methods*, 20(1), 3–31, first published November 30, 2015, 10.1177/1094428115617004.

Erdmier, C., Hatcher, J. and Lee, M. (2016) 'Wearable Device Implications in the Healthcare Industry', *Journal of Medical Engineering & Technology*, 40(4), 141–48, published online March 24, 2016, 10.3109/03091902.2016.1153738.

Godfrey, A., Hetherington, V., Shum, H., Bonato, P., Lovell, N. and Stuart, S. (2018) 'From A to Z: Wearable Technology Explained', *Maturitas*, 113, July 2018, 40–47, 10.1016/j.maturitas.2018.04.012.

Kelly, N. and Gilbert, S. (2018) 'The Wearer, the Device, and Its Use: Advances in Understanding the Social Acceptability of Wearables', *Proceedings of the Human Factors and Ergonomics Society Annual Meeting*, 62(1), 1027–31, 10.1177/1541931218621237.

Majumder, S., Mondal, T. and Deen, M. (2017) 'Wearable Sensors for Remote Health Monitoring', *Sensors (Basel, Switzerland)*, 17(1), 130, 10.3390/s17010130.

Mombers, C., Legako, K. and Gilchrist, A. (2016) 'Identifying Medical Wearables and Sensor Technologies that Deliver Data on Clinical Endpoints', *British Journal of Clinical Pharmacology*, February 2016, 81(2),196–98, first published December 26, 2015, 10.1111/bcp.12818.

Muro-de-la-Herran, A., Garcia-Zapirain, B. and Mendez-Zorrilla, A. (2014) 'Gait Analysis Methods: An Overview of Wearable and Non-Wearable Systems, Highlighting Clinical Applications', *Sensors (Basel, Switzerland)*, 14(2), 3362–94, 10.3390/s140203362.

Pagkalos, I. and Petrou, L. (2016) 'SENHANCE: A Semantic Web Framework for Integrating Social and Hardware Sensors in e-Health', *Health Informatics Journal*, 22(3), 505–22, first published March 10, 2015, 10.1177/1460458215571642.

Pentland, A. (2008) *Honest Signals: How They Shape Our World*. Cambridge, MA: Massachusetts Institution of Technology.

Radhakrishnan, M., Eswaran, S., Misra, A., Chander, D. and Dasgupta, K. (2016) 'IRIS: Tapping Wearable Sensing to Capture In-Store Retail Insights on Shoppers', in *2016 IEEE International Conference on Pervasive Computing and Communications PerCom*. Sydney, March 14–19, 7456526-1-8, retrieved from https://ink.library.smu.edu.sg/sis_research/3238.

Sethumadhavan, A. (2018) 'Designing Wearables That Users Will Wear', *Ergonomics in Design*, 26(1), 29–29, first published December 8, 2017, 10.1177/1064804617747254.

Topol, E., Steinhubl, S. and Torkamani, A. (2015) 'Digital Medical Tools and Sensors', *Journal of the American Medical Association*, 313(4), 353–54, first published January 27, 2015, 10.1001/jama.2014.17125.

CHAPTER 58

Web and mobile analytics

Overview

Web analytics refers to the collection, analysis and reporting of web data for research and/or development. It enables researchers, developers and organisations to understand user behaviour within and across webpages; create, maintain and improve websites; and optimise usage. Mobile analytics (or mobile web analytics) refers to the collection, analysis and reporting of data generated by mobile platforms such as mobile sites and mobile apps, again, for research, development and/or optimisation. Web and mobile analytics enable researchers to collect and analyse information about web behaviour, search behaviour, website or app journey, visitor engagement and location, for example. The two methods use similar tools and techniques and have, therefore, been combined in this chapter to avoid repetition. Web and mobile analytics are covered by the umbrella term of data analytics (Chapter 10) and can be seen to be part of big data analytics (Chapter 3). They can also be used within, and to complement, business analytics (Chapter 4), social media analytics (Chapter 52), HR analytics (Chapter 22) and learning analytics (Chapter 24).

There are different types of web and mobile analytics, examples of which include (in alphabetical order):

Web:

- acquisition analytics to optimise traffic acquisition spending and measure costs and revenue performance for paid advertising campaigns, for example;

- behavioural analytics to discover patterns of behaviour in interaction data;

- keyword analytics that give insight into which words are used to drive people to a site, or which words and phrases are trending at a certain time;

- page analytics to determine whether a site is helping users to meet their needs, including data on how each page on a site is performing;

- session analytics to determine the amount of activity (hits, views or transactions, for example) on a site within a given time;

- traffic source analytics to determine the source of website traffic, including search engine traffic (organic traffic, cost per click traffic or display ads traffic), direct traffic or referral site traffic;

- user analytics that gather data about users, such as location, number and frequency of visits, activity on the site and problems encountered by the user on the site.

Mobile:

- app marketing analytics to monetise apps, increase revenue, attract and keep hold of customers;

- app performance analytics to assess the performance of an app, ascertain how it works on different devices, log crashes and errors, detect when the app is working slowly and make appropriate improvements;

- app session analytics to determine the length and frequency of interaction with an app over a given period of time;

- in-app analytics to find out how an app is used, how people behave within the app and how they navigate around the app;

- mobile location analytics that track and map human mobility (see Chapter 27 for information about location awareness and location tracking).

Web analytics is related closely to webometrics, which is the quantitative study of the World Wide Web (Chapter 59). The two differ in their roots and purpose: web analytics developed from software engineering, computer science and business for research, development and/or commercial purposes, whereas webometrics developed from information sciences for quantification, measurement and/or descriptive purposes. The former concentrates on capturing, analysing, visualising and reporting data and the latter on the application of mathematical and statistical models. Increasingly, however, the two are becoming more closely aligned, with both techniques applied to commercial websites for the purpose of competitive intelligence, to identify and analyse competitors and identify future trends.

Web and mobile analytics are used in a wide variety of disciplines and fields of study, including media and journalism (Tandoc, 2014), politics (Kimball and Payne, 2014), health and medicine (Kirk et al., 2012), international development (Taylor, 2016), mobile learning (Pham et al., 2018) and marketing (Jackson, 2016). Examples of research projects that have used web or mobile analytics, or have considered the use of these analytics for students or in professional practice, include research into how journalists integrate audience feedback from web analytics in their news work (Tandoc, 2014); a project to provide students with hands-on experience in gauging public opinion, using web analytics alongside other methods (Kimball and Payne, 2014); using web analytics to monitor, assess and further develop a free web-based, multimedia genetics-genomics education resource (Kirk et al., 2012); and using mobile analytics to understand participant behaviour on a mobile English language app (Pham et al., 2018).

If you are interested in finding out more about web and mobile analytics, Kirk et al. (2012) provide a good, practical example of how web analytics was used in their research, including visitors and visits, bounce rate, location, traffic source, referring sites, keywords, language and mobile traffic. They also provide an interesting discussion on what web analytics cannot provide (demographics on individual users and occupational role of visitors, for example). Pham et al. (2018) illustrate how their research integrated the Google Firebase analytics tool into an English Practice app to enable them to analyse user data. Jackson (2016) provides in-depth advice for practitioners working in marketing who want to know more about web analytics and other types of data analytics. Clifton (2012) provides a comprehensive guide, including useful case studies and demonstrations, for anyone interested in using Google Analytics. There are a wide variety of digital tools and software packages available if you are interested in web and mobile analytics. Some examples are listed below. It is useful to visit some of these websites as tools, functions and features vary, depending on type and purpose. You may also find it useful to read related chapters on big data analytics (Chapter 3), business analytics (Chapter 4), data analytics (Chapter 10), data mining (Chapter 12), data visualisation (Chapter 13), social media analytics (Chapter 52) and webometrics (Chapter 59).

Questions for reflection

Epistemology, theoretical perspective and methodology

- How might epistemological assumptions (what we accept as knowledge, how we know what we know and how belief is justified) and epistemic beliefs (what we believe about the nature and acquisition of knowledge) shape the type of web or mobile analytics used, how they are used, the outcomes derived from their use and the knowledge generated from their use?

- Do you have a clear understanding of the difference between web analytics and webometrics (Chapter 59)? Which approach is most suitable for your research?

- Are you sure that web and/or mobile analytics is the most appropriate method to answer your research question (this is a clear, concise and complex question around which your research is focussed)? Would a combination of methods enable you to gain deeper insight? If so, which methods/tools are compatible with web and mobile analytics?

- Web and mobile technology are developing continually and rapidly: how can you take account of, and accommodate, change and ensure that your research is up-to-date and relevant?

- How does multiple device ownership affect web and mobile analytics (returning visitors cannot be identified if they use different devices in different locations, for example)?

- How can you avoid inaccurate predictions, misleading results or false conclusions in web and mobile analytics? How can you avoid mistaking correlation with causation? Problems can arise from inaccurate data, incorrect models, flawed algorithms, lack of rigorous validation and incorrect interpretation, for example.

Ethics, morals and legal issues

- How can you ensure that users know and understand what behaviour is being tracked, the depth of tracking and what data are being collected? Data collection processes must be disclosed fully, brought to the user's attention and explained in a way that can be understood easily by users so that they know what is being collected and why it is being collected. They need to be given the opportunity to agree to data being used for research purposes and need to be told how to opt out if they do not agree. Cookies are used to obtain web analytics tracking data. These are small files that hold specific data about a client and website that are stored on a user's computer. Placement of cookies, from organisations and third parties, requires informed consent. How can you ensure that this is the case (checking for a cookie policy information page or an informed consent notification when a website is visited, for example)? Information about rules on the use of cookies in the UK can be obtained from *Guidance on the rules on use of cookies and similar technologies*, produced by the Information Commission's Office, retrieved from https://ico.org.uk/media/1545/cookies_guidance.pdf [accessed January 25, 2019].

- How might web and mobile analytics represent, or fail to represent, data subjects? This point is raised by Taylor (2016) who provides an interesting discussion around this issue in relation to international development.

- How can you minimise, reduce or eliminate bias when undertaking web or mobile analytics? This can include issues such as integrity, reflexivity, sensitivity to context, rigour, transparency, validity and reliability, for example. Issues such as these are addressed in the web analyst code of ethics developed by the Digital Analytics Association (details below).

- Is it possible that users can be manipulated through the harvesting of personal web and mobile information or through the way that harvested information is used to categorise and target web and mobile users? An in-depth and insightful opinion paper on this topic has been produced by the European Data Protection Supervisor: *EDPS Opinion on online manipulation and personal data* (Opinion 3/2018) retrieved from https://edps .europa.eu/sites/edp/files/publication/18-03 19_online_manipulation_en.pdf [accessed January 25, 2019].

- How can you ensure that web and mobile analytics do not lead to discriminatory practice or cause harm to individuals?

Practicalities

- What tools and software are suitable for your project? How will you choose from the wide variety available? This could include decisions about price (open source, free, limited trail or prices based on a number of tiers, for example); features (real-time data points, revenue tracking, funnel development, event tracking, behavioural cohorts and A/B testing, for example); compatibility with your app; hosting services; user-friendliness and ease of use.

- Do you intend to use one tool or multiple tools? Will multiple tools complicate your research or provide further levels of insight? If you intend to use multiple tools, do you have a clear understanding of these tools and possible discrepancies that can arise? For example, web analytics tools that use client-side data to gather information may produce different results from log files that contain server-side information. Log files (or transaction logs) are a list of events that have been recorded, or logged, by a computer or mobile device and these are discussed in Chapter 28.

- If you intend to count number and frequency of visits to a website, are you aware of how results can be misleading or incorrect? For example, web-sites use cookies to differentiate between visitors. They hold a unique visitor ID that is recorded each time a visit is made. However, some users delete their cookie cache when they close their browser, which means that they would hold a new visitor ID when visiting the site at a later date. Users deleting their browser history may also lead to misleading data.

Useful resources

There are a wide variety of digital tools and software packages available at time of writing for those interested in web and mobile analytics. The tools listed (in alphabetical order) have been placed under separate headings, although you will see that some tools and software can be used for both web and mobile analytics.

Web analytics software and tools:

- Clicktale (www.clicktale.com);
- Clicky (https://clicky.com);
- Facebook Analytics (https://analytics.facebook.com);
- Goingup! (www.goingup.com);
- Google Analytics (www.google.com/analytics);
- Kissmetrics (www.kissmetrics.com);
- Linktrack (https://linktrack.info);
- Matomo (https://matomo.org);
- Open Web Analytics (www.openwebanalytics.com);

- Statcounter (https://statcounter.com);

- Woopra (www.woopra.com).

  Mobile analytics software and tools:

- App Analytics (https://developer.apple.com/app-store/app-analytics);

- Appsflyers (www.appsflyer.com);

- Apsalar (https://apsalar.com);

- AskingPoint (www.askingpoint.com);

- Countly (https://count.ly);

- Flurry (www.flurry.com);

- Google Mobile App Analytics (www.google.com/analytics);

- Localytics (www.localytics.com);

- Mixpanel (https://mixpanel.com);

- Taplytics (https://taplytics.com).

The Digital Analytics Association (www.digitalanalyticsassociation.org) was founded in 2004 as the Web Analytics Association. Its name was changed in 2011 to reflect broader approaches in digital analytics. The website contains useful blogs, news and events listings, along with a web analyst code of ethics that covers the topics of privacy, transparency, consumer control, education and accountability.

Key texts

Clifton, B. (2012) *Advanced Web Metrics with Google Analytics*, 3rd edition. Indianapolis, IN: John Wiley & Sons, Inc.

Jackson, S. (2016) *Cult of Analytics: Data Analytics for Marketing*. Abingdon: Routledge.

Jansen, B. (2009) *Understanding User-Web Interactions via Web Analytics (Synthesis Lectures on Information Concepts, Retrieval, and Services)*. San Rafael, CA: Morgan & Claypool.

Kimball, S. and Payne, J. (2014) 'Polling on Public Policy: A Case Study in Engaging Youth Voters in the Public Opinion Process for Effective Civic Discourse', *American Behavioral Scientist*, 58(6), 827–37, first published January 21, 2014, 10.1177/ 0002764213515224.

Kirk, M., Morgan, R., Tonkin, E., McDonald, K. and Skirton, H. (2012) 'An Objective Approach to Evaluating an Internet-Delivered Genetics Education Resource Developed for Nurses: Using Google Analytics to Monitor Global Visitor Engagement', *Journal of Research in Nursing*, 17(6), 557–79, first published October 29, 2012, 10.1177/ 1744987112458669.

Pham, X., Nguyen, T. and Chen, G. (2018) 'Research through the App Store: Understanding Participant Behavior on a Mobile English Learning App', *Journal of Educational Computing Research*, 56(7), 1076–98, first published November 6, 2017, 10.1177/0735633117727599.

Tandoc, E. (2014) 'Journalism Is Twerking? How Web Analytics Is Changing the Process of Gatekeeping', *New Media & Society*, 16(4), 559–75, first published April 11, 2014, 10.1177/1461444814530541.

Taylor, L. (2016) 'No Place to Hide? the Ethics and Analytics of Tracking Mobility Using Mobile Phone Data', *Environment and Planning D: Society and Space*, 34(2), 319–36, first published October 6, 2015, 10.1177/0263775815608851.

CHAPTER 59

Webometrics

Overview

Webometrics is a term that was coined by Almind and Ingwersen (1997) to describe the quantitative study of the World Wide Web. The term cyber-metrics is sometimes used interchangeably, although this technically covers the whole internet and includes non-web based communication and network-ing such as email, newsgroups and instant messaging (the World Wide Web is a way of accessing information over the medium of the internet: it is a service, whereas the internet is the whole computer network). Webometrics has its roots in the discipline of information science, in particular in bibliometrics (the quantitative study of books and other media of communication), sciento-metrics (the quantitative study of scientific activity and the history of science) and informetrics (the quantitative study of information and retrieval). All involve the application of mathematical and statistical methods, which are also used in webometrics to study patterns of use, the number and type of hyperlinks, the structure of the web and social media technologies, for example. Webometrics can involve various techniques and procedures, including (in alphabetical order):

- blog analysis;
- keyword analysis;
- link analysis, link impact analysis and co-link analysis (Chapter 25);
- log file analysis (Chapter 28);
- search engine evaluation;
- social network analysis (Chapter 53);

- video or vlog analysis (Chapter 55);

- web citation analysis;

- web data analysis;

- web data mining (Chapter 12);

- web impact assessment and analysis;

- web page content analysis;

- web technology analysis;

- web usage analysis.

Webometrics is closely related to web analytics (or web metrics), which is discussed in Chapter 58. The two differ in their roots and purpose: webometrics developed from information sciences for quantification, measurement and/or descriptive purposes, whereas web analytics developed from software engineering, computer science and business for research, development and/or commercial purposes. The former concentrates on the application of mathematical and statistical models and the latter on capturing, analysing, visualising and reporting data. Increasingly, however, webometric techniques are applied to commercial websites for the purpose of competitive intelligence, to identify and analyse competitors and identify future trends, thus aligning webometrics more closely with web analytics.

Although webometrics has its roots in information science and librarianship (see, for example, Arakaki and Willett, 2009) and was further developed to measure and quantify academic-related documents on the web (see for example, Figuerola and Berrocal, 2013; Kenekayoro et al., 2014; Ortega and Aguillo, 2008), it is also used as a research method in other disciplines, such as banking and finance (Vaughan and Romero-Frías, 2010), travel and tourism (Ying et al., 2016), politics (Lim and Park, 2013) and business and management studies (Romero-Frías, 2009). Specific examples of research projects that have used, assessed or critiqued webometrics include an investigation into the Web Impact Factor of Arab universities (Elgohary, 2008); a study to 'investigate the relationship between the web visibility network of Korea's National Assembly members and the amount of financial donations they receive from the public' (Lim and Park, 2013); an illustration of how webometric techniques can be applied to business and management studies (Romero-Frías, 2009); an analysis of the websites of UK departments of library and information science (Arakaki and Willett, 2009); and a study to explore 'the structural characteristics of the online networks among tourism stakeholders and their networking behaviors on the Web' (Ying et al., 2016).

If you are interested in undertaking webometrics for your research, a good grounding in the theory, methods and application of informetrics is provided by Qiu et al. (2017) who take time to discuss the relationship between informetrics, biblometrics, scientometrics, webometrics and scientific evaluation, before going on to consider various mathematical models and qualitative analysis techniques. Thelwall (2009) provides comprehensive information for social scientists who wish to use webometrics: the book could do with a little updating, but still raises many pertinent issues covering topics such as automatic search engine searches, web crawling, blog searching, link analysis and web impact assessment. More information about these topics, along with a detailed history of bibliometrics and webometrics can be found in Thelwall (2008). There are various webometrics and bibliometrics software packages and digital tools available and these are listed below. You might also find it useful to obtain more information about data analytics (Chapter 10), data mining (Chapter 12), web and mobile analytics (Chapter 58), information retrieval (Chapter 23) and link analysis (Chapter 25).

Questions for reflection

Epistemology, theoretical perspective and methodology

- Do you have a clear understanding of the difference between webometrics, web analytics, bibliometrics, scientometrics and informetrics? Which approach is most suitable for your research?

- What theoretical and methodological framework do you intend to use to guide your research? Ingwersen and Björneborn (2004) will help you to address this question.

- Thelwall (2008: 616) describes some of the shortcomings of webometrics: the web is not quality controlled, web data are not standardised, it can be impossible to find the publication date of a web page and web data are incomplete. How will you address these shortcomings when planning your research?

- The web and internet are developing continually and rapidly: how can you take account of, and accommodate, change and ensure that your research is up-to-date and relevant?

- Ying et al. (2016: 19) point out that an issue 'related to the quantitative nature of hyperlink network analysis is that it only examines the structure of hyperlink networks but cannot help interpret the content or the motivation of hyperlink creation without content-based web analysis'. How will you take account of this observation when planning your research?

- Lim and Park (2013: 104) point out that while quantitative analyses of online political relationships can count web mentions of politicians, they cannot distinguish whether or not these mentions are favourable. What impact might this observation have on your research? How can you address such problems when planning your research? Could they be overcome through adopting a mixed methods approach, for example?

Ethics, morals and legal issues

- Thelwall (2008: 612–13) points out that 'web coverage is partial' and that 'coverage is biased internationally in favour of countries that were early adopters of the web'. Do you think that this is still the case and, if so, what are the implications for your research?

- What other bias might be present when undertaking webometrics? This could include ranking bias, linguistic bias, search engine bias, geographical bias and sampling or sample selection bias (when choosing URLs, organisations or link structure and format such as clickable logos for example). How can you address, reduce or eliminate bias in your research?

- Vaughan and Romero-Frías (2010: 532) decided to use country-specific versions of search engines as they were concerned that 'search engines from different countries may have databases that favour websites from the host country'. What impact might such an observation have on your research?

- If you intend to rely on data from commercial search engines how can you check that commercial organisations have adhered to all ethical and legal requirements? How can you find out whether commercial organisations are using web data to manipulate, or discriminate against, certain users?

- Is it possible to mitigate unethical attempts to influence website rankings?

- What impact might advertising revenue have on search engine results and rankings?

Practicalities

- Do you have enough understanding of mathematical and statistical methods to undertake webometrics? Qiu et al. (2017) will help you to think more about relevant methods and techniques.

- Do you have a good understanding of digital tools and software packages that are available, and understand which are the most appropriate for your research? Some of these tools are listed below. You might also find it useful to review some of the tools listed in Chapter 58 (web and mobile analytics) and Chapter 25 (link analysis).

- When using search engines for your research, how accurate is your chosen search engine at returning results? What is the level of coverage? How relevant are the results? What impact do accuracy, level of coverage and relevance have on your research?

- If you intend to rely on data from commercial search engines how can you be sure of the reliability and validity of data? This is of particular importance in cases where search engines provide only partial data or limited data. How can you check that commercial organisations have addressed bias that might be present in the data collection and distribution processes (see above)?

Useful resources

The following list provides examples of webometrics and bibliometrics digital tools and software packages that were available at time of writing (in alphabetical order).

- Bing Web Search API (https://azure.microsoft.com/en-gb/services/cognitive-services/bing-web-search-api);
- CitNetExplorer (www.citnetexplorer.nl);
- Issue Crawler (www.govcom.org/Issuecrawler_instructions.htm);
- SocSciBot (http://socscibot.wlv.ac.uk/index.html);
- VOSviewer (www.vosviewer.com);
- Webometric Analyst (http://lexiurl.wlv.ac.uk).

The Ranking Web of Universities (www.webometrics.info/en) is an academic ranking of Higher Education Institutions. It began in 2004 and is carried out every six months by the Cybermetrics Lab, a research group belonging to the Consejo Superior de Investigaciones Científicas (CSIC), which is a public research body in Spain.

Key texts

Almind, T. and Ingwersen, P. (1997) 'Informetric Analyses on the World Wide Web: Methodological Approaches to "Webometrics"', *Journal of Documentation*, 53(4), 404–26, 10.1108/EUM0000000007205.

Arakaki, A. and Willett, P. (2009) 'Webometric Analysis of Departments of Librarianship and Information Science: A Follow-Up Study', *Journal of Information Science*, 35(2), 143–52, first published November 21, 2008, 10.1177/0165551508094051.

Elgohary, A. (2008) 'Arab Universities on the Web: A Webometric Study', *The Electronic Library*, 26(3), 374–86, 10.1108/02640470810879518.

Figuerola, C. and Berrocal, J. (2013) 'Web Link-Based Relationships among Top European Universities', *Journal of Information Science*, 39(5), 629–42, first published April 9, 2013, 10.1177/0165551513480579.

Ingwersen, P. and Björneborn, L. (2004) 'Methodological Issues of Webometric Studies'. In Moed, H., Glänzel, W. and Schmoch, U. (eds.) *Handbook of Quantitative Science and Technology Research*, 339–69. Dordrecht: Springer.

Kenekayoro, P., Buckley, K. and Thelwall, M. (2014) 'Hyperlinks as Inter-University Collaboration Indicators', *Journal of Information Science*, 40(4), 514–22, first published May 13, 2014, 10.1177/0165551514534141.

Lim, Y. and Park, H. (2013) 'The Structural Relationship between Politicians' Web Visibility and Political Finance Networks: A Case Study of South Korea's National Assembly Members', *New Media & Society*, 15(1), 93–108, first published November 1, 2012, 10.1177/1461444812457335.

Ortega, J. and Aguillo, I. (2008) 'Linking Patterns in European Union Countries: Geographical Maps of the European Academic Web Space', *Journal of Information Science*, 34 (5), 705–14, first published April 3, 2008, 10.1177/0165551507086990.

Qiu, J., Zhao, R. and Yang, S. (2017) *Informetrics: Theory, Methods and Applications*. Singapore: Springer.

Romero-Frías, E. (2009) 'Googling Companies – a Webometric Approach to Business Studies', *e-Journal of Business Research Methods*, 7(1), 93–106, open access, retrieved from www.ejbrm.com/issue/download.html?idArticle=206 [accessed January 25, 2019].

Thelwall, M. (2008) 'Bibliometrics to Webometrics', *Journal of Information Science*, 34(4), 605–21, first published June 13, 2008, 10.1177/0165551507087238.

Thelwall, M. (2009) *Introduction to Webometrics: Quantitative Web Research for the Social Sciences* (Synthesis Lectures on Information Concepts, Retrieval, and Services). San Rafael, CA: Morgan and Claypool.

Vaughan, L. and Romero-Frías, E. (2010) 'Web Hyperlink Patterns and the Financial Variables of the Global Banking Industry', *Journal of Information Science*, 36(4), 530–41, first published June 28, 2010, 10.1177/0165551510373961.

Ying, T., Norman, W. and Zhou, Y. (2016) 'Online Networking in the Tourism Industry: A Webometrics and Hyperlink Network Analysis', *Journal of Travel Research*, 55(1), 16–33, first published May 8, 2014, 10.1177/0047287514532371.

CHAPTER 60

Zoning and zone mapping

Overview

Zoning refers to the dividing of land or water into zones, or the assigning of zones onto land or water, for planning and/or management purposes. It is used in land-use planning; urban planning and development; transportation systems; commerce, business and taxation; and in environmental monitoring, management and research, for example. In the UK, zoning is also a method of measuring retail premises to determine and compare their value. Zone mapping is the act of physically mapping out zones or visualising zones or boundaries. Although zoning and zone mapping have a long pre-digital history (see Moga, 2017 for a good example of the history of zone mapping in American cities and Porter and Demeritt, 2012 for a history of Flood Mapping in the UK) it has been included in this book because digitisation and technological developments in visualisation and mapping tools are encouraging and facilitating rapid advances in the field. For example, developments in Geographic Information Systems (GIS) enable researchers to create, store, manage, share, analyse, visualise and publish vast amounts of geographic data (Chapter 21); developments in the Global Positioning System (GPS) enable more accurate and detailed mapping, positioning and navigation; and developments in remote sensing (satellite, aircraft and laser-based technologies) enable researchers to obtain vast amounts of data without the need to visit a particular site (see Chapter 49 for more information about sensor-based methods).

Zoning and zone mapping are used in a number of disciplines and fields of study including human geography, physical geography, cartography, environmental sciences, climatology, geomorphology, urban planning, economic geography and history. Examples of research projects that have used, assessed or critiqued these techniques include research into how zone maps are used to regulate urban development (Moga, 2017); institutional conflicts over the use of Flood Maps in the UK (Porter and Demeritt, 2012); mapping and monitoring land cover and land-use changes in Egypt (Shalaby and Tateishi, 2007); an analysis of how zone maps encourage or hinder development near transport hubs (Schuetz et al., 2018) and research into coastlines that are susceptible to coastal erosion (Sharples et al., 2013).

If you are interested in finding out more about zoning and zone mapping in the urban environment, a useful starting point is Lehavi (2018), which gives a detailed account of the history of zoning and city planning. The book provides perspectives from a number of academic disciplines, with a diverse range of methodologies that will help you to think more about methods and methodology for your research. If you are interested in zoning for marine management, a good starting point is Agardy (2010), who provides a compelling case for improving marine management through large scale ocean zoning, along with some interesting case studies from around the world. You may also find it useful to obtain more information about geospatial analysis (Chapter 21) and spatial analysis and modelling (Chapter 54). It is also important that you become familiar with software and digital tools that are available for zoning and zone mapping. Some of these enable you to create your own maps, using data generated from your own research or from existing data, whereas others enable you to work with existing maps, adding to, altering or highlighting specific points. GIS software, in particular, enables you 'to manipulate the data, layer by layer, and so create specialized maps for specific purposes' Heginbottom (2002: 623). Some software and tools are highly complex (requiring programming) and costly, whereas others are simple to use, user-friendly, free and open source. Some examples are given below: this will enable you to find out about functions, capabilities, purpose and costs. An understanding of data visualisation software and tools is also important and more information about these, including a list of useful digital tools and software packages, can be found in Chapter 13.

Questions for reflection

Epistemology, theoretical perspective and methodology

- What are the ontological and epistemological assumptions underpinning zoning and zone mapping?

- o Can land and water be zoned and mapped objectively?
- o Do zonal maps represent reality, or illustrate an objective truth?
- o Are zonal maps stable and knowable?
- o Are zonal maps social constructions?

- Is zoning and zone mapping static or fluid? Should zonal maps be seen as a representation, process or event, for example?

- In what way can, and should, zoning and zone mapping be contested? Lewinnek (2010) provides an interesting example of how zonal models can be counter-modelled and contested.

- Are zoning and zone mapping the best way to answer your research question and meet your aims and objectives (your 'aim' is a simple and broad statement of intent and the 'objectives' are the means by which you intend to achieve the aim). For example, zoning and zone mapping can help with planning and management; producing an inventory; evaluating or protecting an existing environment; assessing impact of change or development; providing an explanation of existing conditions; forecasting, predicting and modelling. As your research progresses you need to check that zoning and/or zone mapping is helping to answer your research question and meet your aims and objectives.

Ethics, morals and legal issues

- When using existing datasets are you clear about rights, restrictions and obligations of use? Have you read any attached licences or legal statements?

- When zoning or zone mapping, how can you ensure that your representation, visualisation or map does not mislead, provide false information or display incorrect information? How can you ensure that it is understood by your intended audience? More information about perception of, and response to, visualisations is provided in Chapters 13 and 16.

- How can you reduce or eliminate bias that may be introduced into your zones or mapping (language bias, narrative bias or cultural bias, for example)?

- How might zoning and zone mapping be used to maintain or prop up the existing social order or support a particular ideology? Can zoning and zone mapping be used to challenge and transform the existing social order or dominant ideology? Brown et al. (2018: 64) for example, discuss 'participatory mapping methods that engage the general public to explicitly inform zoning decisions'. Their paper goes on to discuss the strengths and weaknesses of such an approach.

- What impact does zoning and/or zone mapping have on individuals and groups? Whittemore (2017: 16) discusses two key areas concerning the impact on racial and ethnic minorities in the US. This first are 'exclusionary effects, resulting from zoning's erection of direct, discriminatory barriers or indirect, economic barriers to geographic mobility'. The second are 'intensive and expulsive effects, resulting from zoning's disproportionate targeting of minority residential neighborhoods for commercial and industrial development'.

Practicalities

- What is to be mapped and zoned? How are zones to be designed and classified? What data are to be collected and analysed? What scale, level of precision, accuracy, colour, map legend (caption, title or explanation) and map key is required for your research?

- Do you have a thorough understanding of the hardware, software and tools that are available to help with zoning and zone mapping? This can include GIS, GPS, remote sensing, LIDAR (Light Detection and Ranging) and mapping tools, for example. What costs are involved? Do you have enough understanding to make the right decisions when choosing equipment?

- If you are mapmaking using GIS, do you understand how the content and detail of the map need to change as the scale changes so that each scale has the appropriate level of generalisation?

- How can you address the problem of non-stationarity of data if using data that have been collected over a long period of time? Such data changes and can be unpredictable: often, non-stationary data are transformed to become stationary.

- How can you mitigate generation error in visualisations? This could be because the format of the files provided by the source is not accepted or files are corrupted at source, for example. More information about data collection and conversion can be found in Chapter 11.

Useful resources

There are a variety of digital tools and software packages available to help you create your own zone maps or enable you to use, analyse and draw on those created by others (or use data provided by others). Examples of tools and software available at time of writing include (in alphabetical order):

- AcrGIS (www.arcgis.com);
- DIVA-GIS (www.diva-gis.org);
- GmapGIS (www.gmapgis.com);
- GPS Visualizer (www.gpsvisualizer.com);
- GRASS: Geographic Resources Analysis Support System (https://grass. osgeo.org);
- gvSIG (www.gvsig.com/en);
- MapMaker Interactive from National Geographic (http://mapmaker. nationalgeographic.org);
- Mapnik (http://mapnik.org);
- MapWindow (www.mapwindow.org);
- QGIS (https://qgis.org/en/site);
- SAGA: System for Automated Geoscientific Analyses (www.saga-gis.org/ en/index.html);
- Scribble Maps (www.scribblemaps.com);
- Whitebox GAT (www.uoguelph.ca/~hydrogeo/Whitebox);
- Zonar (www.zonar.city).

The European Data Portal (www.europeandataportal.eu) provides details of open data (information collected, produced or paid for by public bodies) throughout Europe that is made freely available for re-use for any purpose. Here you can search for datasets relevant to your research. The EU Open Data Portal (http://data.europa.eu/euodp/en/home) provides similar access to open data published by EU institutions and bodies, which are free to use and reuse for commercial or non-commercial purposes (consultation is underway for a possible merger of portals). See Chapter 3 for additional sources of open and big data.

Useful overviews of GIS, along with example images, can be obtained from the US Geological Survey website: https://egsc.usgs.gov/isb//pubs/gis_poster/ #information) and from the National Geographic website: (www.nationalgeo graphic.org/encyclopedia/geographic-information-system-gis).

Maps of India (www.mapsofindia.com/zonal) is described as 'the largest online repository of maps on India since 1998'. On this site you can find examples of zonal maps, one of which divides India 'into six zones based upon climatic, geographical and cultural features'. This site provides useful material for critical analysis of zoning and zone mapping.

Key texts

Agardy, T. (2010) *Ocean Zoning: Making Marine Management More Effective.* London: Earthscan.

Brown, G., Sanders, S. and Reed, P. (2018) 'Using Public Participatory Mapping to Inform General Land Use Planning and Zoning', *Landscape and Urban Planning*, 177, 64–74, September, 2018, 10.1016/j.landurbplan.2018.04.011.

Heginbottom, J. (2002) 'Permafrost Mapping: A Review', *Progress in Physical Geography: Earth and Environment*, 26(4), 623–42, first published December 1, 2002, 10.1191/0309133302pp355ra.

Lehavi, A. (ed.) (2018) *One Hundred Years of Zoning and the Future of Cities.* Cham: Springer.

Lewinnek, E. (2010) 'Mapping Chicago, Imagining Metropolises: Reconsidering the Zonal Model of Urban Growth', *Journal of Urban History*, 36(2), 197–225, first published December 17, 2009, 10.1177/0096144209351105.

Moga, S. (2017) 'The Zoning Map and American City Form', *Journal of Planning Education and Research*, 37(3), 271–85, first published June 30, 2016, 10.1177/0739456X16654277.

Porter, J. and Demeritt, D. (2012) 'Flood-Risk Management, Mapping, and Planning: The Institutional Politics of Decision Support in England', *Environment and Planning A: Economy and Space*, 44(10), 2359–78, first published January 1, 2012, 10.1068/a44660.

Schuetz, J., Giuliano, G. and Shin, E. (2018) 'Does Zoning Help or Hinder Transit-Oriented (Re)Development?', *Urban Studies*, first published June 27, 2017, 10.1177/0042098017700575.

Shalaby, A. and Tateishi, R. (2007) 'Remote Sensing and GIS for Mapping and Monitoring Land Cover and Land-Use Changes in the Northwestern Coastal Zone of Egypt', *Applied Geography*, 27(1), 28–41, 10.1016/j.apgeog.2006.09.004.

Sharples, C., Walford, H. and Roberts, L. (2013) 'Coastal Erosion Susceptibility Zone Mapping for Hazard Band Definition in Tasmania', *Report to the Tasmanian Department of Premier and Cabinet*, October, 2013, retrieved from www.dpac.tas.gov.au/__data/assets/pdf_file/0004/222925/Coastal_Erosion_Susceptibility_Zone_Mapping.pdf [accessed April 26, 2018].

Whittemore, A. (2017) 'The Experience of Racial and Ethnic Minorities with Zoning in the United States', *Journal of Planning Literature*, 32(1), 16–27, first published December 19, 2016, 10.1177/0885412216683671.

Index

acquisition analytics 392
Analytics and Big Data Society 21
anomaly detection 83, 120, 123
API *see* application programming
  interface
app analytics: app marketing analytics
  393; app performance analytics 393;
  app session analytics 393; in-app
  analytics 393; mobile location
  analytics 393
application programming interface 39,
  42, 353
association rules 83
audience response systems *see* live
  audience response
audio methods: acoustic analysis 10;
  audio analysis 10–16, 104; audio-CASI
  see computer-assisted interviewing;
  audio content *see* audio analysis;
  audio diary analysis 10; audio event
  and sound recognition/analysis 10–11;
  multimodal analysis 11; multimodal
  discourse analysis 11; music
  information retrieval 11; music
  retrieval 155; psychoacoustic analysis
  11; semantic audio analysis 11–12, 13;
  sound analysis 12; sound scene and
  event analysis 12; speech analysis 12;
  speech and voice analytics 68; spoken
  content retrieval 155

Bayesian analysis 302, 363
behavioural analytics 392
bibliometrics 169, 398, 400, 402
big data 17, 18, 19, 27, 79

big data analytics 1, 17–23, 66
blog analysis 398
blog retrieval 154
business analytics 24–30, 66, 235

capability analytics 148
capacity analytics 148
classification 83, 120, 123, 197
classroom response systems *see* live
  audience response
cluster methods: biclustering 31; block
  clustering *see* biclustering; cluster
  analysis 31–7; co-clustering *see*
  biclustering; consensus clustering 31;
  density-based spatial clustering 31;
  fuzzy clustering 31–2; graph clustering
  32; hierarchical clustering 32; K-means
  clustering 32; mathematical
  morphology clustering 31;
  model-based clustering 32; nesting
  clustering *see* hierarchical clustering;
  two-dimensional clustering *see*
  biclustering; two-way clustering *see*
  biclustering
coding and retrieval 38–44
co-link analysis *see* link analysis
collaborative virtual environments 376
comparative analysis 60, 101, 104, 349
competency analytics 148
competitive link analysis *see* link
  analysis
computer games: first-person shooter
  games 377; massive multi-player
  online 377; real-time strategy games
  377; role-playing games 377; serious

410

For Product Safety Concerns and Information please contact our EU
representative GPSR@taylorandfrancis.com
Taylor & Francis Verlag GmbH, Kaufingerstraße 24, 80331 München, Germany

www.ingramcontent.com/pod-product-compliance
Lightning Source LLC
Chambersburg PA
CBHW080144060326
40689CB00018B/3843